The Early Modern Human from Tianyuan Cave, China

Hong Shang and Erik Trinkaus

TEXAS A&M UNIVERSITY PRESS
College Station

First edition

This paper meets the requirements of ANSI/NISO Z39.48-1992
(Permanence of Paper).
Binding materials have been chosen for durability.
∞

Library of Congress Cataloging-in-Publication Data
Shang, Hong, 1968–
The early modern human from Tianyuan Cave, China / Hong Shang and Erik Trinkaus.—1st ed.
p. cm.—(Texas A&M University anthropology series ; no. 14)
Includes bibliographical references and index.
ISBN-13: 978-1-60344-177-3 (cloth : alk. paper)
ISBN-10: 1-60344-177-8 (cloth : alk. paper) 1. Tianyuan Man. 2. Cave dwellers—China—Tianyuan Cave. 3. Hunting and gathering societies—China. 4. Tianyuan Cave (China)—Antiquities. 5. Antiquities, Prehistoric—China. I. Trinkaus, Erik. II. Title. III. Series: Texas A & M University anthropology series ; no. 14.
GN286.8.S53 2010
569.9'8—dc22
2009050004

The Early Modern Human from Tianyuan Cave, China

Number Fourteen:
Texas A&M University Anthropology Series
D. Gentry Steele, General Editor

A list of titles in this series appears at the end of this book.

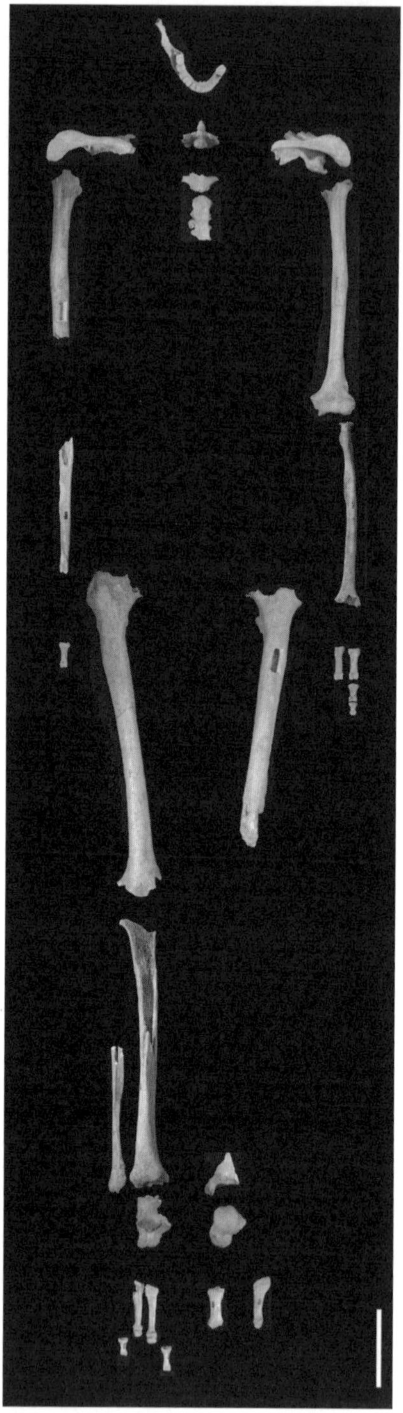

Composite view of the elements preserved for the Tianyuan 1 skeleton.

Contents

To Professor Xinzhi Wu,
member of the Chinese Academy of Sciences,
for his support, enthusiasm, and unfailing interest
in the Pleistocene peoples of China.

Preface

The karstic area around Zhoukoudian has been famous since the discoveries there in the 1920s and 1930s of Middle and Late Pleistocene human remains, abundant faunal remains, and archeological assemblages. This fame, which was justified both by the nature of the discoveries and by the classic monographs written on the sites and their contents during the 1930s and 1940s, was enhanced by the mystery that still surrounds the disappearance of the human fossils during World War II. Subsequent work at the well-known Locality 1, and at various localities surrounding it, has produced additional human remains, archeological materials, and considerable clarification of the geological ages and formation processes of the deposits. But the human fossils have been few and fragmentary compared to the specimens from the early excavations, and the site has become as much a monument to the history of paleoanthropology in China as a locale for scientific investigations.

It is in this context that the discovery of fossil human bones and Pleistocene fauna, in June of 2001 in a small cave (Tianyuandong—Tianyuan Cave) 6 km from Locality 1, takes on significance. Initially investigated by the owners of the Tianyuan Tree Farm and subsequently excavated systematically by a team from the Institute of Vertebrate Paleontology and Paleoanthropology (IVPP) under the direction of Haowen Tong, this little cave has yielded an abundant Late Pleistocene mammalian faunal assemblage, a surprising absence of artifacts, and major portions of an early modern human skeleton. It is this human skeleton, 40,000 years old, that is of concern here.

The Tianyuandong discovery has brought new attention to the paleoanthropological richness of the Zhoukoudian region, and it has occurred during a time when paleoanthropology has been focused on the earliest modern humans to spread across much of the Old World. In the past decade, the Niah Cave human fossil from Sarawak has been confirmed to be 40,000 years old. The field of paleoanthropology has seen the discovery of the oldest modern human remains in Europe, at the Peştera cu Oase in Romania, also 40,000 years old. It has seen the dating to the same time period of the Nazlet Khater skeletons from Egypt. And to this has been added the more recent South African Hofmeyr cranium, as well as direct ages confirming the antiquity of the Mladeč, Muierii, and Cioclovina early modern humans in Europe. In this resurgence of interest in the biology of the earliest modern humans to spread across the Old World, the Tianyuandong human takes its place.

The authors hope that this book, detailing many aspects of the morphology and paleobiology of this partial skeleton from northern China, will help integrate it into our assessments of human evolutionary history in eastern Asia and the processes that were involved in the global establishment of people anatomically similar to ourselves.

This book is principally the fruit of our efforts to describe and analyze these fossil remains in light of current knowledge of Late Pleistocene human biology, with additional contributions by Yolanda Fernández-Jalvo and Peter Andrews on the burial taphonomy of the skeleton and by Yaowu Hu and Michael Richards on its stable isotopes. None of this would have been possible without the encouragement, assistance, support, and enthusiasm of a multitude of individuals.

Our gratitude goes first and foremost to Haowen Tong, of the Institute of Vertebrate Paleontology and Paleoanthropology, for inviting HS to be part of the excavation team, turning the human remains over to her for analysis, and providing support and advice throughout the human paleontological aspects of the project. Without him and his generosity, none of this would have happened. We are also very grateful to Xinzhi Wu (IVPP) for overseeing the whole Tianyuandong project, recognizing its importance, providing continuous support to HS, and being a wonderful host to ET at the IVPP. We also appreciate the constant support of Wu Liu and Xing Gao in IVPP's Department of Paleoanthropology during this project.

During the excavation of Tianyuandong, several people provided invaluable assistance, including Xiumei Tian and her family of the Tianyuan Tree Farm, and Fuyou Chen and Shuangquan Zhang, both of the IVPP. The staff at the Site Museum at Zhoukoudian provided fieldwork assistance, lodging, and laboratory facilities during the excavations. Subsequently at the IVPP in Beijing, Zhongyi Zhao and Haijun Wang made the casts of the human remains; Wei Gao took many of the photographs and provided photographic assistance; and Wending Zhang took the radiographs of the Tianyuan human fossil remains. Yushan Lou provided help in accessing recent human comparative samples. The dating of the Tianyuandong remains, a critical part of the foundation for the interpretation and analysis of the human fossil, was done through the efforts of Zhibang Ma at the Institute of Geology and Geophysics, Chinese Academy of Sciences, Beijing, for uranium-series dating, and of Xiaohong Wu at the School of Archaeology and Museology, Peking University, Beijing, for the accelerator mass spectrometry direct radiocarbon dating of the human fossil and of the faunal remains.

Subsequently, given the multiple abnormalities on the Tianyuandong human skeleton, we consulted several colleagues for their opinions on the etiologies of the lesions, including Arthur C. Aufderheide, Eric Crubézy, Robert Jurmain, Donald J. Ortner, and Richard J. Smith. B. C. Maley performed the multivariate analyses. We also acknowledge, if only generally, the multitude of institutions, curators, and colleagues across the Old World who have provided access to human remains, both Pleistocene and recent. The data available from these human remains have made the comparative analysis of the Tianyuan human fossil possible. Without the openness of these colleagues, much of the comparative data employed here would not have been available.

The archeological and paleontological work at the Tianyuandong has been funded, in both the field and the laboratory, by the National Natural Science Foundation of China (grant 40372015) and the President Fund of the Chinese Academy of Sciences (grant KL203302). The work was also supported by Materials Science and Technology of China (grant 2006CB806400), and the Knowledge Innovation Program of the Chinese Academy of Sciences (grant kzcx2-yw-106). The visiting scholar position of HS at Washington University, which has made this collaboration possible, was funded by the Wenner-Gren Foundation for Anthropological Research. ET's travel to Beijing was made possible by support from Washington University and the IVPP, and previous human paleontological data collection by ET was supported by the National Science Foundation, the Wenner-Gren Foundation, the Leakey Foundation, and his academic institutions.

To all of these individuals, institutions, and foundations, we owe profound gratitude. We hope that our efforts to bring this fossil to the paleoanthropological world justify their confidence and support.

The Early Modern Human from Tianyuan Cave, China

Chapter 1

Introduction

DURING MOST of the past century and into the current century, paleoanthropology has returned time and again to the issue of modern human emergence, querying the when and the where of that evolutionary process and especially of the human behavioral and biological dynamics that were involved. Given the abundance of Paleolithic archeological sites between the Caucasus Mountains and the Atlantic Ocean and the long history of fossil human discoveries in that region, most of the attention of this research has, for better or for worse, been on western Eurasia. This focus has been enhanced by the presence, in that geographical region, of the archetypical premodern humans, the Neandertals. Recent decades have seen this focus spread to the other half of the western Old World, Africa, as paleoanthropological research has increasingly agreed that the earliest anatomically modern humans emerged on that continent.

This attention on modern human emergence in the western Old World has tended to shift interest away from eastern Eurasia and its adjacent islands. But that geographically vast portion of the Old World was equally occupied during the Late Pleistocene, and human populations there underwent an evolutionary transition leading to the establishment of anatomically modern human populations. It is in part to take the focus somewhat away from the western Old World and provide modern human emergence with a more cosmopolitan framework that we present here one of the most complete early modern human fossils from eastern Eurasia, the human remains from Tianyuandong.

Paleontological considerations of the biological nature and phylogenetic position of early modern humans in eastern Eurasia have a long history, dating principally from the discovery in 1933 by W. C. Pei of human remains and abundant Upper Paleolithic archeological remains in the Upper Cave of Zhoukoudian (Weidenreich, 1939a). In addition to describing these human remains and aspects of their biology (see esp. Weidenreich, 1939a,b), Weidenreich (1943, 1947) incorporated the Upper Cave remains into a more global assessment of the patterns of the process of modern human emergence. Since that time, there have been a series of discoveries of the fossils of both late archaic humans and especially early modern humans in eastern Asia (e.g., Woo, 1959; Brothwell, 1960; Sakura, 1981; Suzuki and Hanihara, 1982; Wu and Poirier, 1995; see also Chapter 4).

These discoveries of Late Pleistocene fossils in eastern Eurasia, combined with the ever-present paleoanthropological interest in the phylogenetic emergence of modern humans, have led to a series of attempts by human paleontologists to assess the population dynamics of modern human emergence in the eastern Old World (e.g., Wolpoff et al., 1984; Pope, 1992; Bräuer, 1992; Wu, 1992, 2004, 2006; Trinkaus, 2005a). Many of these paleontological studies have concluded that there was considerable local population continuity between late archaic humans and early modern humans in the region, the studies varying principally in the degree to which population dispersals and/or gene flow were seen as influential in establishing modern human biology in the region. This has occurred despite the limited fossil evidence on either side of the transitional time period and ongoing difficulties in establishing, with limited exceptions (e.g., Barker et al., 2007), the ages of many of the critical fossils (see Brown, 1993; Wu and Poirier, 1995; Trinkaus, 2005a; see also Chapter 4).

At the same time, there has been remarkably little assessment of the paleobiology of these Late Pleistocene fossil human remains. The exceptions are Weidenreich's early attempt (1939b) to assess the paleodemography and paleopathology of the Zhoukoudian–Upper Cave remains and more recent biomechanical analyses of the Yamashita-cho and Minatogawa remains (Kimura and Takahashi, 1992; Trinkaus and Ruff, 1996).

It is in this general context that the 2001 discovery of a human partial skeleton at Tianyuandong (Tianyuan Cave) (see Chapter 2) takes on special importance. Even though Tianyuan 1 lacks the phylogenetically important cranium, it is the first human associated skeleton from eastern Asia that is older than the last glacial maximum and retains substantial portions of the facial skeleton and the limbs. It is the first human skeleton from the region to be directly radiocarbon dated, to ~40,000 years before present (BP). And its geological age places it close to the time period during which modern humans become permanently established across the Old World (between ~50,000 and ~35,000 BP) (Trinkaus et al., 2003; Trinkaus, 2005a; Barker et al., 2007; Grine et al., 2007; Crevecoeur, 2008).

A single skeleton, however interesting and diagnostic, will not solve all the issues of population dynamics or provide a populational framework for assessing paleobiology. However, the small number of human remains bracketing this Late Pleistocene human transition in eastern Eurasia, the often-fragmentary nature of those skeletal remains, and ongoing difficulties with the chronological positions of key samples means that one reasonably complete, well-dated, associated skeleton provides a major addition to our knowledge of early modern human biology. It is in this context that we present here a detailed description and comparative assessment of the human partial skeleton from Tianyuandong, Tianyuan 1. Along with the other scattered Late Pleistocene human fossils from the region, it should provide a new and broader perspective on the emergence of modern humanity in this region of the Old World.

Chapter 2

Tianyuandong: The Cave, Its Context, and Its Contents

The Site of Tianyuandong

TIANYUANDONG, OR Tianyuan Cave, is a small cave situated at the base of a modest exposed limestone cliff, high on the south side of a small valley (Figure 2-1). It is located on the domain of the Tianyuan Tree Farm of Huangshandian village, which is part of Zhoukoudian town, in the Fangshan district of Beijing (39° 39′ 28″ N, 115° 52′ 17″ E) (Figure 2-2). It is approximately 6 km southwest of the classic *Homo erectus* site of Zhoukoudian Locality 1 and its adjacent Late Pleistocene Upper Cave site. Located 174.5 m above sea level, with its opening oriented to the northwest, it is the highest Paleolithic site in the Zhoukoudian district.

The cave has an irregularly shaped surface area; it deviates to the right from the entrance and then curves around to the left in the portion farthest from the entrance (Figure 2-3). The cave is approximately 6 m deep and 6 to 7 m wide, and all portions of it remain within daylight. The small cave is located in the southwest margin of Yanshan Inner Block Tectonic Belt of the North China Platform (Li et al., 2008). It formed within the carbonatite of Jixian System of the Middle Precambrian (Proterozoic). There is a well-developed fold and proximate grade fault in the bedrock, as well as vein intrusion and schistosity, which combine to provide many different forms of passages in the rock. This formation also increases the water permeability, and it forms the basis of the karst system. Tianyuan Cave itself was formed by the fracture water dissolution along the vein and joint fractures, with a vertical vadose zone in the rock. The karstic process continues, such that the back wall has a cascade of calcite formations, and stalactites and stalagmites are still forming in the rainy season. It was this characteristic that led members of the Tianyuan Tree Farm to seek water within the cave.

In front of the cave, there is a small terrace (Figure 2-4). In front of the terrace

Figure 2-1. Panoramic view of the limestone massif rising above the Tianyuan Tree Farm. The location of the Tianyuan Cave (Tianyuandong) is indicated by the arrow.

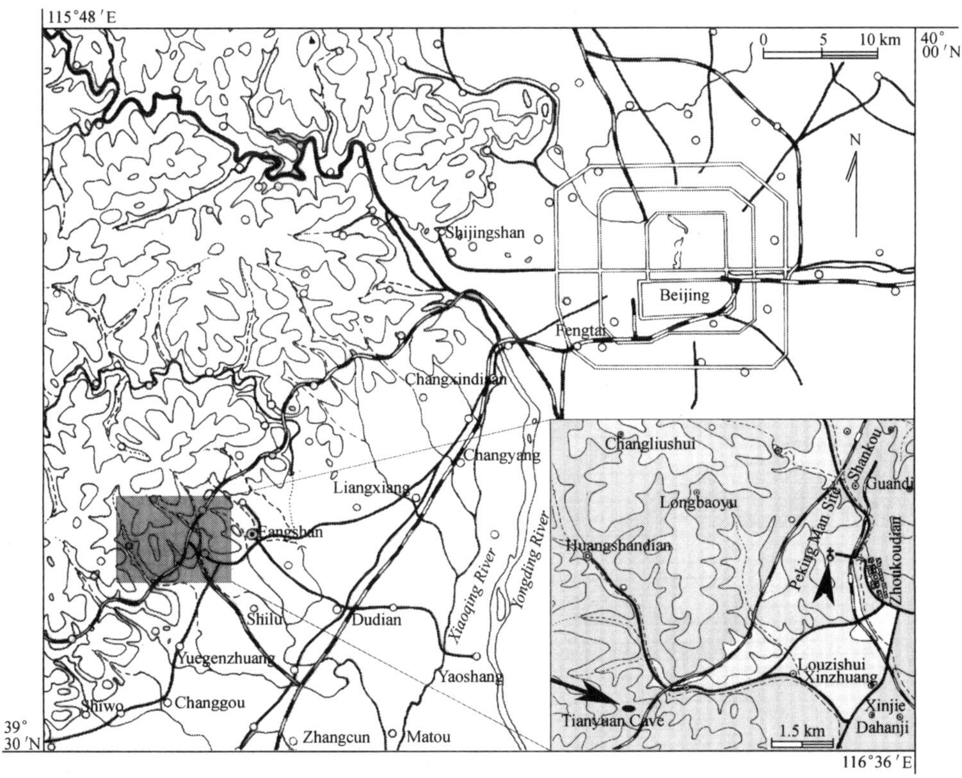

Figure 2-2. The Beijing area, with the Zhoukoudian area in the inset. Location of Tianyuan Cave is indicated by the arrow in lower left of inset; Zhoukoudian Peking Man Site, by the arrow on right. Modified from Tong, Shang, et al. (2004).

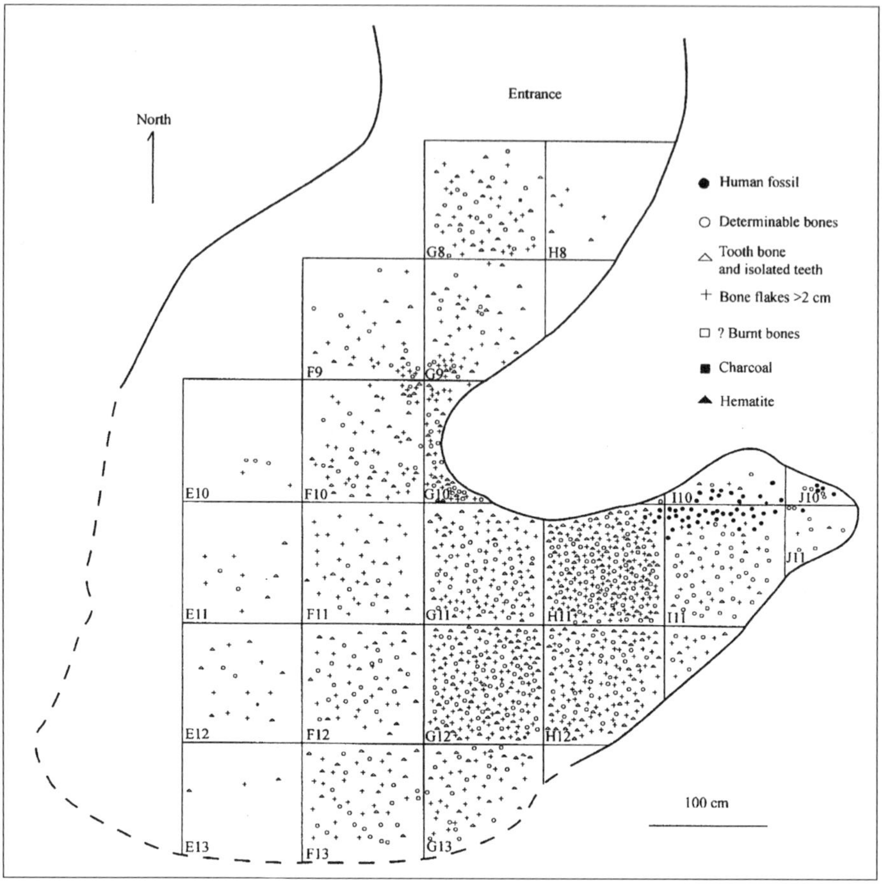

Figure 2-3. Horizontal plan and distribution of bones, teeth, and human remains in Tianyuan Cave. The human remains were clustered in squares I10 and I11 and into the margins of the adjacent squares. Modified from Tong, Shang, et al. (2004).

the modern landscape descends rapidly for a short distance and then slopes more gradually to the valley below (Figure 2-1). Modern access to the cave is along the cliff base to the left of the entrance

Discovery and Excavation

The paleontological nature of the site was first discovered in June 2001, when the owners of the Tianyuan Tree Farm (Xiumei Tian and members of her family) were searching for sources of water in the limestone formations. They found and removed an abundance of bones and teeth, including faunal remains and portions

Figure 2-4. The entrance to the Tianyuan Cave in May 2007. Modern access to the terrace is to the left along the cliff face. Haowen Tong (*left*) and a staff member from the Site Museum at Zhoukoudian prepare to enter the cave.

of a human skeleton. Realizing that the bones and teeth might be of interest, they brought the remains to the Institute of Vertebrate Paleontology and Paleoanthropology (IVPP) of the Chinese Academy of Sciences in Beijing. The IVPP organized a project team to excavate Tianyuan Cave under the direction of Haowen Tong of the IVPP; systematic excavations took place during 2003 and 2004 (Tong, Liu, et al., 2004; Tong, Shang, et al., 2004).

The deposits were excavated in arbitrary levels of 20-cm thickness, which were divided into 5-cm sublevels. After the landowners had dug for water, the remaining deposits varied in thickness. The sediments to bedrock were no more than 40 cm thick in the deep left recess of the cave (squares J10 and J11); they became substantially thicker toward the entrance (Figure 2-5). The excavation at the entrance (square G8) reached a depth of 1 m, but the deposits continued farther in that portion of the cave. The sediment is generally loose and unsorted, with some areas of moderate brecciation.

The human bones, based on the accounts of Xiumei Tian and the discovery of several of the bones in situ during the 2003 excavation season (Figure 2-6), derived from the deposits deep within the cave, squares H11, I10, I11, J10, and J11 (Figure

Figure 2-5. Tianyuan Cave during excavations in 2003 (*above*). View from the interior along the excavation trench to entrance and terrace (*below left*) and from the entrance to the deeper portions of the small cave (*below right*) after the excavations (May 2007).

2-3). The faunal remains were scattered throughout the excavation area, but they were concentrated in 4 squares to the rear of the cave, G10, H11, G12, and H12. The final faunal (and human) assemblages consist of excavated materials, remains screened from the deposits dumped on the terrace by the Tianyuan Tree Farm owners, and the material brought to the IVPP by the landowners.

The discovery of human fossil remains at Tianyuandong was the first such find in the Zhoukoudian district since the discovery of *H. erectus* remains in Locality 1 in 1966 (Qiu et al., 1973) and the first discovery there of Late Pleistocene human remains since the Zhoukoudian–Upper Cave remains found in the 1930s. Understandably there was considerable media interest throughout the 2003 excavation season. The site was designated Zhoukoudian Locality 27; 26 other localities had previously been listed for Zhoukoudian Hill (Figure 2-7).

Figure 2-6. A human metatarsal (indicated by arrow) in situ, as discovered during excavation on June 26, 2003.

Figure 2-7. The stone marker placed below the site. English translation: The Peking Man Site in Zhoukoudian Locality 27 / Tianyuan Cave / August 16, 2003.

Stratigraphy and Radiometric Dating

During excavation, 4 stratigraphic layers were identified within the deposits, varying in thickness across the site. The 4 layers, which are recognized from top to bottom, are Layer I, an interbedding of soil and speleotherms; Layer II, a layer of debris with silt that was stably deposited without sorting; Layer III, a layer of breccia without full cementation that accumulated quickly; and Layer IV, the basal gravel bed (Tong et al., 2006; Li et al., 2008).

Most of the mammalian fossils are from the uppermost Layer I and the stratigraphically deeper Layer III (Table 2-1), but there are a few remains from the intervening Layer II and the basal gravel (Layer IV). The 8 human remains found in situ all derive from Layer III. These bones consist of a mandibular M_3, 4 manual phalanges, a metatarsal, and 2 pedal phalanges; the M_3 fits securely into the mandible, establishing that the mandible was also from Layer III.

The initial assessment of the geological age of the Tianyuan Cave deposits was based on their contained mammalian remains, of which there are 36 securely identified species (including *Homo sapiens*) from Layers I to IV deriving from 32 genera (Table 2-1). If the uncertain genera/species are added to the list, then the total is 41 species from 37 genera. All of the species are extant, but the abundant *Hystrix* (porcupine) remains represent the most recent appearance of that genus in northern China (Tong, 2005, 2008), and the *Aeretes* (Chinese flying squirrel) remains document the earliest appearance of the genus in northern China (Tong, 2007). Several other species are no longer present in the region around Zhoukoudian, including *Ursus tibetanus, Moschus moschiferus,* and *Cervus nippon,* as well as the less certainly identified *Martes* sp., *Crocuta crocuta, Bos* sp., and *Capricornis* sp. (Tong, Liu, et al., 2004).

The faunal remains in Layer I, along with the *Hystrix* remains, exhibit abundant evidence of rodent (porcupine) gnawing (Tong et al., 2008). The warm climate *Macaca mulatta* and *Paguma larvata* remains are solely from the upper Layer I, whereas the faunal profile of the deeper Layer III closely resembles that of the Late Pleistocene levels of Zhoukoudian–Upper Cave (Tong et al., 2006).

An initial attempt was made to date the site by uranium series dating at the Institute of Geology and Geophysics, Chinese Academy of Sciences. It provided an age of 25,300 ± 1,500 cal BP on a cervid tooth from Layer III (U-ppm: 7.098 ± 0.230; $^{234}U/^{238}U$: 1.372 ± 0.041; $^{230}Th/^{232}Th$: 10.771; $^{230}Th/^{234}U$: 0.208 ± 0.009). Given potential uncertainties in the uranium series dating of tooth enamel, primarily whether the tooth is a closed system after deposition and/or whether it has reached equilibrium at some point in the past, it was unclear whether this date represented a minimum age or the actual age of the deposit. As a result, a series of accelerator mass spectrometry (AMS) radiocarbon dates were run on the fauna

Figure 1. Views of Tianyuan Cave (Tianyuandong). Upper left: the excavation team; from left to right, Shuang-quan Zhang, Fu-you Chen, Hao-wen Tong, Hong Shang. Upper right: Erik Trinkaus. Middle: students excavating. Lower left: looking up at Tianyuandong. Lower right: the view from Tianyuandong.

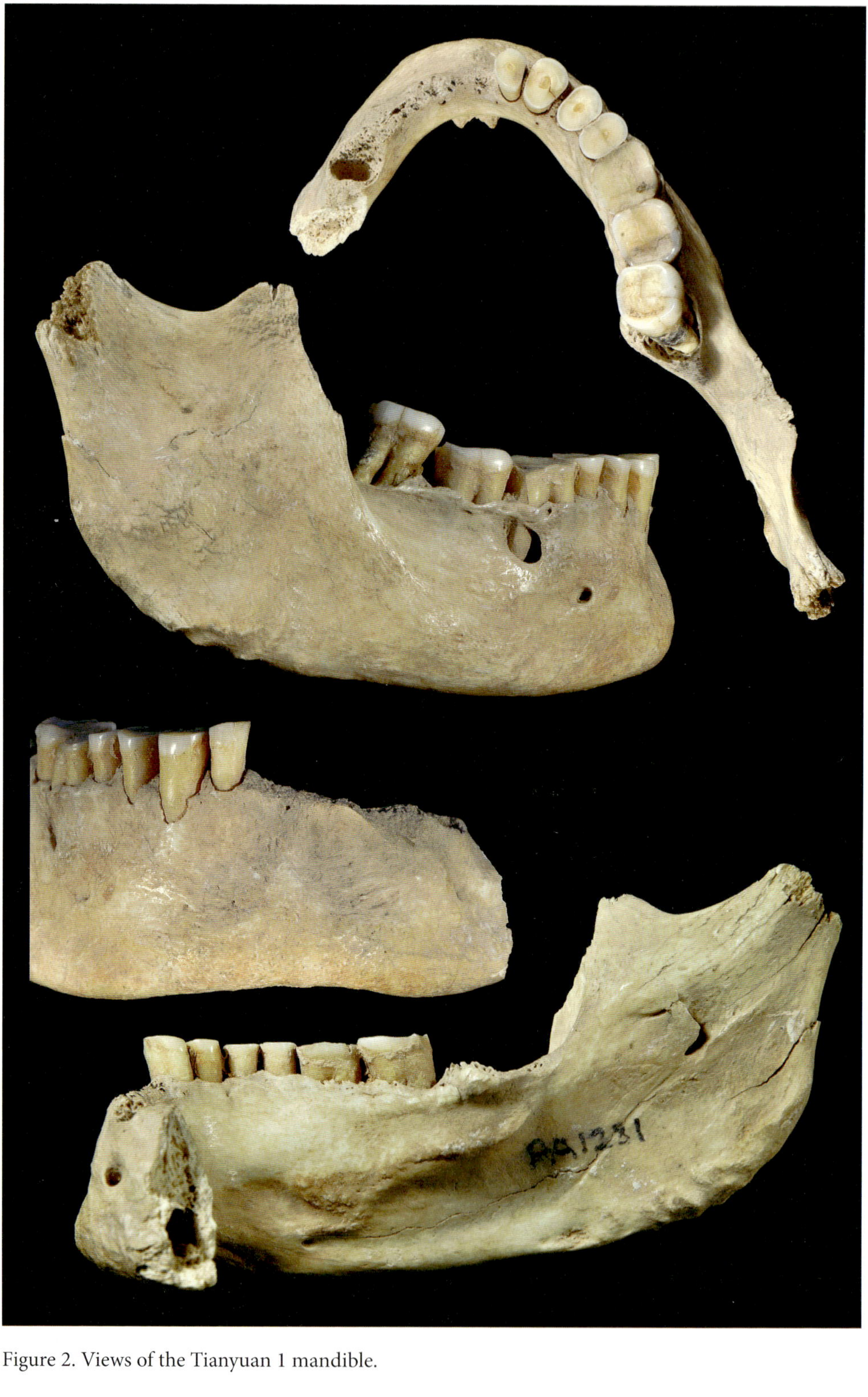

Figure 2. Views of the Tianyuan 1 mandible.

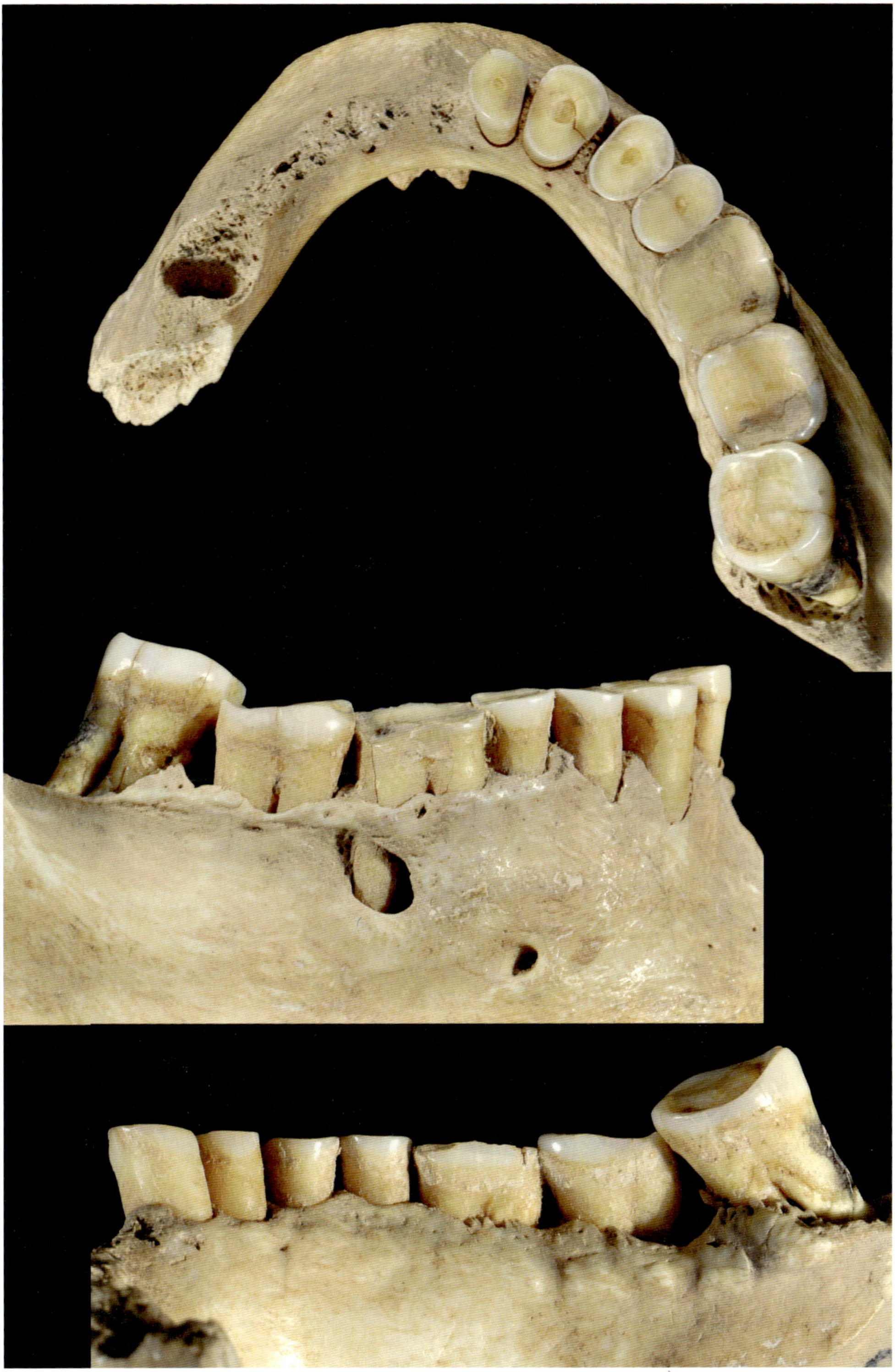

Figure 3. Views of the Tianyuan 1 mandibular dentition.

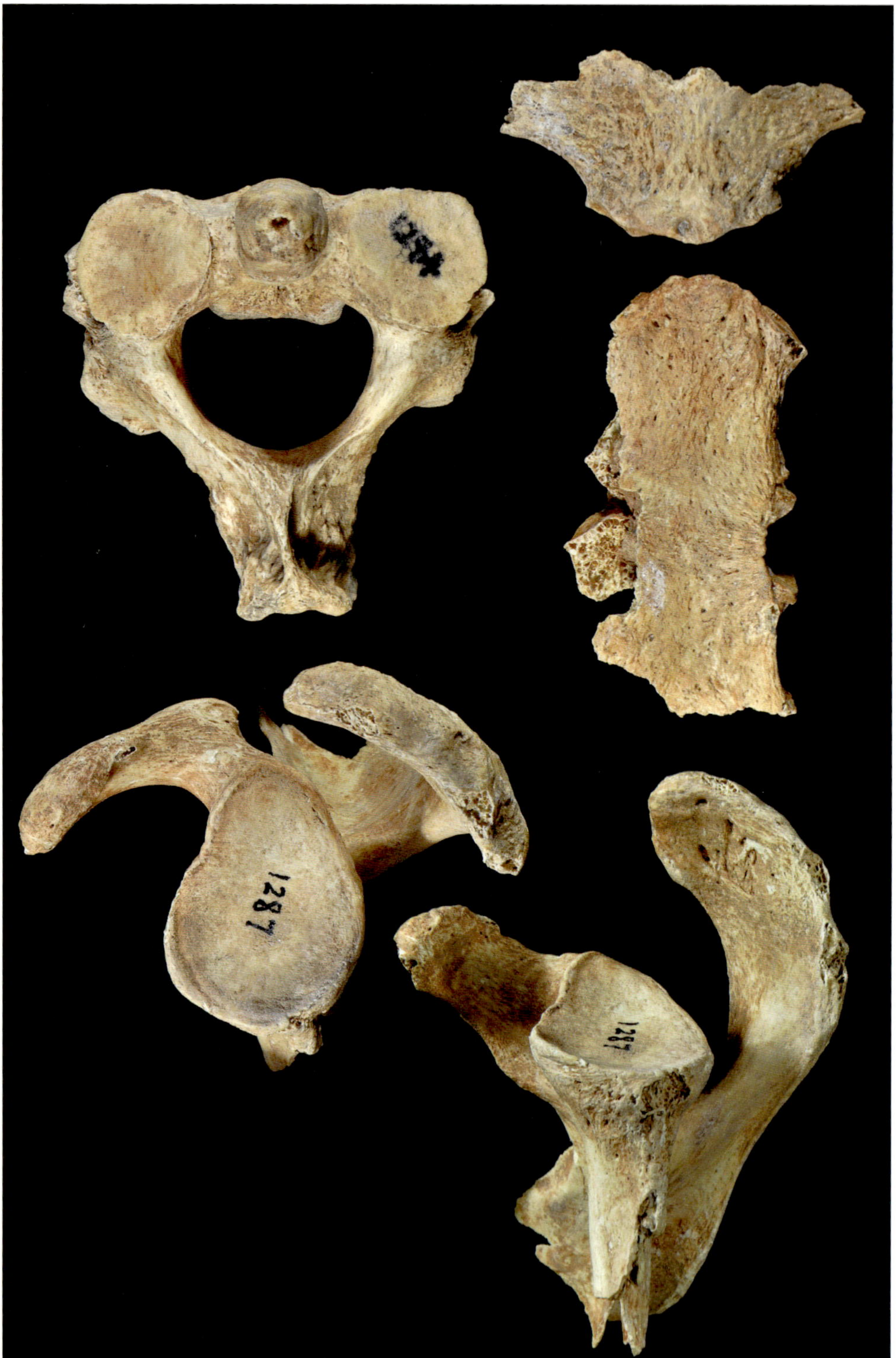

Figure 4. The Tianyuan 1 axis and sternum (above) and left scapula (below).

Figure 5. The Tianyuan 1 humeri, right ulna and left radius.

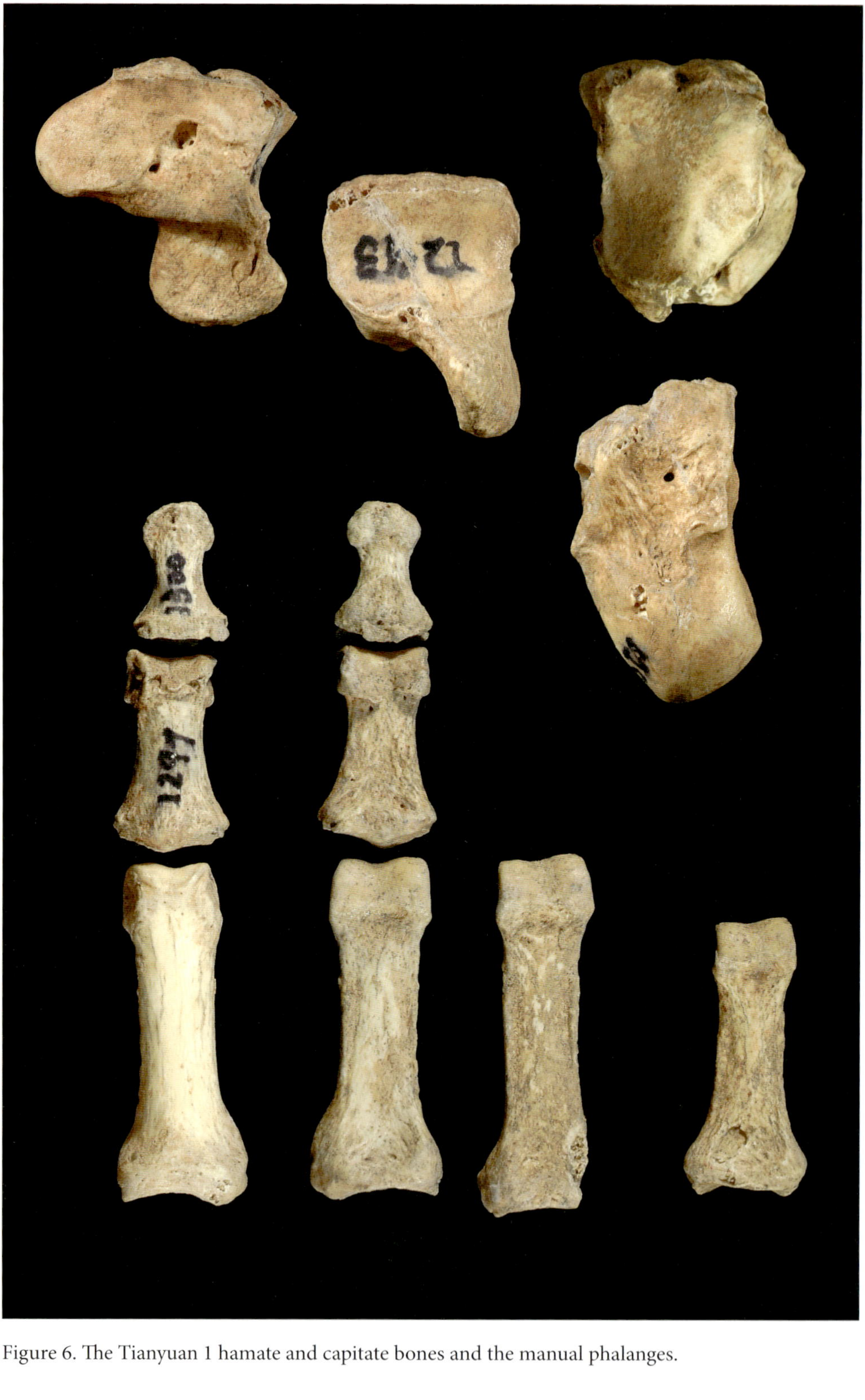

Figure 6. The Tianyuan 1 hamate and capitate bones and the manual phalanges.

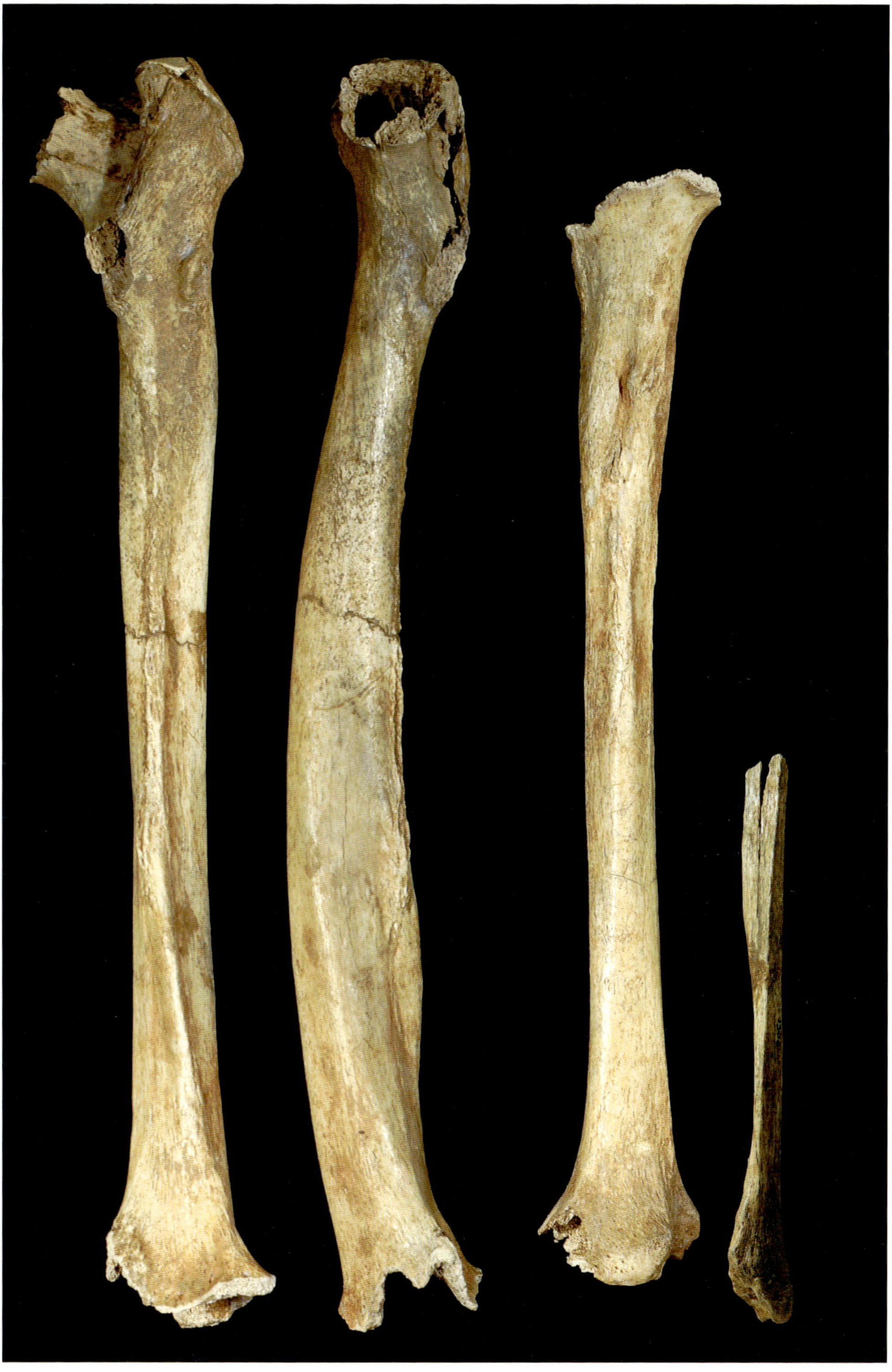

Figure 7. The right femur in posterior and medial views, and the right tibia and fibula in posterior view.

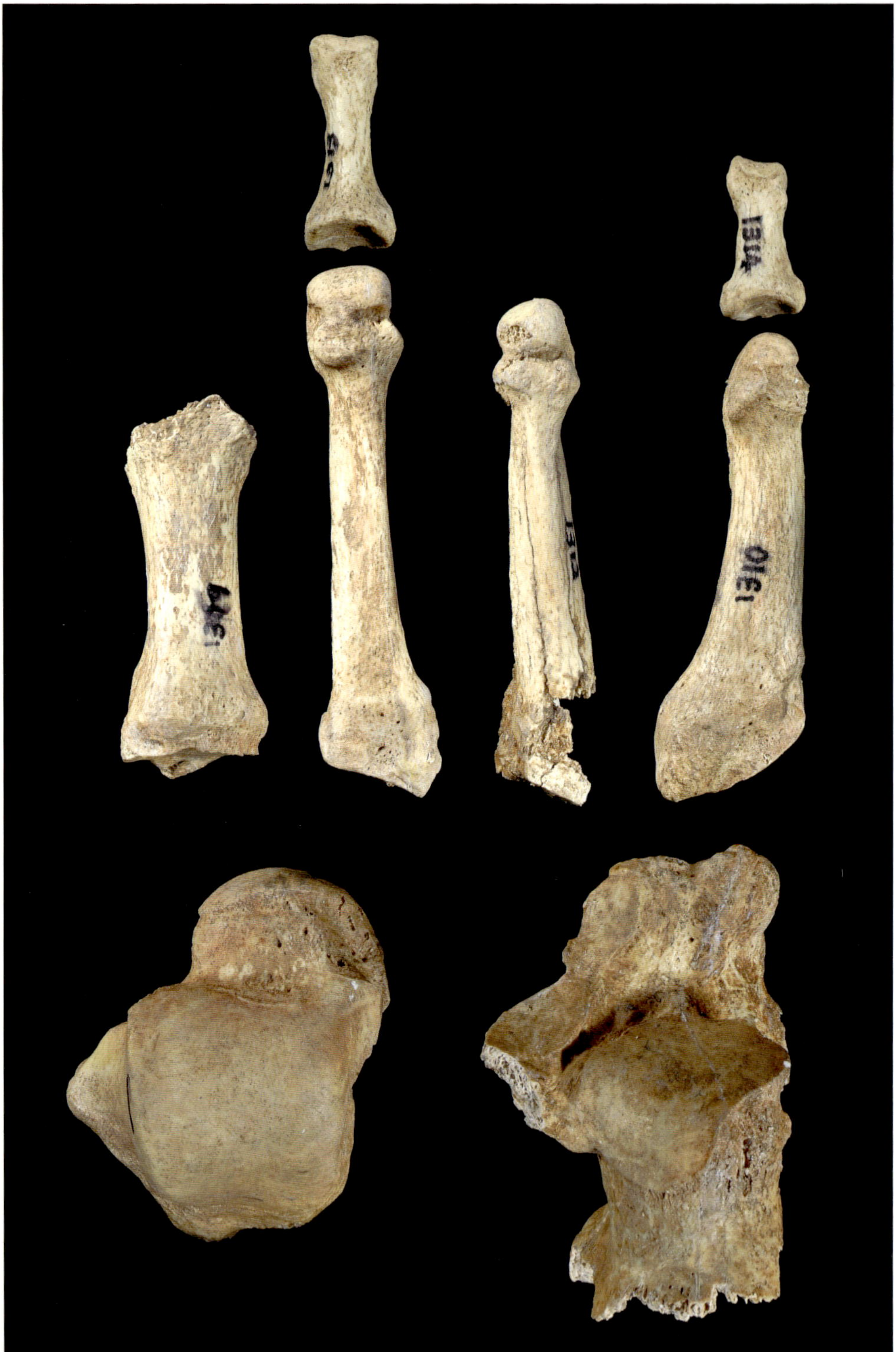

Figure 8. The Tianyuan 1 foot bones.

Table 2-1. Distribution of genera and species in Layers I to IV of Tianyuan Cave

		I	*II*	*III*	*IV*
Primates					
Homo sapiens	human			+	
Macaca mulatta	rhesus macaque	+			
Carnivora					
Arctonyx collaris	hog badger	+			
Canis lupus	wolf			+	
Felis microtis	small-eared cat	+			
Nyctereutes procyonides	raccoon dog	+			
Paguma larvata	palm civet	+	+		
Panthera pardus	leopard	+			
Ursus tibetanus	Asiatic black bear	+			
Vulpes vulpes	fox	+		+	
Crocuta crocuta ultima	Asian spotted hyena			?	
Martes sp.	marten			?	
Mustela altaica	mountain weasel			?	
Artiodactyla					
Capricornis sp.	serow	+			
Cervus elaphus	red deer			+	
Cervus nippon	sika deer	+		+	+
Moschus moschiferus	Siberian musk deer	+		+	
Ovis sp.	sheep			+	
Sus scrofa	boar	+			
Bos sp.	cattle	?			
Capreolus capreolus	roe deer			?	
Rodentia					
Aeretes melanopterus	Chinese flying squirrel	+		+	
Apodemus sylvaticus	wood mouse	+	+	+	
Cricetulus triton	greater long-tailed hamster	+		+	
Cricetulus barabensis	Chinese striped hamster			+	
Cricetulus griseus	Chinese hamster			+	
Cricetulus longicaudatus	long-tailed hamster			+	
Crocidura suaveolens	lesser white-toothed shrew	+			
Erinaceus europaeus	hedgehog	+		+	
Eutamias sibiricus	Siberian chipmunk			+	
Hystrix subcristata	porcupine	+	+	+	
Microtus sp.	vole			+	
Mus musculus	house mouse		+		
Myospalax sp.	Siberian zokor	+	+	+	
Ochotona dauurica	Daurian pika	+	+	+	
Rattus norvegicus	Norway rat	+			
Scaptochirus sp.	short-faced mole	+			
Sciurotamias davidianus	Père David's rock squirrel	+		+	+
Lagomorpha					
Lepus capensis	cape hare	+	+	+	
Chiroptera					
Rhinolophus ferrumequinum	greater horseshoe bat	+	+	+	
Eptesicus serotinus	serotine bat	+			

Source: Modified from Tong et al. (2006).

Table 2-2. AMS radiocarbon results for faunal remains and the Tianyuan 1 human skeleton from Tianyuan Cave

Sample no.	*Square*	*Lab no.*	*Layer*	*Sample material*	*Age (^{14}C years BP)*	*Age (calendar years BP)*
2		BA-04245	I	bone flake with rodent gnawing	1135 ± 30	1037 ± 39
03-4-27-06		BA-03228	I	animal bone	4670 ± 60	5430 ± 87
1		BA-04244	I	bone flake with rodent gnawing	6025 ± 20	6868 ± 39
111-(11)-B	111	BA-03226	III	animal bone	30,500 ± 370	34,755 ± 403
111-(11)	111	BA-04247	III	bone fragment	31,115 ± 190	35,120 ± 347
J11-(11)	J11	BA-03225	III	animal bone	33,970 ± 540	39,353 ± 1,115
110-(11)	110	BA-03227	III	animal bone	34,990 ± 400	40,027 ± 872
J10-(11)	J10	BA-03224	III	animal bone	39,430 ± 680	43,458 ± 665
		BA-03222	III	human femur	34,430 ± 510	39,713 ± 882
G12-(12)	G12	BA-04248	IV	bone fragment	37,785 ± 250	42,292 ± 330

Note: Calendrical years calculated using CalPal v. 1.5 2007 (www.calpal.de/).

and the human skeleton by Xiaohong Wu at the School of Archaeology and Museology, Peking University, Beijing (Table 2-2).

The radiocarbon dates for Layer I are all Holocene in age, including a relatively recent (~900 A.D.) date on a rodent gnawed piece of bone. This is in agreement with the warm climate nature of some of the fauna from Layer I.

Of more interest are the dates for the fauna from Layer III. For this purpose, a series of bone pieces of sufficient size were selected from the abundant but extremely fragmented diaphyseal remains of herbivores (Appendix 1). Most of them probably represent remains of *Cervus nippon,* the most common artiodactyl in the deposits. All of the resultant determinations are between ~30,000 and ~40,000 ^{14}C BP (~34,500 and ~44,000 cal BP) [all conversions of radiocarbon ages (^{14}C BP) to calendar ages (cal BP) are done using CalPal (quickcal2007 ver.1.5) (Weninger et al., 2007)].

These dates place Layer III securely in the middle of marine (oxygen) isotope stage (MIS/OIS) 3. One sample from the underlying Layer IV falls toward the earlier portion of that range, slightly younger than the oldest date from Layer III, suggesting that there might have been modest mixing of bone fragments between Layers III and IV and/or little chronological separation between the 2 adjacent deeper levels of Tianyuandong. At 2 standard deviations the 2 dates, BA-03224 and BA-04248, are not statistically different from each other.

To establish the age of the Tianyuan 1 human skeleton, a sample of bone was removed from the proximoanterior diaphysis of the left femur, 34.0 mm long and 14.6 mm wide (see Figure 3-1). It produced an AMS date of 34,430 ± 510 ^{14}C BP, which converts to a calendar age of ~40,000 cal BP. This date places it securely in the middle of the range of dates for Layer III, likewise in the middle of MIS 3. This direct radiocarbon date therefore establishes the age of the Tianyuan 1 human partial skeleton, despite major portions of it (including the femur) having been disturbed by the owners of the Tianyuan Tree Farm and further supports the association of the disturbed human remains with those found in situ in Layer III.

The direct date on the Tianyuan 1 femur makes the human skeleton among the oldest MIS 3 anatomically modern human remains. It is approximately the same geological age as the early modern human remains from Niah Cave, Sarawak (Borneo) [35,690 ± 280 and 35,000 ± 400 ^{14}C BP by association (40,625 ± 782 and 40,033 ± 873 cal BP) (Barker et al., 2007)], Peştera cu Oase, Romania [34,950 +990 –890 ^{14}C BP direct date (39,872 +1152 –1077 cal BP) (Trinkaus et al., 2003)], Nazlet Khater, Egypt [37,570 +350 –310 ^{14}C BP by association (42,161 +355 –342 cal BP) and 38,000 ± 6,000 cal BP direct by ESR (Vermeersch, 2002; Crevecoeur, 2008)], and Hofmeyr, South Africa [36,200 ± 3,300 cal BP direct by OSL and U-series (Grine et al., 2007)] [as with the Tianyuan dates, all radiocarbon calibrations are done using CalPal v. 1.5 (2007) for consistency]. Older diagnostically anatomically modern humans are east African from MIS 6 to 4 and occurred briefly in extreme southwest Asia in MIS 5 (Trinkaus, 2005a; see Chapter 4). Subsequent specimens of early modern humans confidently dated to the dozen millennia after Tianyuan 1 derive from Yamashita-cho, Zhoukoudian–Upper Cave, Pinza Abu, and Moh Khiew in eastern Asia; Fa Hein and Batadomba lena in south Asia; and Brassempouy, Cioclovina, La Crouzade, Mladeč, Muierii, La Quina Aval, Les Rois, and others in Europe (Trinkaus, 2005a; see Chapter 4).

Cave Contents

The contents of the cave's sediments consist almost entirely of bones and teeth, most of the faunal remains being heavily fragmented. There are no artifacts. The bones and teeth are well preserved structurally, not heavily fossilized, and variably weathered (see Appendix 1).

Faunal Remains

The Late Pleistocene MIS 3 remains from Layer III are heavily fragmented, consisting principally of bone fragments and isolated teeth. The faunal specimens are dominated by artiodactyls, primarily deer (*Moschus moschiferus, Cervus elaphus,* and especially *C. nippon*). Analysis of the age-at-death profile of the *C. nippon* remains, based principally on tooth eruption and attrition, shows that young and middle-aged adults dominate the assemblage (Li, 2007). A prime-age mortality profile for the deer remains might reflect human selective predation, but it is also common in the prey age-at-death profiles for large carnivores that primarily ambush hunt (Stiner, 1994). There is no carnivore damage to the Layer III faunal remains (Appendix 1), and the remains of carnivores are extremely rare, consisting of one tooth of *Canis lupus,* remains of *Vulpes vulpes,* and fragments tentatively attributed to *Martes, Mustela* and *Crocuta.* This dearth of carnivores in the assemblage makes it unlikely that the site served as a carnivore den (at least with respect to larger denning carnivores) (Klein and Cruz-Uribe, 1984).

At the same time, there is clear evidence of human processing of the deer remains in the form of multiple percussion marks on their surfaces and impact marks on their edges, indicating that the Layer III assemblage is humanly derived. (See Appendix 1 for details and rationale.) Despite this damage pattern on the faunal remains, there is a complete absence of humanly modified materials in the site. There are no lithic implements. There is one piece of hematite and one piece of charcoal, both from square G8 near the cave entrance, but it is unclear what their origins might be. The site is unusual in presenting faunal damage patterns characteristic of human bones processing, plus a human skeleton, but an absence of artifacts.

Human Remains

The ultimate origin of the human remains in the cave is uncertain (Appendix 1). Based on the nonduplication of skeletal and dental elements (Chapter 3), the bilateral symmetry of the few elements preserved for both sides (despite the upper limb functional asymmetry; Chapter 9), and the anatomical compatibility of all the elements, there is little reason to doubt that all of the human remains derive from one individual. Based on the pieces found in situ and the information provided by the initial local discoverers of the site, all the human bones came from one small area in the deep recess of the cave.

The human bones are relatively solid, but they lack the hardness or density of the associated animal remains. This contrast in material preservation is frequently encountered among Late Pleistocene human remains. It usually represents the accumulation of cleaned (butchered) animal remains, whereas the human remains retained their soft tissue, which promoted the bacterial decay of the skeletal tissue (Jans et al., 2004; Nielsen-Marsh et al., 2007). The pattern of skeletal preservation therefore suggests that the human bones were largely covered in flesh, as in an intact cadaver, rather than separated and cleaned, through either human processing or carnivore scavenging. In any case, there are no carnivore tooth marks or cut marks on the human bones (Appendix 1).

It is therefore highly probable that the human bones and teeth represent one individual, whose intact body came to rest in or on the sediments of the back of the cave. Since the original in situ juxtaposition of the bones is not known, it is not known whether they were in at least partial anatomical connection or merely a jumble of bones. As detailed in Appendix 1, it remains unclear whether there was an intentional primary burial of Tianyuan 1 in a shallow grave, the body was placed on the surface of the cave, or the individual died there alone.

Curiously, even though intentional human burials are well known among western Eurasian Middle Paleolithic late archaic humans (Vandermeersch et al., 2008) and early modern humans (McCown and Keith, 1939; Vandermeersch, 1981; Tillier, 1999) and among Mid Upper Paleolithic (~32,500—~24,000 cal BP) modern humans across Eurasia (Zilhão and Trinkaus, 2002; Matsumura and Pookajorn, 2005; Zilhão, 2005; Einwögerer et al., 2006), they are rare among modern humans in the time period of Tianyuan 1. The only definite burials from this period of modern human emergence across the Old World are those of Nazlet Khater 1 and 2 (Vermeersch, 2002). A probable infant burial from Kostenki XII (Kostenki 7) may date to this time period (Sinitsyn, 2004), and it has been suggested that the Mladeč human remains were intentionally deposited within the Mladeč cave system (Svoboda, 2000). If the Willandra Lakes (Lake Mungo) 1 and 3 skeletons date to this time period (see Habgood and Franklin, 2008), they would add to this meager list of burials. At the same time, isolated human remains occur within archeological sites throughout the Late Pleistocene, with or without intentional burials in the same sites, and human remains for this important time period are scarce, making it difficult to assess the nature of human burial practices more generally for the time period between the last of the late archaic humans and the Mid Upper Paleolithic modern humans.

Summary

From these considerations, it is apparent that the Tianyuan Cave, near the well-known sites of Zhoukoudian, was a small cavity that was formed within limestone, well above the valley floor, and then filled with sediment (including bones and teeth) initially during the middle of the last glacial (middle of MIS 3), and then subsequently during the Holocene. In the latter phase, it appears to have served principally as a porcupine den, although various carnivores may have occupied it as well. During the earlier MIS 3 period, it accumulated a rich but highly fragmented assemblage of ungulate remains, plus micromammals, and a major portion of a human skeleton. It appears that humans accumulated the Layer III faunal remains in the cave and may have intentionally laid to rest the human individual there.

Chapter 3

Preservation of the Tianyuan 1 Human Fossil Remains

THE PRESERVED remains of the partly disturbed Tianyuan 1 skeleton retain portions of nearly all aspects of the skeleton except the cranium. It has 8 teeth, a mandible, 3 axial elements, 15 upper limb bones, and 13 lower limb bones (Table 3-1; Figures 3-1 and 3-2). Of this total of 32 bones, 4 are midsagittal; 14 are right; 11 are left; and 3 are indeterminate as to side. Of the 8 teeth, the left maxillary molar is isolated; the right mandibular right I_2 to M_2 are securely in the mandibular corpus; and the right M_3 fits into the right mandible but was found separately.

These remains are inventoried here, and their preservation is described. Photographs of the specimens are in the chapters discussing their morphological and paleobiological aspects. For the few manual and pedal elements for which digital allocation is uncertain, the criteria for their assignment are provided. The details differ slightly from the list provided in Shang et al. (2007), since a couple of elements or portions of elements were identified among the isolated bone fragments subsequent to that publication. The Institute of Vertebrate Paleontology and Paleoanthropology (IVPP) catalog numbers (between PA1281 to PA1319) are provided for each element.

Dentition

The Tianyuan 1 teeth all experienced heavy occlusal attrition, but the identification of the right mandibular teeth is secure since all of them are in or fit into the mandibular corpus. The left maxillary molar exhibits mesial and distal interproximal facets, and therefore it is a first or second molar.

The right I_2 has an enamel chip gone from the mesiobuccal corner, but it is otherwise intact (I_2 to M_2: PA1281). The right C_1 has a crack around the labial secondary dentin to the distal interproximal facet. The right P_3 is intact with a mesiodistal midline crack. There is a similar mesiodistal midline crack on the otherwise

Table 3-1. Skeletal and dental elements preserved for Tianyuan 1

Left		*Right*
	Mandible symphysis, corpus, and right ramus (f)	
M^{1-2} (ff)		I_2—M_3 (i)
	Cervical vertebra 2 (axis) (i)	
	Manubrium (f), sternal segments 1 and 2 (f)	
Scapula (f)		Scapula (ff)
Humerus (f)		Humerus (ff)
		Ulna (ff)
Radius (f)		Radius (ff)
Hamate (i)		Lunate (i), Capitate (i),
Proximal hand phalanges 2 (i) and 3 (i)		Proximal hand phalanx 5 (i)
	Middle hand phalanx 2? (i)	
	Distal hand phalanx 2? (i)	
Femur (ff)		Femur (f)
Tibia (ff)		Tibia (f)
		Fibula (f)
Talus (i)		Calcaneus (f)
Metatarsals 1 (f) and 5 (i)		Metatarsals 3 (i) and 4 (i)
	Proximal pedal phalanx 2 (i)	
		Proximal pedal phalanx 5 (i)

Note: Following the *Catalogue of Fossil Hominids* (Oakley et al., 1975), (i) indicates an essentially complete element with little or no marginal damage; (f) indicates a largely preserved piece but lacking a portion or having substantial damage; and (ff) identifies a piece preserving a minority of the original element, with damage. Elements with a question mark are ones for which the digit is probable but not conclusive. Bones in the middle are either midsagittal elements or ones indeterminate as to side.

complete right P_4. The heavily worn right M_1 exhibits very fine linear cracks on buccal side of the occlusal crown. The right M_2 has similar very fine line cracks, but they are across the distal half of the crown.

The right M_3 (PA1283) was found separately and in situ, and it fits cleanly into the altered socket for the tooth on the right mandibular corpus. There are radial fine line cracks on the occlusal surface. The mesial edge of the mesial root was split postmortem, and the distal root was broken and reglued. As discussed in Chapters 5 and 12, this tooth was markedly supererupted and therefore the intact socket for it receives only the mesial root and the apical portion of the distal root. However, the fit is clean despite the postmortem alteration of the mesial root. The tooth is a right M_3; its wear pattern, both occlusal and interproximal, matches those of the other Tianyuan 1 teeth; and minimal adjustment for the mesial root damage indicates that it belongs to Tianyuan 1.

The other isolated tooth is a maxillary molar (PA1282). It is one of the elements found in situ. Its crown is missing most of the buccal rim. The distobuccal root was broken and reglued at a slight angle. It has mesial and distal inter-

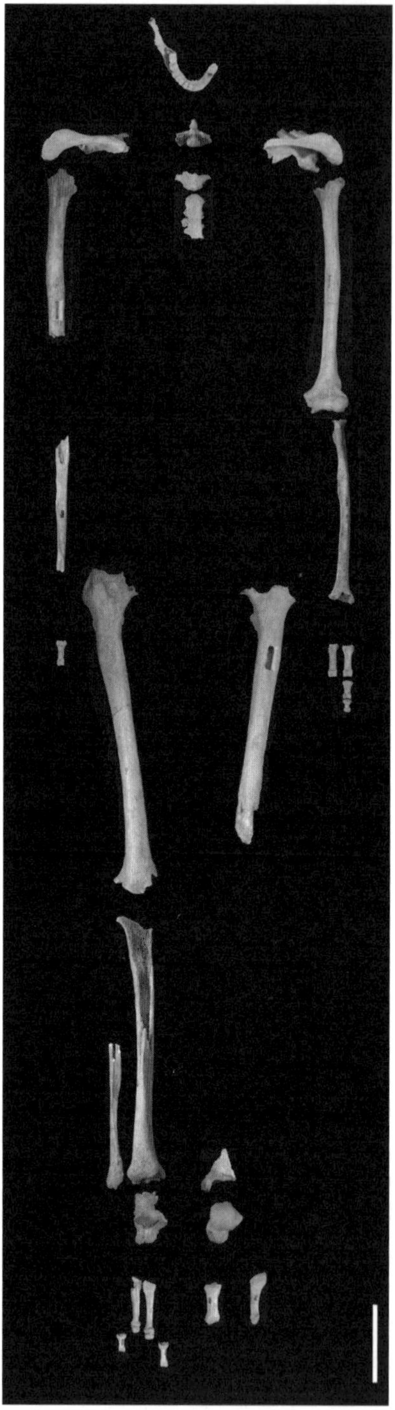

proximal facets, and it is a 3-rooted tooth. It is left based on the assumption of a distal curvature to the buccal root apices. Its occlusal attrition matches that of the right mandibular molars, in particular that of the M_2 (see Table 5-1). It is therefore left and either an M^1 or an M^2; based on attrition it is more likely to be an M^2 assuming that the antemortem tooth loss of the individual (Chapter 12) had little effect on the occlusal attrition. Since it provides little morphological or paleobiological data and no morphometrics, its attribution to M^1 or M^2 would add little to its paleontological contribution.

Mandible

The Tianyuan 1 mandible (PA1281) consists of the entire right corpus, the anterior half of the left corpus, and most of the right ramus. The piece was ex situ among the bones dug up by the Tianyuan Tree Farm members, but the discovery of the right M_3 in situ, which fits cleanly into the mandibular corpus once it is corrected for damage, establishes the Layer III provenance of the mandible.

The individual sustained significant antemortem tooth loss, especially from the right I_1 to the left P_3, with complete resorption

Figure 3-1. Composite view of the elements preserved for the Tianyuan 1 skeleton. Very small elements (maxillary molar and carpal bones) are not shown. The mandible is in superior view; the axial remains in ventral view; the scapulae in superior view; the long bones in anterior view; the manual phalanges in palmar view; and the pedal bones in dorsal view. The scale bar is 10 cm.

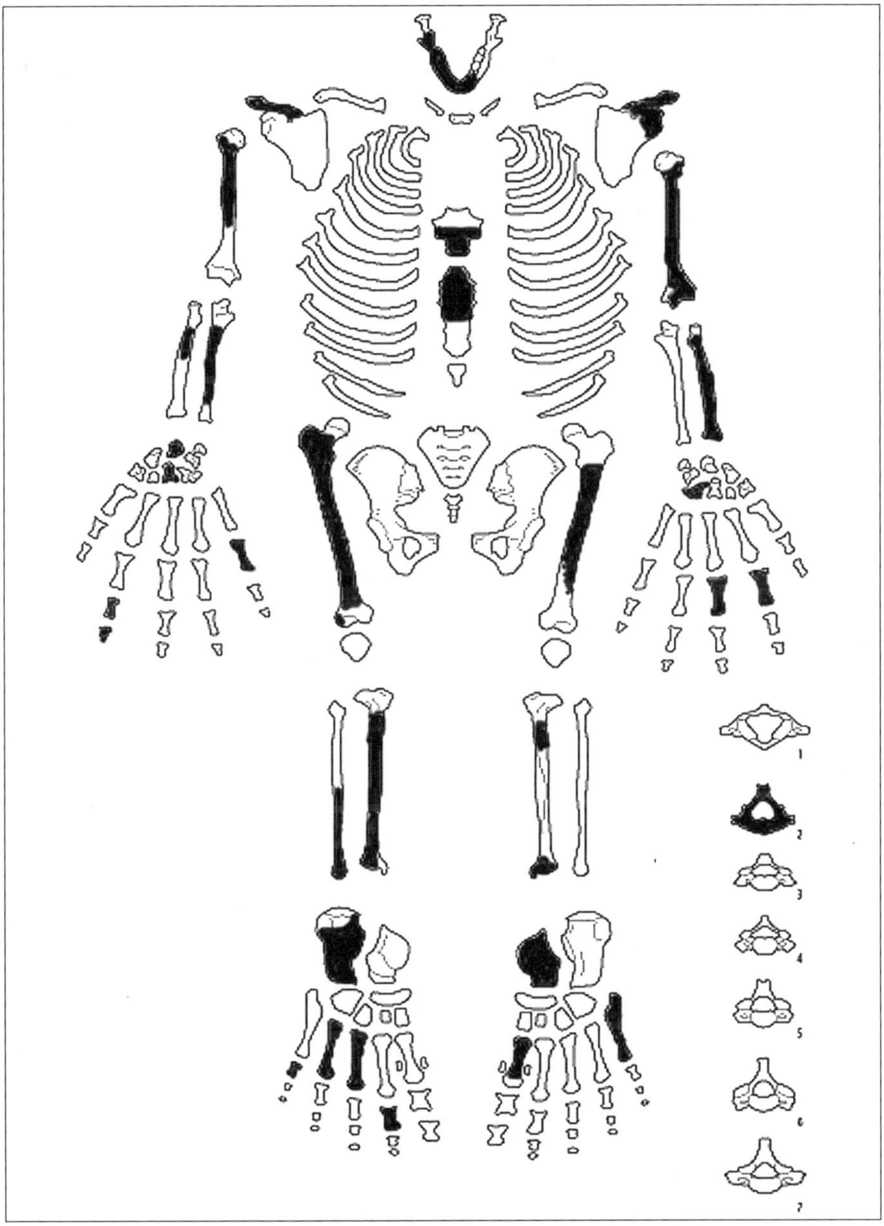

Figure 3-2. Human skeleton diagram with bones or portions thereof preserved for Tianyuan 1 shaded in. No cranial remains nor thoracic or lumbar vertebrae were found. The attributions of the middle and distal hand phalanges and the pedal proximal phalanx 2 to side are uncertain. Diagram modified from Courtaud (1996).

in that area of the alveolar bone (Chapter 12). There has also been marked dental occlusal and interproximal attrition, which has resulted in supereruption of the teeth and shortening of the dental arcade (Chapter 5). These age-related and pathological changes have altered the alveolar process of the mandible, especially at the symphysis, but they do not appear to have markedly altered the size or shape of the remainder of the mandible.

The right corpus is complete from the symphyseal midline to the ramus. Buccally there has been chipping of the alveolar margin for 3.5 mm down at the C_1 and 2.5 mm down at the P_3. Otherwise, the remainder of the buccal alveoli and all of the lingual alveoli are intact, with no damage. The M_3 alveolus is pathologically altered but undamaged. On the lingual corpus, there is a small crack from below the M_1 and 7.0 mm above the basal margin to the mylohyoid groove; it continues up the mylohyoid groove with little to no expansion of the bone.

The left corpus is complete superiorly to the distal P_4 socket, but it consists of the resorbed I_1 to P_3 alveoli and then the altered P_4 alveolus. The corpus below these alveoli is intact, and it continues 17 mm more distally than does the alveolar margin. The distal oblique break of the bone is modern, with a fresh bone break from just distal of P_4 socket to the mandibular base. There is no sign of the M_1 mesial socket in the bone exposed at the break, only trabeculae with a cortical bone rim.

The right ramus retains most of the medial and lateral surfaces but none of its superior extensions. The tip of the coronoid process is absent, and its anterior margin is sheared off. The anterior break begins vertically 9.5 mm above the alveolar plane. The condyle is completely absent, and the neck consists of eroded bone with some of the neck core bone. One cannot therefore confirm the spatial relationship between the notch crest and the condyle. The middle and inferior portions of the mandibular notch are preserved, from the anterior margin of the condylar neck to the posterior margin of the coronoid process, for a chord length of 19.5 mm through the middle of the notch.

Medially the mandibular foramen has a complete lingula with all of the original edges and no abrasion. There is a fine line crack in the inferior two-thirds of the mylohyoid groove that extends onto the lingual corpus. The gonial area is continuous and intact, with a crack on the posterior superior edge that extends anteroinferiorly ~16–17 mm internally and externally. There is a slight posterior displacement (~0.6 mm) of the inferior side of the crack at its posterosuperior end. The masseteric area has an anteroposterior crack across the superior portion of the rugosity, which minimally displaces surface bone laterally.

Axial Remains

Sternum

The Tianyuan sternum preserves a partial manubrium (PA1285) and the two proximal elements of the sternal body (the first two sternebrae; PA1286). The manubrium and the first sternebra were completely unfused, but the two cranial sternebrae are fully fused between the costal cartilage surfaces.

The manubrium has most of the caudal ventral surface, the full sternal synchondrosis, the caudal lateral edges, and a small portion of the caudal left dorsal surface. The facets for the first costal cartilages are absent, although the bone approaches their caudal extents. All of the cranial portions for the clavicles and the jugular notch are absent. The surface for the caudal synchondrosis with the first sternebra is complete, along with the cranial demifacets for the second costal cartilages. Maximum preserved breadth: 48.8 mm; maximum preserved height: 25.5 mm.

The sternebrae are present from the proximal synchondrosis to the mid-distal portion of the second sternebra. They are complete on the left side, including the third costal cartilage facets, but the right side is largely absent. Only cranial and caudal portions of the third costal cartilage facets remain of the right side. The caudal break is irregular, but the bone extends farther ventrally than dorsally. Maximum preserved height: 59.5 mm; maximum preserved breadth, at the third costal cartilage facets: 30.9 mm.

Second Cervical Vertebra (Axis)

There is only minimal damage to this otherwise complete bone (PA1284). There is abrasion on the lateral edges of the transverse processes, minimal crushing to the caudal tip of the left end of the spinous process, and modest abrasion to the lateral third of the right caudal facet and the ventral right facet base including the edge of the facet ventrally. Maximum transverse diameter: 51.5 mm; maximum dorsoventral diameter: 52.4 mm. None of the other vertebrae were recovered.

Upper Limb Remains

Scapulae

The right scapula (PA1288) is the less complete of the two. The acromion is complete with a small amount of abrasion to the mid dorsolateral edge and the dorsal tip. The spine is preserved lateral of the glenoid fossa to the medial narrowing of the dorsal edge, clearly lateral of the vertebral margin. There is minimal supraspinatus surface present and up to ~32 mm of the infraspinatus surface preserved, with the associated subscapularis surface. The glenoid fossa is absent, with preservation only of the beginning of the curve below the spine for the glenoid fossa. Maximum preserved mediolateral dimension (orientation approximate, given preservation): 121.5 mm.

The left scapula (PA1287) is more complete and retains the morphologically important glenoid fossa. The acromion is complete with minimal abrasion to the dorsolateral edge. The spine is complete from its lateral extent to the medial portion of the middle swelling of its dorsal edge. The coracoid is present with abrasion to the ventral edge of the tip. The glenoid fossa is complete with slight abrasion to the glenoid labrum attachment just dorsal of the triceps brachii tuberosity. The axillary border retains ~24 mm of the triceps brachii tuberosity and ~36 mm of the ventral axillary ridge or bar. Maximum mediolateral dimension on the spine: 105.0 mm.

Humeri

The right humerus (PA1290) retains the diaphysis from the base of the proximal epiphysis to the mid-distal shaft. Most of the lesser tubercle is present, but none of the greater tubercle remains. There is a midshaft oblique break with a clean rejoin; there is no separation or bone loss. The distal break is close to the distal minimum circumference. A sample for bone chemistry, 26.1 mm long and 12.0 mm wide, was removed between the midshaft and the mid-distal break on the medial portion of the anterior surface (see Appendix 2). Maximum preserved length: 216.0 mm.

The more complete left humerus (PA1289) consists of the diaphysis and the distal epiphysis from the distal margin of the proximal epiphysis to the distal articulations. Proximally there is the flare for the tubercles only, with just a small portion of the greater tubercle present medioposteriorly. The diaphysis had a midshaft break with a clean join of the sections, but there was the loss of a chip of bone posterolaterally, 15.7 mm long and 12.3 mm wide. Distally the subchondral bone is intact with abrasion to the medial edge of the trochlea. Only the base of the medial epicondyle remains. The lateral epicondyle is absent, and there was loss of bone for 61 mm proximally along the supracondylar crest. Most of the lateral pillar is absent, but the medial one is present. There are sufficient landmarks present on this bone for an accurate estimation of its original standardized lengths (Chapter 9). Maximum preserved length: 305.0 mm.

Right Ulna

The bone preserves a damaged diaphysis (PA1291). Proximally it begins just distal of the coronoid process dorsally, with an oblique break proximoposterior to distoanterior. Distally there is an oblique break, with bone absent especially anteriorly, that is in the middle of the pronator quadratus tuberosity. There is a chip of bone 28 mm long missing from the interosseus crest near the maximum extent of the crest. Maximum preserved length: 178.0 mm.

Radii

The right bone preserves a proximal diaphyseal section, with ~22.5 mm of the interosseus crest and 72.7 mm of the anterolateral surface. The proximal break, which is proximolateral to distomedial, is old, but the distal break is recent. The distal break is close to, if not at, the maximum extent of the interosseus crest. Maximum preserved length: 72.7 mm.

The more complete left radius (PA1292) has a complete diaphysis with portions of the epiphyses but none of the articular subchondral bone. Proximally it retains all of the neck but none of the head. Distally it extends to the proximal end of the extensor tendon sulci, especially laterally. It is sufficient to provide standardized length estimates (Chapter 9). Maximum preserved length: 239.0.

Hand Remains

Right Lunate

This is a complete bone with minor damage to the distodorsal corner (PA1294). Maximum depth: 18.3 mm.

Right Capitate

The bone is complete with damage to the radiopalmar base between the metacarpal and scaphoid surfaces (PA1295). Maximum length: 27.5 mm.

Left Hamate

A complete bone with a maximum length of 23.3 mm (PA1293).

Left Proximal Manual Phalanx 2

A complete bone (PA1298) found in situ. It is assigned to the 2nd left ray based on the differential development of its dorsal interosseus tendon insertions on the base. Maximum length: 41.7 mm.

Left Proximal Manual Phalanx 3

The complete bone, with loss of the proximoulnar corner (PA1299), found in situ. Assignment to ray and side is based on its radial base tubercle and length relative to the 2nd proximal phalanx. Maximum length: 44.2 mm.

Right Proximal Manual Phalanx 5

A complete bone with a postmortem depression on the right diaphysis near the base (PA1296). It is assigned to the right side based on the larger right tubercle on the base, for the abductor digiti minimi muscle, and the right deviation of the head trochlea. Maximum length: 34.5 mm.

A Middle Manual Phalanx 2?—Side Indeterminate

A complete bone with osteoarthritic degenerations on the trochlea (PA1297), found in situ. Visual comparisons of the base morphology and breadth to those of the proximal phalangeal heads make it likely that it derives from ray 2 as opposed to ray 3. A metric comparison of middle phalanx proximal maximum breadth to proximal phalanx distal breadth for a recent European sample of known ray for the phalanges (N = 44) (Musgrave, 1970) indicates that the Tianyuan 1 middle phalanx has a relatively wide middle phalangeal base compared to its distal proximal phalanges 2 and 3 (14.3 mm, versus predicted values of 13.8 and 13.5 mm respectively) and that the predicted value for it as a second ray phalanx is closer to the actual measurement. Based on these considerations, it is tentatively assigned to the second ray, but of indeterminate side. It is possible that its larger than predicted base breadth reflects bilateral asymmetry and that it is right. Maximum length: 24.9 mm.

A Distal Manual Phalanx 2?—Side Indeterminate

A complete bone (PA1300) found in situ. There are osteoarthritic degenerations on the base which match those on the head of the middle phalanx (Chapters 9 and 12), indicating that they derive from the same digit. On the basis of the assignment of the middle phalanx tentatively to the second ray, this distal phalanx is similarly provisionally assigned to the second ray. Maximum length: 17.2 mm.

Lower Limb Remains

Femora

The majority of the preserved right femur (PA1302) consists of the neck and trochanteric region plus the diaphysis. The diaphysis is complete from the distal lesser trochanter to the supracondylar area, with pathological alterations of the mid-distal diaphysis (Chapters 10 and 12). There is an oblique break just proximal of midshaft, but the join is clean with minimal bone loss. Distally the bone extends posterolaterally close to the supracondylar area, posteromedially to the adductor tubercle, and anterolaterally adjacent to the patellar surface. Proximally, the bone retains the base of the lesser trochanter but none of the trochanter surface. The full circumference of the neck is present close to the head, with the beginning of the flare for the head, but none of the internal trabeculae nor any of the head subchondral bone present. Most of the greater trochanter is present, but there is a proximal and posterior break that resulted in the posteromedial side being pushed in slightly, especially just posterior of the digital fossa. The presence of the proximal neck landmark and several distal metaphyseal region landmarks permits the estimation of standardized lengths for this bone (Chapter 10). Maximum preserved length: 417.0 mm.

Preserved separately, and without contact with the diaphysis, is the lateral portion of the lateral condyle. It retains the distal surface and lateral margin of the condylar surface from the anterior patellar surface to the posterior condylar surface, but none of the medial (intercondylar fossa) margin remains. Up to ~17 mm of the lateral surface with the distal corner of the lateral epicondyle is present. The posterior surface retains the full convexity of the condylar surface, but the anterior edge is very close to, but possibly trivially short of, the full anterior extent of the lateral patellar margin. Maximum preserved depth (anteroposterior): 74.6 mm.

The left femur (PA1301) is a diaphyseal section from the distal lesser trochanter to approximately midshaft. Proximally it retains the anterior, medial and lateral surfaces to the edges of the trochanters, plus the medial base of the neck. The anterior diaphysis extends to about midshaft, but the posterior margin with the linea aspera and the associated pathological crest extends farther distally. In the mid-distal diaphysis the lateral side is mostly absent, but the medial side is present with an oblique proximoanterior to distoposterior break. A radiocarbon dating sample was removed from the proximoanterior diaphysis, 34.0 mm long and 14.6 mm wide (Chapter 2). Maximum preserved length: 334.0 mm.

Tibiae

The right tibia (PA1305) retains much of the diaphysis and of the distal epiphysis. The distal half of the diaphysis is complete. Proximoposteriorly, the diaphysis is largely complete to the edge of the proximal capsular attachment posterolaterally. Anteriorly, only the distal half is present, and a lateral expansion crack may have moved the anterior margin slightly anteriorly near midshaft. Laterally, only the posterior half of the diaphysis is present on the proximal half of the diaphysis, and medially the posterior two-thirds is present up to the soleal line level. The distal trochlear surface is present, but the medial malleolus and the posteromedial corner of the epiphysis are absent. Maximum preserved length: 359.0 mm

The left tibia (PA1303) retains two pieces. There is a section of proximoposterior diaphysis with the middle of the soleal line and up to the posterior flare for the epiphysis. Its maximum preserved length is 89.6 mm. There is an eroded distal epiphysis, with the trochlear surface lacking the lateral and posterior margins, the anterior surface with the squatting facet area, and the lateral and distal surfaces of the medial malleolus. Maximum proximodistal dimension: 50.0 mm; maximum preserved breadth: 46.5 mm; maximum preserved depth: 35.0 mm.

Right Fibula

The bone (PA1306) retains the complete distal epiphysis and the distal approximately half of the diaphysis. The diaphysis has longitudinal cracks anteriorly and posteriorly as it approaches midshaft, and it lacks the anterior corner of the midshaft bone due to an oblique break. Maximum preserved length: 189.5 mm.

Foot Bones

Left Talus

It is a complete bone (PA1308) with abrasion to the posterolateral trochlear corner. Maximum length to the lateral tubercle: 60.9 mm.

Right Calcaneus

The bone (PA1307) retains the complete talar and cuboid facets, the sustentaculum tali with abrasion posteromedially, the dorsal and medial body walls, and the medial two-thirds of the plantar body. The tuberosity, the lateral body wall, and most of the internal body trabeculae are absent. Maximum preserved length: 71.0 mm.

Left Metatarsal 1

The bone (PA1319) preserves the proximal epiphysis and the diaphysis, lacking the head. Maximum preserved length: 53.9 mm.

Right Metatarsal 3

It is a complete bone with minor damage to the proximoplantar corner (PA1311). Maximum preserved length: 78.0 mm.

Right Metatarsal 4

The bone retains the head, the diaphysis with longitudinal cracks, and the base with damage to the medial side and loss of the metatarsal 3 facets (PA1312). It was found in situ. Maximum preserved length: 74.9 mm.

Left Metatarsal 5

A complete bone (PA1310). Maximum length: 69.8 mm.

Proximal Pedal Phalanx 2—Side Indeterminate

Complete bone (PA1313), found in situ. In order to assess from which ray this phalanx derives, the metatarsal 2 articular length was estimated to be 76.6 mm from its metatarsal 3 and 4 lengths using a linear multiple regression from a pooled recent Euroamerican and Amerindian sample (N = 75) (Trinkaus and Hilton, 1996). The Tianyuan 1 proximal phalanx articular lengths were predicted from its metatarsal articular lengths as 27.9, 25.8 and 24.7 mm respectively for the second to fourth rays. Given an articular length of 29.2 mm for the Tianyuan 1 phalanx, it is most likely to derive from the second digit. The head trochlea deviates 2° to the right, which may indicate that it is a right phalanx, but the deviation is insufficient to be diagnostic. Maximum length: 32.5 mm.

Right Proximal Pedal Phalanx 5

A complete bone (PA1314), found in situ. The bone has a strongly developed proximal right tubercle, for the abductor digiti minimi, and its distal trochlea deviates 11° to the right; both of these indicate that it is a right bone. Maximum length: 25.0 mm.

Chapter 4

Comparative Materials and Methods

THIS BOOK is concerned with the comparative morphology and paleobiology of the early modern human partial skeleton from Tianyuan Cave. However, all paleontological description is by its nature comparative. Therefore, it is necessary to provide a human paleontological framework to assess the continuous and discrete morphological pattern of this individual and to place its paleobiological aspects in context. It is also desirable to employ standardized criteria for this assessment to the extent possible. This chapter defines these comparative samples and, in general terms, the methods employed to quantify the morphological and pathological aspects of Tianyuan 1.

Materials

The comparative samples consist principally of those that bracket the Tianyuan skeleton in time and space across the Old World. Ideally, these samples would consist of late archaic and early modern human remains from the eastern Old World, given the chronological position of Tianyuan 1 as one of the earliest known modern humans from eastern Asia. However, securely dated human remains from this time period east of ~80° E longitude are rare, fragmentary, and limited in their anatomical completeness. Moreover, since the earliest modern humans known are currently Middle Paleolithic marine isotope stages (MIS) 6–5 remains from eastern Africa and extreme southwestern Asia (Valladas et al., 1988; White et al., 2003; McDougall et al., 2005), and since it is likely that the pan–Old World emergence of the modern human morphological pattern involved population dispersal and/or gene flow from those earliest modern humans (Wu, 2004, 2006; Trinkaus, 2005a; Bräuer, 2008; Cartmill and Smith, 2009), it is relevant to include comparative samples from western Eurasia and at least neighboring areas of Africa. The comparative samples therefore include geographical samples of late archaic humans plus chronological and regional samples of early modern humans. The samples are described below. As appropriate to provide a broader context for aspects of the Tianyuan 1 remains, data from recent (late Holocene) human samples are included.

The data employed are derived primarily from personal analysis of the original remains, the primary publications of the fossil remains, and/or personal communications from colleagues who have derived data from the original remains. In some cases, additional data for discrete traits or dental macroscopic wear stages have been derived from high-quality casts of the specimens or high resolution photographs in which the configuration is unambiguous. This has been necessary particularly for the now lost Zhoukoudian–Upper Cave remains (Weidenreich, 1939a, 1941; Wu, 1961) and the Předmostí fossils (Matiegka, 1934, 1938; Veleminská and Brůžek, 2008).

Eastern Asian Late Archaic Humans

Late Pleistocene (MIS 5–3) late archaic humans from the eastern Old World are rare and consist solely of cranial and dental remains (Wu and Poirier, 1995; Shpakova, 2005). (For location of human fossil sites, see Figure 4-1.) The most complete is the calotte and superior facial skeleton from Maba (Mapa) (Woo, 1959). There is also a series of cranial fragments, isolated teeth, and an immature maxilla from Xujiayo (Hsuchiayao) (Chia et al., 1979; Wei, 1979); a partial maxilla and a lower premolar from Changyang (Chia, 1957); isolated teeth and partial parietal bone from Dingcun (Woo, 1958; Wu and Poirier, 1995); a tooth from Zhoukoudian Locality 4 (Gu, 1978); and isolated teeth from several other sites (Wu and Poirier, 1995), all in China. In addition, an archaic human calotte of probable Late Pleistocene age is known from Salkhit, Mongolia (Coppens et al., 2008), and there are isolated teeth from the Altai Mousterian caves of Denisova and Okladnikov in southern Siberia (Shpakova, 2005). Of these remains, only the Changyang P_4, the Dingcun M_2 and the Okladnikov P_3, M_1, and M_3 match anatomical regions preserved on the Tianyuan 1 partial skeleton. Moreover, given the extensive occlusal attrition of the Tianyuan 1 teeth (Chapter 5), the only comparisons possible are dental crown metrics.

Eastern Asian Early Modern Humans

There is a relatively long list of early modern human remains from eastern Asia, as well as a few from south and southeast Asia (including adjacent islands), but many of them have uncertain associations with otherwise well-dated Late Pleistocene deposits. Because of this uncertainty, the securely dated comparative remains for early modern humans in the eastern Old World are limited. The time range for the comparative sample is therefore relatively broad, including remains

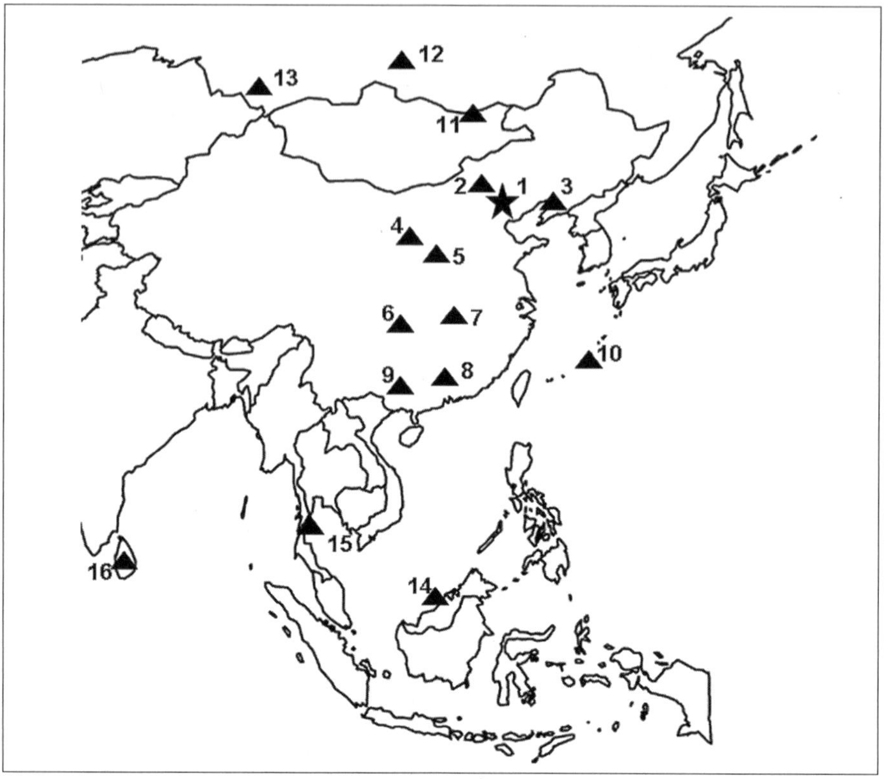

Figure 4-1. Map of eastern Asia showing the locations of Tianyuandong, relevant Late Pleistocene human fossil sites, and the late Middle Pleistocene Jinniushan site. *1:* Zhoukoudian localities, including Tianyuandong and Zhoukoudian–Upper Cave; *2:* Xujiayao; *3:* Jinniushan; *4:* Salawusu; *5:* Dingcun; *6:* Ziyang; *7:* Changyang; *8:* Mapa; *9:* Liujiang; *10:* Okinawan sites of Yamashita-cho, Pinza-Abu, and Minatogawa; *11:* Salkhit, Mongolia; *12:* Malta, Siberia; *13:* Altai sites of Denisova and Okladnikov caves; *14:* Niah Cave, Sarawak; *15:* Moh Khiew, Thailand; *16:* Sri Lankan sites of Fa Hien, and Batadomba lena.

from the same age as Tianyuan 1 (e.g., Niah 1) to remains ~18 ka ^{14}C BP (~21.5 ka cal BP) (e.g., Minatogawa).

The partial skeleton from Niah Cave, Sarawak, is securely dated, directly and by stratigraphic association, to ~40 ka cal BP (Barker et al., 2007), but only the cranium has been described (Brothwell, 1960). The next oldest are the immature femur and tibia from Yamashita-cho, Okinawa, at ~32 ka ^{14}C BP (~36 ka cal BP) (Suzuki, 1983; Trinkaus and Ruff, 1996), followed by the isolated remains from Pinza-Abu, Okinawa, and the partial skeleton from Moh Khiew, Thailand, both ~26 ka ^{14}C BP (~31 ka cal BP) (Sakura, 1981, 1985; Matsumura and Pookajorn, 2005).

Probably at approximately the same age as these samples are the now lost

Zhoukoudian–Upper Cave (ZKD-UC) human remains (Weidenreich, 1939a, 1941; Wu, 1961). The age of those human remains is uncertain, given a range of radiometric dates from ~10 ka BP to ~34 ka BP for Layer 4, which produced the three human crania and most of the other human remains, and a range of dates from ~18 ka BP to ~34 ka BP for the Lower Recess (Chen et al., 1992; Norton and Gao, 2008). Many scholars assume that the human remains from Layer 4 date to ~24–29 ka ^{14}C BP (~29–33.5 ka cal BP) (see Cartmill and Smith, 2009), and that is probably a reasonable estimate. Further dating of faunal remains that were securely associated with the lost human remains is needed (see Norton and Gao, 2008). It will be assumed that the ZKD-UC humans are close in age to, but slightly younger than, Tianyuan 1.

The ZKD-UC human crania have been the subject of a multitude of analyses, many of them craniometric and mostly focusing on ZKD-UC 101 (e.g., Weidenreich, 1939a; Wu, 1961; Kamminga, 1992; Bräuer and Mímisson, 2004). The portions of the skeleton of relevance here (mandible, dentition, and postcrania) have received far less attention. The data available for the Zhoukoudian–Upper Cave remains derives from the summary data provided by Weidenreich plus the original series of plaster casts of the crania, mandibles, contained teeth, and a few postcranial elements preserved in the Institute of Vertebrate Paleontology and Paleoanthropology; metric and discrete data from them are used, but it remains uncertain to what extent they have been modified from the original specimens. Weidenreich (1941) provided only limited and generally summary data on the ZKD-UC postcranial remains, but additional data can be gleaned from the few photographs and drawings of the ZKD-UC femora published in his monograph on the Zhoukoudian Locality 1 postcrania.

More recent than these remains are two samples from western and eastern margins of this region. The first is the child burials from Malta, Siberia (Gerasimov, 1935; Shpakova, 2005), radiocarbon dated to ~20 ka ^{14}C BP (~24 ka cal BP) (Kuzmin and Orlova, 1998; Richards et al., 2001). They provide only one dental dimension for comparison with Tianyuan 1. The second, and much more complete, sample is the ~18 ka ^{14}C BP (~21.5 ka cal BP) Minatogawa sample from Okinawa of four partial skeletons and isolated remains (Suzuki, 1982; Hanihara and Ueda, 1982; Baba and Endo, 1982). Both of these samples are included in the eastern Asian early modern human sample.

From geographically farther away in Asia are the fragmentary remains from Fa Hien and Batadomba lena, Sri Lanka, ~30 ka ^{14}C BP (~34 ka cal BP) and ~28.5 ka ^{14}C BP (~33 ka cal BP) respectively (Kennedy and Deraniyagala, 1989; Kennedy and Zahorsky, 1995). Late Pleistocene human remains from Australia probably all postdate ~20 ka BP, even though some were buried into substantially older geological levels (see reviews in Trinkaus, 2005a; Zilhão, 2005); in any case, relevant comparative data for most of them are largely unavailable.

There are additional early modern human remains from three sites or localities that deserve mention, Liujiang, Ziyang, and Salawusu. Each one presents difficulties. The southern Chinese Liujiang partial skeleton may predate 60 ka BP, based on extensive dating of the site after the discovery of the skeleton, but questions remain as to its original stratigraphic context (Shen et al., 2002). The Ziyang 1 cranium may be associated with fauna dated to 32–44 ka ^{14}C BP (36–47.5 ka cal BP) (Pei and Woo, 1957; Wu and Poirier, 1995). However, its original stratigraphic context is poorly documented, and a multivariate assessment of its neurocranial proportions (Bräuer and Mímisson, 2004) places it among recent Chinese and separate from the Late Pleistocene specimens such as the ZKD-UC crania.

The association of the Salawusu (Ordos) human bones, which include cranial and postcranial remains, with well-dated but substantially older geological deposits is currently debated (Yuan et al., 1986; Shang et al., 2006). The only one of the Salawusu human remains that has been directly dated is a surface find, and it yielded a late Holocene date (Shang et al., 2006). However, that date may not relate to the other human remains from Salawusu, some are surface finds and others were found in geological levels that date to various phases of the Late Pleistocene. Many uncertainties remain regarding the provenance of the Salawusu human remains, and it is not feasible to use them currently as indicating Late Pleistocene human morphology in the region.

Except for the Salawusu sample, the human remains from these three sites preserve few of the same skeletal elements as Tianyuan 1. They are therefore not included in the comparative samples.

Western Eurasian Late Archaic Humans

The largest comparative sample of late archaic humans consists of fossils from western Eurasia, the Neandertals of Europe and western Asia. It is apparent that the Neandertals exhibit a series of uniquely derived characteristics relative to earlier (Middle and Early Pleistocene) archaic *Homo* and early modern humans (Trinkaus, 2006a), but it remains unclear to what extent these aspects separate them from Eurasian archaic *Homo* generally. A couple of their configurations, such as prominent nasal bridges and suprainiac fossae, are not present in the few available eastern Asian late archaic fossils that preserve the same regions (e.g., Maba and Xujiayo). However, Maba 1, at least in the anatomical regions preserved, exhibits the ovoid sagittal profile ("forme en bombe") of the Neandertals. Another purportedly Neandertal feature (Smith, 1978; Jidoi et al., 2000; Trinkaus, 2006a), the bridging of the mandibular foramen, is absent from earlier eastern Asian *Homo* [e.g., Chenjiawo (Lantian) 1 and Zhoukoudian Locality 1 (Weidenreich, 1936; Woo, 1964)] but occurs in the ZKD-UC 104 early modern human mandible (Chapter 6).

The Neandertals consist of an abundance of partial skeletons and isolated remains, and they are here limited to remains dated to MIS 3, 4, and late 5. The earlier MIS 5 samples from sites such as Krapina and Tabun are not included, except for a couple of cases in which later Neandertal remains are scarce. The sites that have yielded data for one or more anatomical regions include Amud, Arcy-Hyène, Arcy-Renne, Artenac, Aubesier, Banyoles, Biscegli, Bisitun, Bombarral, Caminero, Châteauneuf-sur-Charente, La Chapelle-aux-Saints, Combe Grenal, La Crouzade, Dederiyeh, Fate, Feldhofer, Fenera, La Ferrassie, Fond-de-Forêt, Fossellone, Genay, Guattari, Hortus, Kebara, Kiik-Koba, Lakonis, Lezetxiki, Macassargues, Maglie, Malarnaud, Merveilles, Monsempron, Montgaudier, Le Moustier, La Naulette, Neussing, Oliveira, La Quina, Palomas, Petit-Puymoyen, Peyrards, Portel, Pradelles, Regourdou, Rochers de Villeneuve, Saccopastore, Saint Brelade, Saint-Césaire, Shanidar, El Sidrón, Šipka, Spy, Stadel, Subalyuk, Švédův stůl, Taddeo, Taubach, Valdegoba, Vaufrey, Vergisson, Vindija, Zafarraya, and Zaskalnaya.

All of these Neandertals are Middle Paleolithic except for Arcy-Renne, Lakonis, Saint-Césaire, and Spy, which are Initial Upper Paleolithic. Given differences in diaphyseal cross-sectional shapes for at least the Saint-Césaire 1 femur (Trinkaus, Churchill, et al., 1999), it is highlighted in some of the comparisons.

Middle Paleolithic Modern Humans

The earliest modern human remains are now clearly those from the late Middle Pleistocene (MIS 6) eastern African sites of Omo-Kibish (Day, 1969; Day et al., 1991; Pearson, Royer, et al., 2008; Pearson, Fleagle, et al., 2008) and Herto (White et al., 2003). These earlier sites are followed in time by the larger samples of associated skeletons from the sites of Qafzeh and Skhul in extreme southwestern Asia (McCown and Keith, 1939; Vandermeersch, 1981; Tillier, 1999), and the various Middle Stone Age–Middle Paleolithic fragmentary remains from Aduma and Bouri in eastern Africa (Hailie-Selassie et al., 2004) and Haua Fteah in northeastern Africa (Tobias, 1967). For comparison with the Tianyuan 1 remains, it is principally the Qafzeh and Skhul samples that provide data. Some relevant mandibular and postcranial data are available for earlier Omo-Kibish 1, and mandibular and dental data are available for Haua Fteah 1 and 2. All of these remains are Middle Stone Age–Middle Paleolithic associated.

Middle Paleolithic African Late Archaic Humans

It should be noted that there are additional Middle Paleolithic–Middle Stone Age human remains from Africa that have been variously referred to as early modern humans. One sample of these "modern" humans consists of the Aterian (Middle Paleolithic) remains from the northwest African sites of Dar-es-Soltan,

El Harhoura, and Témara (Vallois and Roche, 1958; Ferembach, 1976, 1998; Ménard, 1998; Hublin, 1992; Bräuer, 2008). Although clearly not Neandertals, they exhibit a complex mix of archaic and modern (sensu Trinkaus, 2006a) cranial, dental, and especially mandibular features.

The other sample consists of Middle Stone Age remains from south and southeast Africa, from the sites of Blombos, Equus, Klasies River Mouth, Die Kelders, Mumba, Pinnacle Point, and Sea Harvest (Singer and Wymer, 1982; Grine and Klein, 1985, 1993; Bräuer and Mehlman, 1988; Churchill et al., 1996; Lam et al., 1996; Pearson and Grine, 1997; Grine, 1998; Marean et al., 2004). They similarly exhibit a long list of archaic cranial, mandibular, and postcranial characteristics and only a few apparently derived modern human features. They are distinctive primarily in having small teeth and hence small dental arcades; small teeth are not diagnostic of archaic versus modern human anatomy (see Trinkaus 2005a for a review), especially since the earliest definite modern human from southern Africa, the Hofmeyr 1 cranium, appears to have had rather large teeth (Grine et al., 2007).

These two samples, given their mix of archaic and modern morphological configurations and their geographical distance from eastern Asia, are only considered peripherally in the assessment of the Tianyuan 1 skeletal remains.

A number of later Middle Pleistocene human remains from Africa, including the northwest African Irhoud crania, mandible, and humerus; the east African Laetoli 18, Eliye Springs ES-11693, and Ileret ER-3884 crania; and the south African Florisbad 1 cranium, have been variously included within early modern versus late archaic humans (see Bräuer, 2008 and references therein). We follow Bräuer (2008) in keeping these specimens within late archaic humans, albeit probably representing the ancestral population to the Middle Stone Age African early modern humans, at least in eastern Africa. Given their temporal, geographical, and phylogenetic distance from Tianyuan 1, they are not considered further here.

Western Old World Early Upper Paleolithic Modern Humans

The most relevant sample for comparison with Tianyuan 1 is the small sample of approximately contemporaneous early modern human remains. The eastern Old World Niah Cave and Yamashita-cho fossils are mentioned above, even though neither one provides direct comparisons for Tianyuan 1. The situation is similar for the ~36 ka cal BP isolated cranium from Hofmeyr, South Africa (Grine et al., 2007). However, the ~38 ka cal BP Nazlet Khater 2 skeleton from Egypt provides an abundance of data, even though it evinces developmental and degenerative abnormalities (Crevecoeur and Trinkaus, 2004; Crevecoeur and Villotte, 2006; Crevecoeur, 2008). In particular, Nazlet Khater 2 has developmental abnormalities of the lower limb that resulted, minimally, in relatively short femora with high neck shaft angles; lower limb considerations of it employ those features that appear to be least affected by its abnormalities, but it should be kept in mind that its lower limb (especially femoral) size and proportions are anomalous. The ~35 ka ^{14}C BP (~40 ka cal BP) mandible from the Peştera cu Oase, Romania (Trinkaus et al., 2003) provides mandibular and molar data.

Slightly more recent from Europe are the Peştera Muierii remains from Romania (Soficaru et al., 2006), the Mladeč sample from the Czech Republic (Frayer et al., 2006; Trinkaus, Smith, et al., 2006), and isolated specimens from a series of western European (all French) sites (Abeille, Brassempouy, Castanet, La Crouzade, La Ferrassie, Fond-de-Gaume, Fontéchevade, Le Piage, La Quina Aval, Vachons, and Les Rois). The fragmentary and purportedly Aurignacian remains from Fontana Nuova and Fossellone are not included since their stratigraphic proveniences are uncertain (Mallegni and Segre-Naldini, 1992; Chilardi et al., 1996). All of the specimens included in this sample are >28 ^{14}C BP (>32.5 ka cal BP) and most of them are associated with Aurignacian assemblages. Hofmeyr 1, La Crouzade 6, Nazlet Khater 2, Oase 1, Muierii 1, and the Mladeč sample have direct dates on the human remains (Trinkaus et al., 2003; Wild et el., 2005; Soficaru et al., 2006; Grine et al., 2007; Crevecoeur, 2008; Henry-Gambier and Sacchi, 2008). However, except for some individual dental crown diameters, sample sizes for this group are extremely small.

Western Old World Mid Upper Paleolithic Modern Humans

To provide a larger comparative sample of Late Pleistocene Upper Paleolithic early modern humans, data are also included for western Old World (primarily European) Mid Upper Paleolithic (Gravettian sensu lato) remains. They date from ~19 ka ^{14}C BP (~24 ka cal BP) to ~28 ka ^{14}C BP (~32.5 ka cal BP). They include associated skeletons from the sites of Arene Candide, Barma Grande, Bausu da Ture, Brno, Caviglione, Cro-Magnon, La Crouzade, Dolní Věstonice, Grotte des Enfants (Fanciulli), Kostenki, Lagar Velho, Nahal-Ein-Gev, Ohalo, Paglicci, Pataud, Paviland, Pavlov, Předmostí, La Rochette, Sunghir, and Veneri (Parabita). Additional data (mostly dental dimensions) are included for isolated remains from these sites plus the sites of Blanchard, Caldeirão, Castanet, Labuttut, Miesslingtal, Silická Brezová, and Willendorf.

Early and Middle Pleistocene Archaic *Homo*

Even though they are more remotely relevant to the assessment of the Tianyuan 1 skeletal remains, occasional reference is made to more ancient (preterminal Middle Pleistocene) Early and Middle Pleistocene *Homo* fossils. This occurs particularly with respect to aspects of the postcrania for which they might provide some additional perspective.

The Middle Pleistocene fossil of particular relevance is the late Middle Pleistocene partial skeleton of an archaic human from Jinniushan, China (Lü, 1989, 1990; Rosenberg et al., 2006). In particular, Jinniushan 1 retains postcranial elements and indications of both size and proportions for which no other eastern Eurasian late archaic specimen is known. Of interest also, primarily for the mandible and some of the postcrania, are the Chenjiawo (Lantian) 1 mandible (Woo, 1963) and the Zhoukoudian Locality 1 remains (Weidenreich, 1936, 1941, 1943) Other mainland eastern Asian Early and Middle Pleistocene human remains, for example, from Gongwangling (Lantian), Hulu (Nanjing), Hexian, and Dali, are principally cranial and dental (Woo, 1965; Wu, 1981; Huang et al., 1982; Wu et al., 2002; Wu and Poirier, 1995) and therefore less directly relevant to morphological comparisons with the Tianyuan 1 remains.

In the western Old World, comparative reference is appropriate for postcranial remains and body proportions to these earlier *Homo* fossils. These include in particular the Middle Pleistocene Arago, Atapuerca-SH (or AT-SH) and Boxgrove remains (Sigmon, 1982; Arsuaga et al., 1999; Trinkaus, Stringer et al., 1999; Carretero et al., 2004), the probably Middle Pleistocene African remains from Berg Aukas and Broken Hill (Pycraft, 1928; Grine et al., 1995; Trinkaus, Ruff, and Conroy, 1999; Trinkaus, 2009), and the Early Pleistocene KNM-WT 15000 skeleton (Walker and Leakey, 1993).

Methods

The protocols for osteometrics, discrete traits, and cross-sectional geometry are provided here, since they apply across the remains. Methodological issues that relate to specific bones are discussed in the relevant chapters.

Comparisons consist principally of tables of means ± standard deviations for osteometrics and frequencies for discrete traits, box plots for individual measurements, and scatter plots for proportional comparisons. Although many of the comparisons involve bivariate plots, and despite small sample sizes and the difficulties inherent in the use of ratios (or indices), there are a number of comparisons of indices (especially when comparative data are only available as such) and angles. These data are presented in tables as means and standard deviations when sample sizes are ≥4 and as individual measurements for sample sizes <4. Assessments of degrees of difference across the comparative samples are done using nonparametric Kruskal-Wallis tests, since the sample sizes are generally too small to assess normality. All calculations are done with NCSS (Hintze, 2007).

In the scatter plots, Tianyuan 1 and the few relevant Early Upper Paleolithic specimens (from Mladeč, Nazlet Khater, and Oase) are individually labeled, as well as the Initial Upper Paleolithic Saint-Césaire 1 remains for a few functionally relevant postcranial aspects.

For two sets of measurements in which multivariate proportions are relevant, dental buccolingual diameters along the mandibular tooth row and talar osteometrics, the Tianyuan 1 and sufficiently complete and available comparative specimens were compared using principal components analysis (PCA). The PCA was initially done for the dental dimensions using the correlation matrix of the raw linear measurements for both data sets, combined with discriminate function analysis to determine the posterior probabilities of the specimens belonging to the reference samples using cross-validation. Since size is relevant to dental crown assessment, the dental data were not "size adjusted." For the talus, the data were "size adjusted" by dividing the individual measurements by the specimen's geometric mean, following Darroch and Mosimann (1985; see also Jungers et al., 1995). For each analysis, given variable preservation, different variable sets were analyzed to maximize either the numbers of variables or the sample sizes. The computations were done using SPSS.

Osteometrics

The overwhelming majority of the quantified data provided for the Tianyuan 1 skeletal and dental remains follows standard landmarks and measurements, the Martin system updated by Bräuer (1988). All such measurements are identified in the tables by: M-##. When additional measurements are provided, they are defined in the notes to the tables and references are provided to other sources if such exist.

Long Bone Lengths

None of the Tianyuan 1 long bones is sufficiently complete to provide standard length measurements, since each one lacks at least one epiphysis (Chapter 3). However, the left humerus and radius and the right femur and tibia preserve sufficient landmarks proximally and distally to permit the reliable estimation of their standard lengths. The Tianyuan 1 lengths were therefore estimated using least squares regressions based on samples of recent human long bones, regressing the desired length against the distance between the average longitudinal positions of the proximal and distal landmarks. The results and formulae are in the osteometric tables for the 4 bones (see Tables 9-2, 9-7, 10-1, and 10-4). All of the standard errors of the estimates are $<2.0\%$ of the estimated values, and most of them are $\leq 1.4\%$ of the estimated lengths. Given low levels of long bone length asymmetry (Jolicoeur, 1963; Trinkaus et al., 1994; Čuk et al., 2001; Auerbach and Ruff, 2006), it is assumed that the calculated lengths can be applied bilaterally for size scaling.

Discrete Traits

A series of discrete traits of the mandible and postcranium are provided; if not readily apparent from the characterization, they are described in the text and/or tables. Mandibular lateral corpus and ramal discrete traits follow Stefan and Trinkaus (1998a,b). Mandibular mentum osseum development follows Dobson and Trinkaus (2002). Humeral and femoral discrete traits are standard ones. Radial tuberosity orientation follows Trinkaus and Churchill (1988). Talocrural and pedal articular facet variations follow Trinkaus (1975a,b) and references therein (especially Barnett, 1954; Singh, 1959).

Cross-Sectional Geometry

These data have been supplemented by cross-sectional geometric data for the mandibular symphysis and for humeral, femoral, and tibial diaphyses at standard percentages of biomechanical lengths. For the long bone diaphyses, these cross sections are taken at 20%, 35%, 50%, 65%, and 80% of biomechanical length (approximately articular length for all but the femora), in which 0% is distal, following Ruff and Hayes (1983a). The mandibular symphyseal data model the symphysis as a solid beam and scale it to appropriate mandibular dimensions (Dobson and Trinkaus, 2002). The diaphyseal cross sections model them as hollow tubes of uniform cortical bone density and scale them to bone length and/or bone length times estimated body mass (the latter especially for the weight-bearing diaphyses) (Ruff et al., 1993; Ruff, 2000; Polk et al., 2000).

The majority of the cross sections, including those of Tianyuan 1, were generated by using polysiloxane dental putty (Optosil [Bayer] or Cuttersil Putty Plus [Heraus Kulzer]) to transcribe the subperiosteal contour (for all), and parallax-corrected biplanar radiography to interpolate the endosteal contour (for diaphyses) (Trinkaus and Ruff, 1989; O'Neill and Ruff, 2004). In a few cases in the comparative samples, scaled photographs of natural fossilization breaks of diaphyses were used if the break was sufficiently perpendicular to the diaphyseal axis to minimize distortion. The subsequent cross sections were primarily digitized using a Summagraphic 1812 tablet, and the cross-sectional parameters were computed using a PC version (SLCOMM: Eschman, 1992) of SLICE (Nagurka and Hayes, 1980). Additional data for some Upper Paleolithic femora and tibiae were generated by B. Holt using an NIH macro version of SLICE (Holt, 1999) and by I. Crevecoeur for Nazlet Khater 2 using an Image-J macro of SLICE by C. B. Ruff (www.hopkinsmedicine.org/FAE/mmacro.htm) from CT slice digital data (Crevecoeur, 2008). The resultant data consist of total subperiosteal area (TA), cortical area (for diaphyses) (CA), anatomically oriented second moments of area (I_x (anteroposterior) and I_y (mediolateral), maximum and minimum second moments of area (I_{max} and I_{min}), and the orientation of the maximum second moment of area (theta). The polar second moment of area (J, or I_p) is the sum of any two perpendicular second moments of area and provides an overall measure of bending and torsional rigidity. Areas are in mm^2, and second moments of area are in mm^4.

Even though the polar moment of area may become a poor measure of torsional rigidity when the cross section deviates markedly from circularity (Daegling, 2002) and second moments of area may poorly reflect resistance to habitual strain distributions at low levels of activity (Demes et al., 2001; Lieberman et al., 2004), they conform more closely to strains at higher activity levels (Demes et al., 2001), reflect long-term adaptation of bone to habitual strains (Shaw and Stock, 2009a,b), and are a generally useful and appropriate measure of bone rigidity for comparative purposes (Ruff et al., 2006). Moreover, they are a more accurate and scalable measure of diaphyses than tradition external diameters, all of which model diaphyses as rectangles and have uncertain biomechanical relationships to bone length and body mass.

Bilateral Asymmetry

For bones and teeth that normally exhibit little bilateral asymmetry, right and left values, when available for an individual, were averaged prior to the calculation of sample statistics. Moreover, the comparative samples pool right and left values for those largely symmetrical aspects. For those dimensions that are frequently asymmetrical, such as reflections of both diaphyseal and articular hypertrophy in the upper limb (Trinkaus et al., 1994; Churchill and Formicola, 1997; Čuk et al., 2001; Auerbach and Ruff, 2006), the right and left sides were compared separately. Long bone lengths tend to be largely symmetrical, as do many articular dimensions (Trinkaus et al., 1994; Auerbach and Ruff, 2006); as a result, the estimated long bone lengths obtained for the left humerus and radius (see Tables 9-2 and 9-7) and for the right femur and tibia (see Tables 10-1 and 10-4) are considered to be appropriate for the right and left of each bone. Similarly, the estimates for the right femoral head diameters are considered to apply generally to the individual.

The assessments of asymmetry are computed by dividing the difference between the right and left measurements by the average of the right and left measurements, expressed as a percentage [((right—left)/((right + left)/2)) x 100]. This is the percent directional asymmetry (%DA) value of Steele and Mays (1995) and Auerbach and Ruff (2006), and it is different from the percent asymmetry values of Trinkaus et al. (1994) and Churchill and Formicola (1997). As such, individuals with larger right measures have positive asymmetry values, and those with larger left side measurements have negative asymmetry values.

Age and Sex

Tianyuan 1 represents a fully mature but not geriatric individual (Chapter 11). Therefore, its skeletal osteometrics are compared solely to mature or almost mature (late adolescent) remains. However, some of the discrete traits appear to be stable during development, and for those aspects, a minority of immature individuals is included in the comparative samples. The sex of Tianyuan 1 is unknown and can be assessed primarily based on its overall body size, as are the sexes of the majority of the Late Pleistocene comparative samples. The known males and females, as well as the fossils of unknown sex, are therefore pooled together in the comparative samples. It is unlikely that this will bias the results, since any sexual dimorphism, other than in overall size, of the elements preserved for Tianyuan 1 is likely to be small. Moreover, many of the purported proportional differences between males and females, especially in features not directly related to reproduction and secondary sexual characteristics, disappear when the proportions are assessed with appropriate bivariate plots and scaling (see Ruff et al., 1993).

Chapter 5

The Tianyuan 1 Dentition

THE DENTITION of Tianyuan 1 was markedly reduced, antemortem and postmortem. Prior to death, the individual sustained substantial tooth loss, which is reflected in the altered alveolar process of the mandible and the super-eruption of the right M_3 (Chapter 12). There is also advanced occlusal and interproximal wear on the remaining teeth. After death and through fossilization and recovery, any left mandibular teeth that might have been there and all but one of the remaining maxillary teeth were lost. However, the remaining eight teeth (right I_2 to M_3 and left $M^{2?}$) and their alveolar processes provide considerable data on the dentition of this individual.

Occlusal and Interproximal Attrition

Maxillary Molar

The single maxillary molar, a left M^1 or M^2, has an incomplete crown (Figure 5-1). It is still possible to determine that its lingual occlusal enamel was largely lost antemortem through attrition, leaving only an enamel ring around the dentin basin. Buccally, the crown edge was lost postmortem, but there are the remains of the enamel from the sulcus between the paracone and the metacone. The wear most closely approximates Molnar's (1971) wear stage 5. This buccolingual wear differential is in agreement with a normal maxillary overbite, despite antemortem tooth loss. Its degree of wear, with some remaining occlusal enamel, is less than the wear on the M_1 and more closely approximates that of the M_2 (Table 5-1). This attrition suggests that the tooth is most likely an M^2. It is not possible to determine the extent of the interproximal wear facets given the buccal wear damage. However, they planed off the mesial and distal crown surfaces, leaving only thin borders of enamel between the exposed occlusal dentin and the interproximal surfaces.

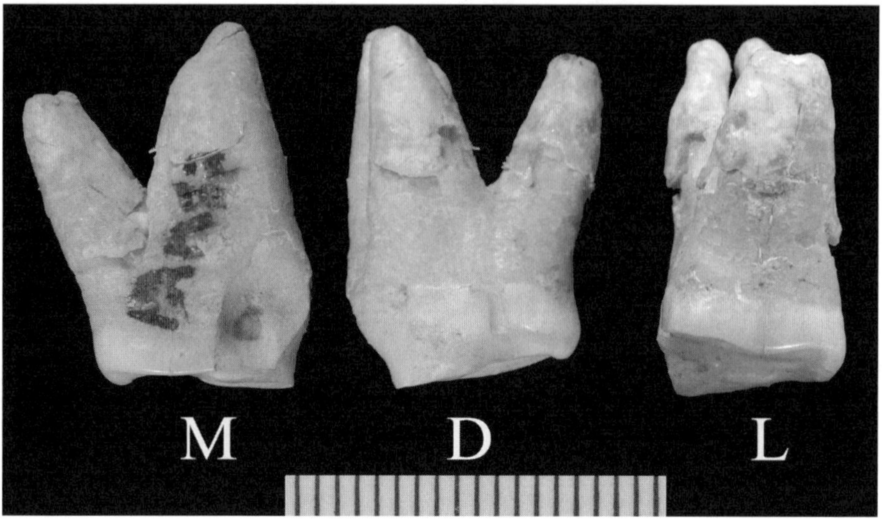

Figure 5-1. Views of the maxillary molar (M^1 or M^2). *M:* mesial; *D:* distal; *L:* lingual. Scale in millimeters.

Incisor, Canine, and Premolars

The I_2, the C_1 and both lower right premolars had their occlusal surfaces planed off to horizontal dentin surfaces with thin rings of enamel (Figure 5-2). Their buccal crown heights are slightly greater than their lingual ones (Table 5-1), but their largely flat occlusal surfaces remain buccolingually essentially horizontal relative to the alveolar plane of the mandible. The C_1 and the P_3 occlusal surfaces slope downward slightly to their interproximal contact area, and the P_4 surface slopes slightly distally toward the M_1. All four teeth match Molnar's wear stage 6.

In combination with their marked occlusal attrition, these four teeth have high cervix to alveolar crest (C/AC) heights; only the buccal P_4 value is noticeably below 3 mm. In healthy adults, a C/AC measurement of 1–2 mm is considered normal (Wood and Goaz, 1997). The values are not likely to have been influenced by the minor damage to the anterior buccal alveoli. Since there is no evidence of alveolar degeneration, especially mesially, these elevated cervix to alveolar crest measurements reflect normal dental supereruption in response to the pronounced reductions in crown heights of these teeth. Supereruption of the teeth is sometimes confused with alveolar recession due to periodontal disease (Whittaker et al., 1982; Costa, 1982; Newman, 1998). Given the absence of diagnostic periodontal indicators, such as interdental bony pockets, on the Tianyuan 1 mesial alveoli, it is likely that the increased C/AC heights are due to normal supereruption.

Table 5-1. Crown height and C/AC measurements for the Tianyuan 1 mandibular dentition, in millimeters

	Crown height		*C/AC (cervix to alveolar crest)*		
	Buccal	*Lingual*	*Buccal*	*Lingual*	*Wear stage*[1]
I_2	3.1	2.5	2.8	3.0	6
C_1	2.5	2.3	2.7	3.8	6
P_3	2.0	1.1	3.4	3.3	6
P_4	2.5	1.9	2.3	3.7	6
M_1 mesial	1.1	2.1	4.2	3.0	6/7
M_1 distal	0.2	2.3	—	3.6	
M_2 mesial	3.6	2.2	4.8	4.0	5
M_2 distal	2.4	2.3	4.9	5.1	
M_3 mesial	4.8	2.4	—	—	6
M_3 distal	4.3	2.2	—	—	
M^{1-2}	—	3.3	—	—	5

[1] Molnar (1971) occlusal wear stages for the teeth.

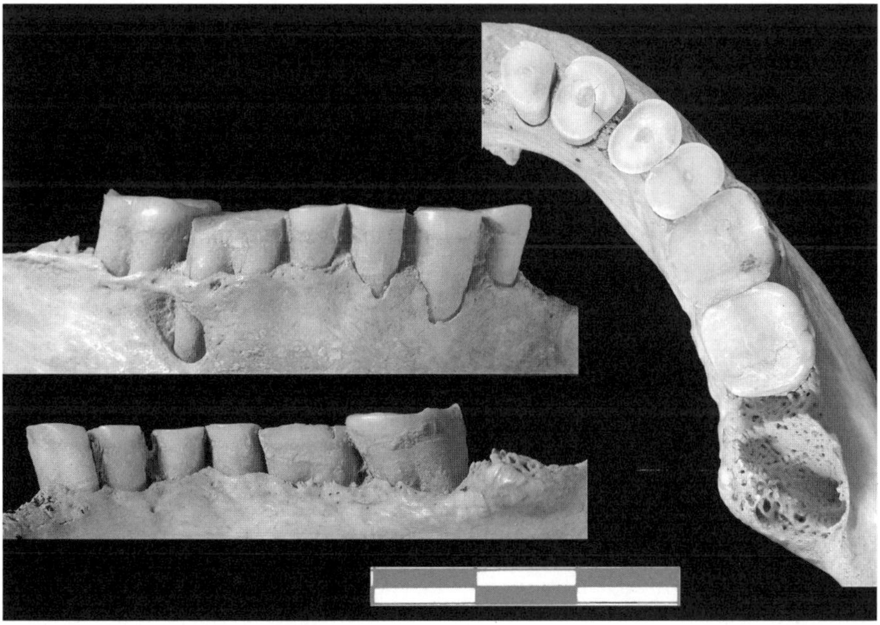

Figure 5-2. Views of the mandibular I_2 to M_2 dentition. *Right:* occlusal; *upper left:* buccal; *lower left:* lingual. Scale in centimeters.

Molars

The right M_1 is more extensively and irregularly worn (Figure 5-2). There is a complete ring of enamel around the exposed dentin, but it narrows to a tiny edge on the distobuccal corner of the crown. It is therefore scored as 6/7 (or 6.5) in the Molnar wear scale. The mesial half of the occlusal surface is largely flat to minimally concave, and it slopes distally downward. On the distal half of the crown, there is a broad and smooth sulcus in the exposed dentin that goes from midcrown to the distal margin; the sulcus slopes downward buccally, leading to the lowest margin being at the distobuccal corner of the crown. The distal enamel margin then rises to meet that of the M_2 interproximally. There is also a low peak of the buccal margin enamel between the original positions of the two buccal cusps. The short buccal margins of the crown are reflected in its crown heights (Table 5-1), in which the lingual ones are similar to those of the more mesial teeth but its buccal ones are markedly less. These low buccal crown height values fit with the higher values for the M_1 C/AC height.

The M_2 has less wear than the M_1 overall, with a complete ring of enamel around the occlusal dentin basin and only a trivial tongue of enamel extending into the middle of the surface in the middle of the buccal margin (Figure 5-2). It corresponds best to Molnar's wear stage 6. The surface undulates mesiodistally. It rises up from the M_1 distal transverse groove to be mesiodistally convex on its mesial half, and then it becomes mesiodistally concave on its distal half. The distal enamel edge then rises up to meet the M_3. Despite its lower amount of occlusal wear and greater crown heights buccally (but not lingually), the M_2 has greater C/AC heights than the M_1 (Table 5-1). This may represent normal irregularities and/or some alveolar recession, or it may be related to the marked supereruption of the M_3 (Chapter 12). If the latter, then it implies reduced occlusal opposition for the M_2, sufficiently before death to permit the contour of the distal M_1 and mesial M_2 crowns to be one continuous curving surface.

The mesial M_2 interproximal facet with the M_1 is transversely straight, 6.8 mm wide, and spans most of the two crown breadths. The distal one for the M_3 is flat, 5.5 mm wide, and it is both across the distal M_2 and beveled for the abnormal position of the M_3. It slopes superomesially and mesiolingually on the lingual two-thirds of the distal M_2 crown, and it would closely match the surface of the mesial M_3 cervix were the mesial M_3 root undamaged and permitted the tooth to sit fully within its socket. The mesial M_3 crown has a normal facet on the midmesial crown 5.5 mm wide. Clearly, at one point the M_3 was in normal occlusion and in line with the M_2, but it became supererupted long enough before death to produce the abnormal facet on the more lingual distal M_2 crown.

The right M_3 was markedly supererupted for some time before death, the details of which are provided in Chapter 12 as part of the presentation of the ante-

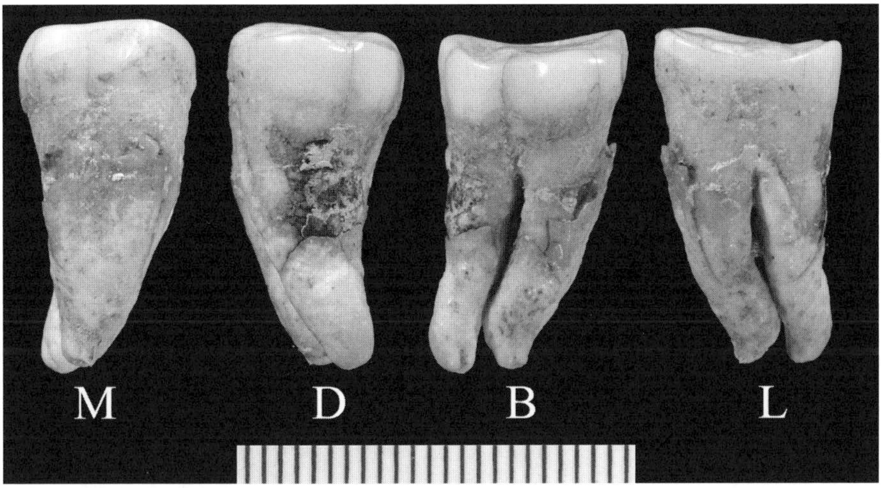

Figure 5-3. Views of the mandibular M_3. *M:* mesial; *D:* distal; *B:* buccal; *L:* lingual. Scale in millimeters.

mortem tooth loss. The tooth by itself, however, shows no evidence of its abnormal position at death (Figure 5-3). The occlusal wear is largely flat, dipping downward slightly distolingually. Most of the occlusal surface is dentin. The enamel ring is continuous, and there is an enamel extension buccally into the surface. This conforms to Molnar's wear stage 5.

Differential Molar Wear

Given these wear stages for the Tianyuan 1 molars, it is possible to assess the wear gradient of these teeth. This assumes that the degree of wear on the M_3 (and M_2) is close to what it would have been had the teeth remained in normal occlusion until the death of the individual. The differences between the wear stages of the M_1 versus the M_2 and the M_2 versus the M_3 are in Table 5-2.

For the M_1 to M_2 comparison, the wear gradient of Tianyuan 1 is relatively modest. Among the comparative samples, the Tianyuan 1 wear differential most closely matches the two Middle Paleolithic samples, both early modern and late archaic. Among more recent Upper Paleolithic modern humans, a similarly small contrast in M_1/M_2 wear is found only on Dolní Věstonice 16 and Minatogawa 5. If the Tianyuan 1 M_2 had become nonocclusal as a result of the M_3 supereruption and maxillary tooth loss, one would expect the differential to be even greater than the minimal amount evident on the teeth. The M_3 to M_2 comparison provides a slightly greater differential, one that matches the highest percentages of the larger comparative samples. There is therefore nothing unusual in the differential molar

Table 5-2. Differential molar wear for Tianyuan 1 and comparative samples. % except N

	0	*0.5*	*1*	*1.5*	*2*	*2.5*	*N*
M_1 vs. M_2							
Tianyuan 1		X					
East Asia EMH		16.7	50.0	16.7		16.7	6
West MUP	6.3		62.5	12.5	12.5	6.3	16
West EUP			50.0			50.0	2
MPMH	42.9	28.6	14.3		14.3		7
Neandertals	18.8	25.0	50.0	6.3			16
M_2 vs. M_3							
Tianyuan 1			X				
East Asia EMH	25.0	25.0	25.0	25.0			4
West MUP		11.1	44.4		33.3	11.1	9
West EUP		100					1
MPMH	14.3	28.6	42.9		14.7		7
Neandertals	12.5	31.3	56.3				16

Note: EMH: early modern humans; *MUP:* Mid Upper Paleolithic; *EUP:* Early Upper Paleolithic; *MPMH:* Middle Paleolithic modern humans. Using the Molnar (1971) wear scale, the distributions reflect the differences in wear scores between the adjacent molars.

wear of Tianyuan 1, even though the M_1/M_2 differential is slightly less than in the majority of the earlier Upper Paleolithic modern human remains, from both ends of Eurasia.

Dental Discrete Morphology

Given the extensive attrition of the Tianyuan 1 teeth, there is little of note that remains concerning discrete dental traits. It is not possible to discern general crown shape for more than the M_3, which is distally bulbous, and not even cusp number can be reliably inferred. Based on the high positions of the mesial and distal root bifurcations of the three mandibular molars, they are non-taurodont.

Dental Crown Size and Proportions

Crown Diameters

Given the extensive occlusal and interproximal wear on the Tianyuan mandibular teeth, only buccolingual diameters are provided for the six mesial teeth. It is not appropriate, even continuing contours of the remaining crowns, to reliably interpolate the original mesiodistal diameters of the I_2 to M_2 crowns. The M_3, however, does provide a reliable mesiodistal diameter that was little, if at all, reduced by attrition. The maxillary molar is too damaged to provide a buccolingual breadth and too worn to provide a mesiodistal one; it appears to be in proportion to the mandibular molars.

The crowns of the six mesial teeth also appear to have had their buccolingual diameters reduced, since the wear extends below the normal lingual and buccal convexities of the crowns. The M_3 buccolingual measurement is unaffected by wear. To compensate for the more mesial tooth crown reductions, buccolingual crown and cervix diameters were measured on a series of original teeth and high resolution epoxy casts of Late Pleistocene and later Holocene teeth (late archaic and modern humans). From these paired measurements, least squares regressions were computed to estimate the original Tianyuan 1 I_2 to M_2 buccolingual crown diameters from its cervical diameters (Table 5-3); regression sample sizes range from 30 to 42 per tooth (total N = 223) and r^2 values range from 0.608 (M_1) to 0.927 (I_2) (mean r^2: 0.806). The standard errors of the estimates are moderately large (Table 5-3), but most of the mean estimated values appear to provide more realistic values for the original crown breadths. The original diameter of the I_2 is slightly larger than the mean estimated value and hence is employed. The resultant crown diameters used in the comparisons are in Tables 5-4 and 5-5.

Univariate and Bivariate Comparisons

The only one of these teeth that shows a significant difference across the Late Pleistocene samples is the I_2, the difference driven mostly by the moderately larger incisors of the Neandertals relative to the later Eurasian Upper Paleolithic humans. The Tianyuan 1 I_2 labiolingual diameter is unexceptional for an early modern human, although it is moderately small for a Neandertal. There are nonsignificant differences between the means for the P_3 and M_1 crown breadths, with the Middle Paleolithic humans having slightly larger P_3s and the Neandertal and European Upper Paleolithic samples having smaller M_1s. In the eastern Eurasian samples, the early modern human dental dimensions are unexceptional, but the

Table 5-3. Crown and cervical bucco(labio)lingual (BL) diameters of the Tianyuan I mandibular teeth, in millimeters

	Cervical BL diameter	*Preserved crown BL diameter*	*Estimated crown BL diameter*
I_2	6.8	7.1	7.0 ± 0.2
C_1	8.7	8.7	8.8 ± 0.3
P_3	7.7	8.1	8.6 ± 0.4
P_4	8.0	8.4	8.9 ± 0.3
M_1	10.3	(11.1)	11.4 ± 0.4
M_2	9.5	10.2	10.7 ± 0.4
M_3	9.8	10.7	

Note: The estimated crown diameters were derived from buccolingual cervical diameters using least squares regressions based on Late Pleistocene (late archaic and early modern) and later Holocene human teeth: predicted value ± standard error of the estimate (SE_{est}).

Table 5-4. Mandibular incisor, canine, and premolar crown buccolingual (BL) diameters for Tianyuan 1 and Late Pleistocene comparative samples [mean ± SD (N)]

	I_2 *BL*	C_1 *BL*	P_3 *BL*	P_4 *BL*
Tianyuan 1	7.1	8.8	8.6	8.9
East Asia EMH	6.9	8.2 ± 0.5 (6)	8.3 ± 0.1 (5)	8.7 ± 0.3 (6)
East Asia LAH	—	—	8.6	10.6
West MUP	6.8 ± 0.5 (23)	8.6 ± 0.7 (20)	8.5 ± 0.5 (19)	8.7 ± 0.6 (20)
West EUP	7.2 ± 0.5 (5)	8.7 ± 0.8 (6)	8.4, 9.0	8.7, 8.9
MPMH	7.2 ± 0.2 (10)	8.3 ± 0.8 (10)	8.8 ± 0.5 (8)	8.9 ± 0.6 (8)
Neandertals	7.7 ± 0.5 (30)	8.9 ± 0.8 (37)	9.0 ± 0.7 (36)	8.9 ± 0.8 (38)
K-W P-value	<0.001*	0.123	0.030	0.722

Note: EMH: early modern humans; *LAH:* late archaic human; *MUP:* Mid Upper Paleolithic; *EUP:* Early Upper Paleolithic; *MPMH:* Middle Paleolithic modern humans. Kruskal-Wallis (K-W) P-values across the comparative samples are provided; * $P < 0.05$ using a multiple comparison correction. The Tianyuan 1 values are the estimated values (Table 5-3) except for the I_2, for which the 0.1 mm larger preserved diameter is used.

Table 5-5. Mandibular molar buccolingual (BL) crown diameters and M_3 mesiodistal (MD) diameter for Tianyuan 1 and Late Pleistocene comparative samples [mean ± SD (N)]

	M_1 *BL*	M_2 *BL*	M_3 *BL*	M_3 *MD*	M_3 *"Area"*
Tianyuan 1	11.4	10.7	10.7	11.5	123.1
East Asia EMH	11.2 ± 0.6 (10)	11.0 ± 0.5 (8)	10.6 ± 0.2 (5)	10.8, 11.2, 11.6	111.2, 121.0, 130.0
East Asia LAH	10.3	10.1	10.4	—	—
West MUP	11.1 ± 0.7 (35)	10.9 ± 0.7 (31)	10.8 ± 0.9 (19)	11.3 ± 0.9 (17)	122.5 ± 18.4 (17)
West EUP	11.0 ± 0.7 (12)	10.9 ± 0.9 (13)	10.9 ± 0.9 (4)	11.8 ± 1.7 (4)	130.0 ± 29.7 (4)
MPMH	11.4 ± 0.6 (15)	11.0 ± 0.7 (11)	10.8 ± 0.5 (10)	11.7 ± 0.8 (9)	126.8 ± 14.1 (9)
Neandertals	10.9 ± 0.6 (52)	10.9 ± 0.8 (41)	11.0 ± 0.8 (44)	11.7 ± 0.6 (42)	128.7 ± 14.0 (42)
K-W P-value	0.099	0.927	0.510	0.271	0.350

Note: EMH: early modern humans; *LAH:* late archaic human; *MUP:* Mid Upper Paleolithic; *EUP:* Early Upper Paleolithic; *MPMH:* Middle Paleolithic modern humans. Crown "areas" (MD x BL) are provided for the M_3, the only Tianyuan 1 tooth to provide a reliable mesiodistal diameter. The high standard deviation for the western EUP M_3 mesiodistal diameter reflects the exceptional dimensions of the Oase 1 M_3s. Kruskal-Wallis (K-W) P-values across the comparative samples (not including the East Asian LAH) are provided; none is significant.

late archaic Changyang P_4 is relatively large, whereas the Dingcun and Okladnikov molars are fairly small. In all of these crown diameters, the Tianyuan 1 values are unexceptional. The same applies to the M_3 mesiodistal length and to the resultant M_3 crown "area" (Table 5-5).

At the same time, it is principally the dental crown proportions along the dental arcade that differentiate at least western Eurasian samples, rather than individual tooth dimensions (Stefan and Trinkaus, 1998b). This is partly reflected in a comparison of summed anterior (I_2 to P_3) to posterior (P_4 to M_2 or P_4 to M_3) crown breadths (Figure 5-4A). The comparisons are done both with and without the M_3, given the high level of variation in its size in Late Pleistocene samples (Trinkaus, 2004). The plots of summed anterior teeth to posterior teeth provide a largely complete separation of the Neandertals from the early modern humans, with only two of the early modern human specimens falling among the Neandertals with relatively smaller anterior teeth. Only in the comparison using the M_3 (Figure 5-4B) is there a specimen completely with the others, in this case the MUP (Mid Upper Paleolithic) Arene Candide 1 with the Neandertals. Tianyuan 1 falls in both comparisons along the narrow overlap zone between the two larger groups. The EUP (Early Upper Paleolithic) Mladeč 54 is even farther from the middle of the early modern human distribution and among the Neandertals in the I_2–M_2 comparison; the other EUP specimen, Nazlet Khater 2, is distant from the Neandertals. The pattern remains similar if only the sum of the I_2 and the C_1 diameters is compared to those of the postcanine teeth. The difference with the previous preliminary results, in which the Tianyuan 1 dental proportions appeared more distinctly Neandertal-like (Shang et al., 2007) derives from the larger diameters used for the postcanine teeth given the current correction for wear.

Multivariate Comparisons

An alternative assessment of dental crown proportions along the tooth row can be obtained through the use of principal components analysis and discriminant function analysis of the buccolingual diameters across these samples (Chapter 4). This was done using the full set of seven crown diameters (I_2–M_3) available for the Tianyuan 1 mandibular teeth (Figure 5-5A; Table 5-6). Since the resultant comparative samples for all seven teeth are limited, the same analyses were run on more limited sets of tooth crown diameters, I_2–M_2 (Figure 5-5B), C_1–M_3, and C_1–M_2. Since absolute size, as well as relative size, is an important component of Late Pleistocene dental variation, the analyses were done without size standardization.

The bivariate plot of principal components 1 and 2 for the I_2–M_3 analysis (total variance explained: 79.8%) largely repeats the pattern evident in the comparisons of summed anterior to summed posterior crown diameters (Figure 5-5A). The Neandertals cluster in one diagonal half of the distribution and all but two of the early modern humans are within the other diagonal half. Tianyuan 1 falls in the middle of the distribution, along the interface between the Neandertal and early modern human distributions, adjacent to Qafzeh 27. The two MUP specimens within the Neandertal distribution are Arene Candide 1 and Dolní Věstonice 13. None of the eastern Asian early modern humans provide all seven crown diameters, but the contemporaneous Nazlet Khater 2 dentition is far from Tianyuan 1 and from the Neandertals.

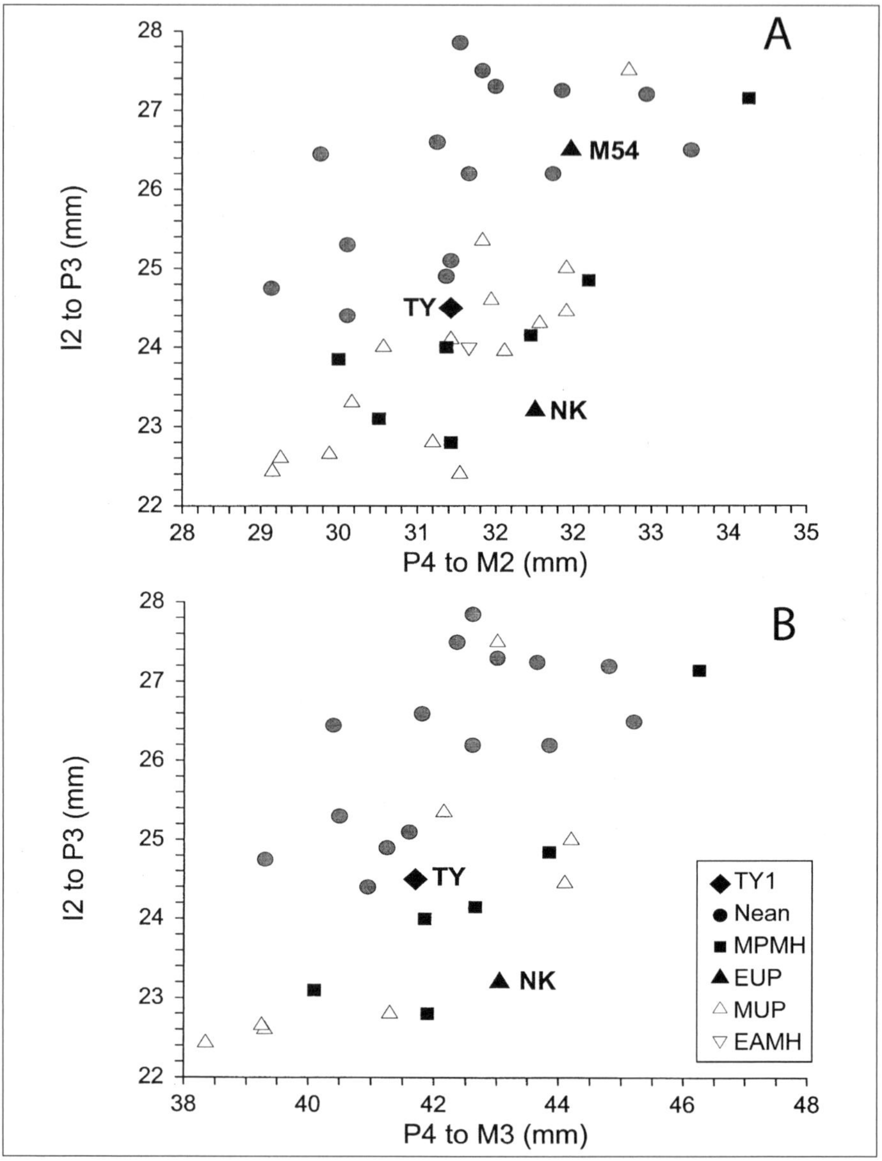

Figure 5-4. Bivariate plot of (*A*) summed anterior (I_2 to P_3) to posterior (P_4 to M_2) crown breadths and (*B*) I_2 to P_3 versus P_4 to M_3 crown breadths. *TY:* Tianyuan 1; *Nean:* Neandertals; *MPMH:* Middle Paleolithic modern humans; *EUP:* western Early Upper Paleolithic; *MUP:* western Eurasian Mid Upper Paleolithic; *EAMH:* eastern Asian modern humans; *NK:* Nazlet Khater 2; *M54:* Mladeč 54.

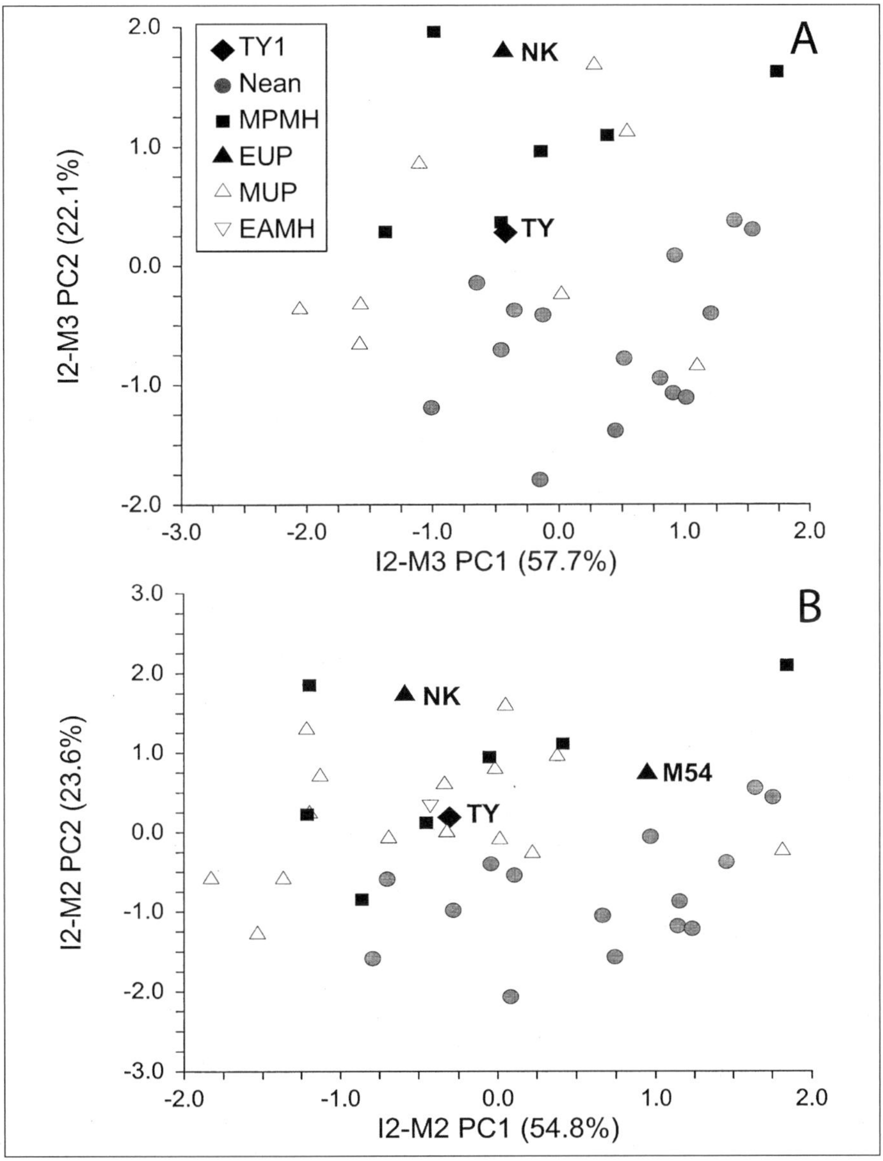

Figure 5-5. (*A*) PC2 versus PC1 for I_2 to M_3; (*B*) PC2 versus PC1 for I_2 to M_2. *TY:* Tianyuan 1; *Nean:* Neandertals; *MPMH:* Middle Paleolithic modern humans; *EUP:* western Early Upper Paleolithic; *MUP:* western Eurasian Mid Upper Paleolithic; *EAMH:* eastern Asian modern humans; *NK:* Nazlet Khater 2; *M54:* Mladeč 54.

Table 5-6. Discriminant function affinities and posterior probabilities for the Tianyuan 1 mandibular dentition using buccolingual diameters

	I_2—M_3	I_2—M_2	C_1—M_3	C_1—M_2
Tianyuan 1 1st	MPMH (0.872)[1]	EAMH (0.487)	MPMH (0.606)	MPMH (0.538)
Tianyuan 1 2nd	MUP (0.121)	MPMH (0.412)	EAMH (0.197)	EAMH (0.314)
% Correct—All	66.7	40.0	54.5	47.8
% Correct—LAH/EMH	96.7	87.5	87.9	82.6
Comparative N	30	40	33	46

Note: MPMH: Middle Paleolithic modern humans; *EAMH:* eastern Asian modern humans; *MUP:* Mid Upper Paleolithic; *LAH:* late archaic humans; *EMH:* early modern humans. Includes correct classification rates for the comparative samples across all 5 samples and between the late archaic humans (Neandertals) versus the pooled early modern human samples (LAH/EMH). Only the first two posterior probabilities are provided, since the others are markedly smaller.

[1] For each set of tooth measurements, the group with the highest and then second highest posterior probability for Tianyuan 1 is provided, followed in parentheses by the posterior probability of Tianyuan 1 for that sample.

The resultant posterior probabilities, which provide a 97% correct distinction between the Neandertals and the early modern humans, place Tianyuan 1 primarily with the Middle Paleolithic modern humans (MPMH) sample and a distant second with the MUP sample.

Removing the M_3 from consideration increases the sample size, but it does not change the pattern (Figure 5-5B). With modest overlap with the MUP sample, the Neandertals remain within their own diagonal half of the distribution, separate from most of the early modern humans. The posterior probabilities show slightly less correct classification between the Neandertals and the early modern humans (87.5%), given the overlap of a couple of MUP specimens and the EUP Mladeč 54 with the Neandertal distribution, although the latter has its posterior probability affinities primarily with the MUP sample (0.544) and secondarily with the MPMH sample (0.285). Tianyuan 1 has essentially equal affinities to the MPMH and the eastern Asian EMH samples (Table 5-6), although the latter is represented by only one individual, Moh Khiew 1.

The deletion of the I_2 from the measurements, with or without the M_3, reinforces the affinities of the Tianyuan 1 dental size and proportions with the MPMH sample, and very secondarily with the eastern Asian EMH sample (Table 5-6). For these comparisons, the C_1–M_3 analysis includes ZKD-UC 101 and Minatogawa 1, and Minatogawa 5 and 6 are added in the C_1–M_2 analysis.

Summary

The partial dentition of Tianyuan 1 is notable principally for its extensive occlusal and interproximal attrition, which removed most or all of the occlusal enamel from the individual teeth. This was associated with extensive antemortem tooth loss (Chapter 12). The overall wear gradient of the molars, however, was relatively modest and similar to most Late Pleistocene humans between the M_2 and M_3 and slightly below the average between the M_1 and M_2.

There is little discrete dental morphology preserved on the teeth. The estimated crown buccolingual diameters for the I_2 to M_2, plus the mesiodistal and buccolingual diameters for the M_3, place the Tianyuan 1 dentition in the middle of Late Pleistocene human dental metric variation in size and proportions along the arcade. Although its dental size and proportions fall among the early modern human samples, especially the earlier MPMH sample, the slightly younger EUP Mladeč 54, and the more recent eastern Asian early modern humans, its positions in both comparisons of summed anterior to posterior teeth and principal components analysis are close to the modest overlap zone between the late archaic Neandertals and the various early modern human samples.

Chapter 6

The Tianyuan 1 Mandible

THE TIANYUAN 1 mandible is largely complete from the left P_4M_1 region, around the symphysis and the right corpus, to the mandibular notch (Figures 6-1 and 6-2). As such, it provides data on the symphyseal region, the lateral corpus, and all of the ramus except the coronoid tip and the condyle.

The mandible had been altered as a result of antemortem tooth loss (Chapters 5 and 12), but it appears that the associated pathological alterations were limited to the alveolar process. The changes primarily affected the left corpus, and as a result the morphological assessment of the corpus employs only the symphyseal region and the right side. Both I_1s were lost antemortem, and their sockets completely closed over; therefore in the assessment of the symphysis, the original alveolar bone was estimated using the preserved labial and lingual contours and the level of the alveolar plane at the right I_1I_2 and I_2C_1 interdental septa.

The alveolar bone changes associated with the extensive wear of the molars and the supereruption of the right molars, and especially the right M_3 (Chapters 5 and 12), appear to have affected only the alveolar margin and not the underlying corpus, and those alveolar changes are taken into account in the assessment of the lateral corpus and anterior ramus. The extensive interproximal wear of the teeth almost certainly produced mesial drift of the dentition, such that the more distal teeth may well be farther mesial than they would have been when they were all first erupted in the late 2nd decade of life. This would have mainly affected the perceived positions of the mental foramen relative to the dentition, the possibility of a retromolar space, and to a lesser extent the exact positions of corpus heights and breadths taken relative to the dentition. The overall effect of these changes should be minimal.

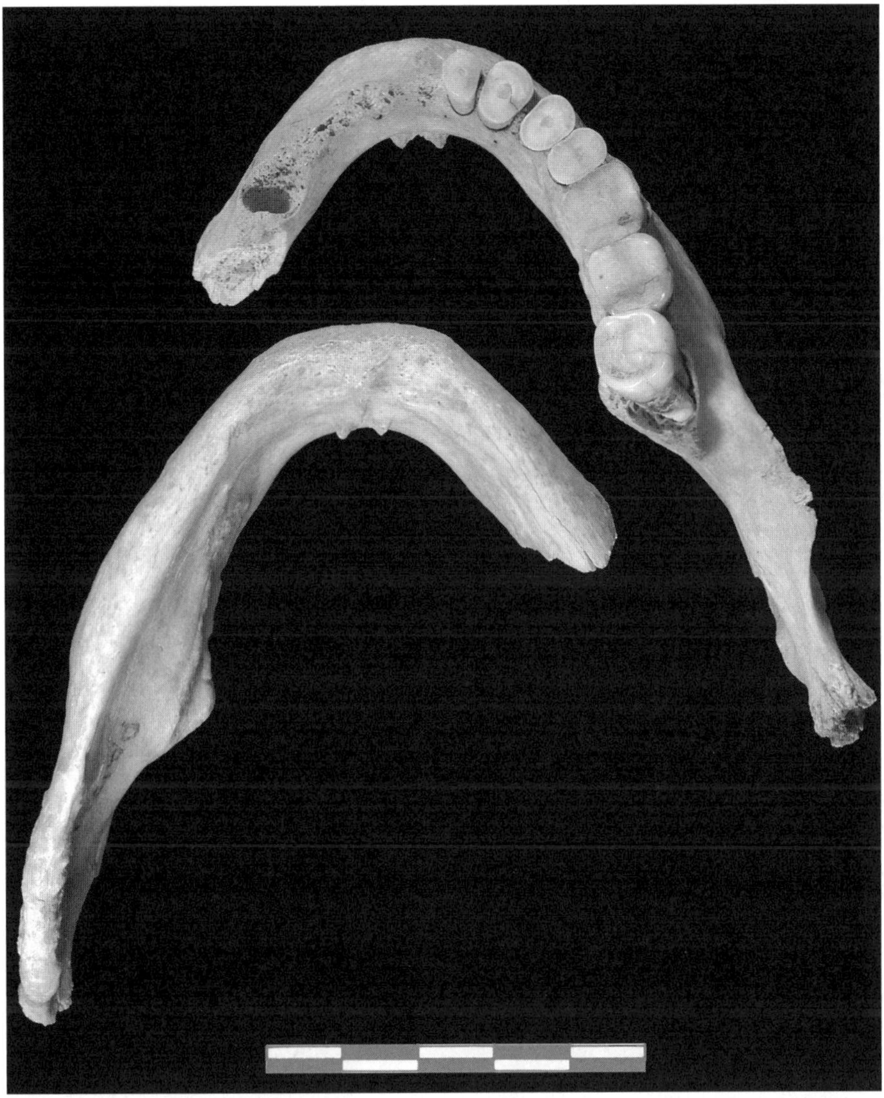

Figure 6-1. Superior (*above*) and inferior (*below*) views of the Tianyuan 1 mandible. Scale in centimeters.

Overall Size and Dimensions

Despite the postmortem damage to the Tianyuan 1 mandible, it is possible to estimate reliably its overall length and, through mirror imaging, its dental arcade breadth. Mandibular breadths across the condyles and gonion are less reliable, given the complete absence of the condyle and lack of knowledge of its posi-

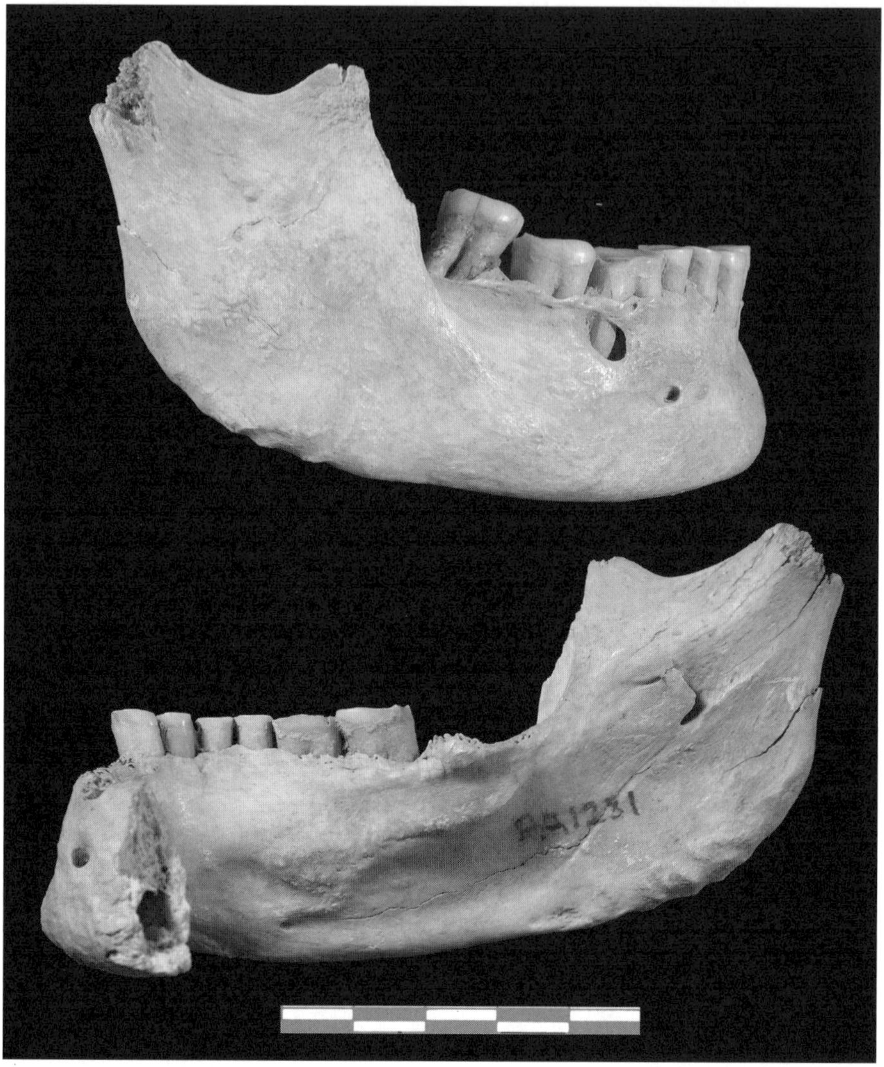

Figure 6-2. Lateral (*top*) and medial (*bottom*) views of the Tianyuan 1 mandible. Lateral view is in *norma lateralis* (perpendicular to the sagittal plane); medial view is an oblique one perpendicular to the lateral corpus. Scale in centimeters.

tion relative to the mandibular notch crest and given the trigonometric effects of being farther from the symphysis. Moreover, reliable bigonial and/or bicondylar breadths are rare in the fossil record, given ramal damage and/or distortion along the mandible.

The estimated superior mandible length (midsagittal distance from the mid-

Table 6-1. Mandibular osteometrics of Tianyuan 1, in millimeters

	Right	*Left*
Mandible superior length[1]	(98.0)	
Dental arcade external M_2 breadth	(64.5)	
Digastric fossae		
Maximum length chord	18.2	—
Main fossa length chord	13.8	13.4
Maximum fossa breadth	7.1	7.1
Mental foramen position		
Pogonion to midforamen	30.0	28.5
Basal margin to midforamen	13.8	13.7
Minimum ramus breadth (M-71a)	39.5	—
Distal M_3—Mandibular foramen	25.7	—

Note: Estimated values are in parentheses.
[1] Direct midline distance from the middle of each mandibular condyles to infradentale (Twiesselmann and Brabant 1967; Twiesselmann, 1973).

condyles to infradentale) of Tianyuan 1 (Table 6-1) of ~98 mm is relatively small (Table 6-2). It is slightly below the smallest Neandertal value (Regourdou 1 at 100 mm; although La Naulette 1 was probably smaller), and it is below the three available Middle Paleolithic modern human (MPMH) values and the two western Early Upper Paleolithic (EUP) lengths. It is nonetheless well within the ranges of variation of the western Mid Upper Paleolithic (MUP) sample and the eastern Asian early modern human sample, falling slightly below Minatogawa 1 and ZKD-UC 101 and slightly above ZKD-UC 104. It is in the middle of recent human sample distributions (Trinkaus, 2003).

At the same time, the dental arcade breadth of ~64.5 mm is modest but unexceptional for a Late Pleistocene human. It is essentially on the means of the western early modern human samples, but it is moderately low for a Neandertal or for an eastern Asian early modern human (Table 6-2). It remains within the ranges of variation of all of the comparative samples.

The minimum ramus breadth of Tianyuan 1 of 39.5 mm (Tables 6-1 and 6-2) is small compared to the two Middle Paleolithic samples and especially relative to the very high values for the EUP Nazlet Khater 2 and Oase 1 mandibles. Nonetheless, three of the Neandertals have small ramus breadths; it is small for Muierii 1 (Soficaru et al., 2006), and they were probably small for Mladeč 1 and 2 given their facial configurations (Wolpoff et al., 2006). It is in the middle of the western Eurasian MUP sample and the eastern Asian early modern human sample. When plotted against mandibular superior length (Figure 6-3), Tianyuan 1 falls in the middle of the generally smaller more recent human ranges of variation. But most of the Late Pleistocene specimens have similar ramus breadth to mandible length proportions; Oase 1 and especially Nazlet Khater 2 are high outliers.

Table 6-2. Overall mandibular and corpus measurements of Tianyuan 1 and comparative samples, in millimeters

	Superior length	*Dental arcade breadth*	*Minimum ramus breadth*	*Symphyseal height*	*Mental foramen height*	*Mental foramen breadth*	*M_1M_2 height*	*M_1M_2 breadth*
Tianyuan 1	(98.0)	(64.5)	39.5	(27.5)	29.7	12.1	29.6	11.7
East Asia EMH	96.1, 99.1, 104.4	66.9 ± 2.2 (4)	37.9 ± 4.1 (4)	32.3 ± 2.4 (5)	30.6 ± 1.6 (5)	12.1 ± 0.8 (5)	28.2 ± 2.5 (5)	13.2 ± 1.1 (4)
West MUP	101.7 ± 5.9 (12)	64.5 ± 3.8 (11)	38.9 ± 2.5 (12)	32.6 ± 3.8 (15)	31.7 ± 4.4 (12)	12.4 ± 1.4 (11)	29.9 ± 3.3 (11)	14.3 ± 2.0 (11)
West EUP	103.4, 107.6	63.5, 65.5	35.1, 46.2, 51.0	36.5 ± 1.9 (4)	27.5, 33.2, 35.3	11.6, 12.2, 15.7	32.0 ± 3.0 (5)	14.7 ± 1.7 (4)
MPMH	109.0, 118.0, 126.0	65.0	42.5, 43.0, 44.0	37.3 ± 3.2 (4)	35.0, 36.0, 40.5	13.2, 15.0, 16.6	31.0, 33.0, 35.5	12.0, 19.5, 20.0
Neandertals	109.9 ± 6.6 (14)	67.5 ± 3.4 (16)	41.8 ± 2.6 (14)	34.7 ± 4.0 (22)	32.4 ± 3.5 (26)	15.5 ± 1.8 (26)	31.5 ± 3.4 (24)	16.2 ± 1.6 (24)
K-W P-value	0.004*	0.066	0.027	0.079	0.269	<0.001*	0.123	0.017

Note: EMH: early modern humans; *MUP:* Mid Upper Paleolithic; *EUP:* Early Upper Paleolithic; *MPMH:* Middle Paleolithic modern humans. Mean ± standard deviation (N) are provided for samples >3. Kruskal-Wallis (K-W) P-values across the comparative samples are provided; * $P < 0.05$ using a sequentially reductive multiple comparison correction.

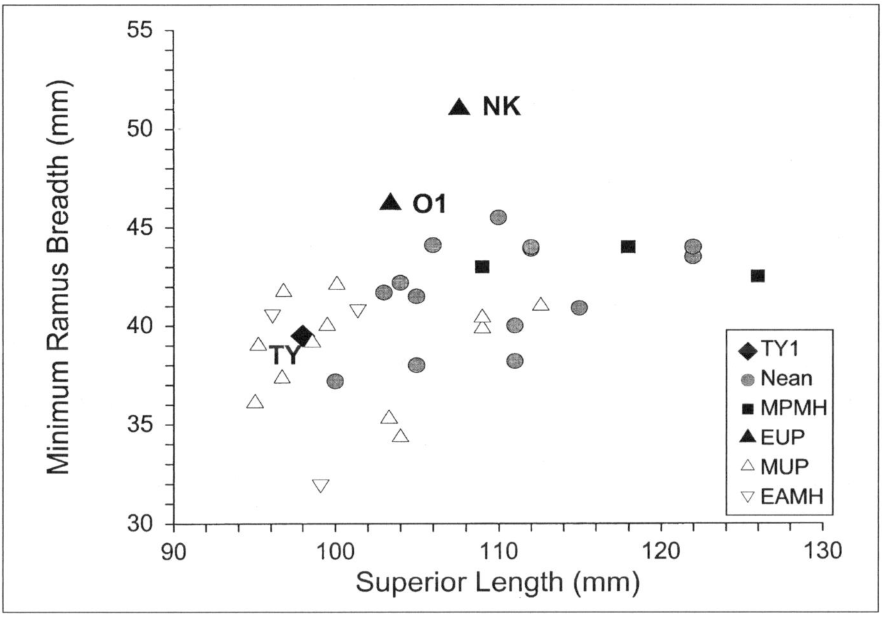

Figure 6-3. Bivariate plot of minimum ramus breadth versus mandible superior length for Tianyuan 1 (*TY*) and Late Pleistocene comparative samples. *TY:* Tianyuan 1; *Nean:* Neandertals; *MPMH:* Middle Paleolithic modern humans; *EUP:* western Early Upper Paleolithic; *MUP:* western Eurasian Mid Upper Paleolithic; *EAMH:* eastern Asian modern humans; *NK:* Nazlet Khater 2; *O1:* Oase 1.

The Symphysis

Morphology

There is a slight concavity of the right anterior symphysis just below the alveolar plane, which gives the impression of an anterior buccal mandibular depression (sensu Arensburg et al., 1989) (Figure 6-4). Since the alveolar margin was altered by resorption after the right I_1 to left C_1 antemortem tooth loss, it is not clear whether this represents a definite depression of this form or merely an alteration of the bone at that level. The two sides of the superior labial symphysis are nonetheless divided by a distinct but low crest rising from the tuber symphyseos to the alveolar margin.

The tuber symphyseos is a prominent triangular bulge occupying the inferior half of the anterior symphysis. It narrows superiorly into the sharp crest that rises up to infradentale. There are depressions superolaterally of both sides of the tuber symphyseos, but they are gently concave and their margins are not sharply delimited. The middle of the tuber symphyseos is raised, but there is no clear center.

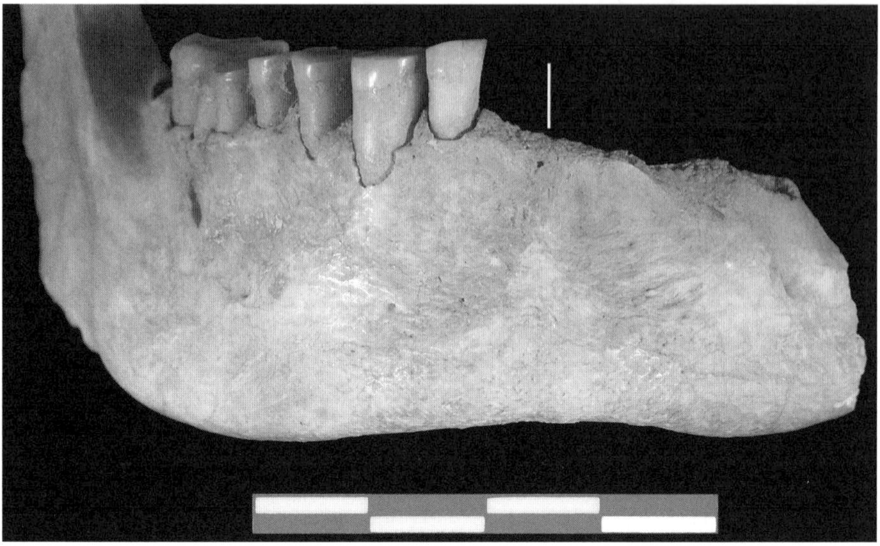

Figure 6-4. Anterior view of the Tianyuan 1 mandibular symphysis. The white line indicates the symphyseal midline. Scale in centimeters.

Pogonion is therefore identified only as the most projecting point on the midline. The resultant position of pogonion is 19.0 mm below infradentale, or 69% of the estimated symphyseal height of 27.5 mm (Table 6-2).

The lateral tubercles are rounded angles, rounding onto the lateral and inferior corpus, below the C_1 on the right side. Given their indistinct margins, the bilateral breadth of ~28 mm is approximate.

The mentum osseum rank of the Tianyuan 1 mandible is therefore category 4 of Dobson and Trinkaus (2002); projecting tuber symphyseos but little or no projection of the lateral tubercles. This is the most common pattern among early modern humans, even though the east (and south) Asian sample is evenly split between categories 4 and 5 (the latter form also exhibiting prominent lateral tubercles), with ZKD-UC 101 being intermediate between the two categories (Table 6-3). Among the early modern humans, only the immature Qafzeh 10 and Předmostí 2 have little or no projection of the tuber symphyseos. Category 4 is especially the dominant pattern among western EUP mandibles, with only the Mladeč 52 mandible having prominent lateral tubercles [based on the photographs (Frayer et al., 2006) of the now lost mandible]. None of the Neandertals, or Aterian fossils, have a projecting tuber symphyseos, and only a minority of the South African Middle Stone Age mandibles exhibits it.

The digastric fossae (Table 6-1) are moderately rugose with clear margins. They face posteroinferiorly and are separated by a central rounded area 3–4 mm wide.

Table 6-3. Distributions of mentum osseum ranks for Tianyuan I and Late Pleistocene human comparative samples

	Category 1, %	*Category 2, %*	*Category 3, %*	*Category 4, %*	*Category 5, %*	*N*
Tianyuan I				X		
East/South Asia EMH				50.0	50.0	7
West MUP			3.2	67.7	29.0	31
West EUP				85.7	14.3	7
MPMH			12.5	87.5		8
Neandertals	11.5	57.7	30.8			26
South Africa MSA			62.5	37.5		6
Aterian		50.0	50.0			2

Note: EMH: early modern humans; *MUP:* Mid Upper Paleolithic; *EUP:* Early Upper Paleolithic; *MPMH:* Middle Paleolithic modern humans; *MSA:* Middle Stone Age. Categories based on Dobson and Trinkaus (2002), in which categories 4 and 5 have a prominent "chin," with both categories having a prominent tuber symphyseos and category 5 also having prominent lateral tubercles.

The main rugose areas are similar bilaterally, but the right digastric fossa has a continuous rugose area that extends 4.4 mm farther posterolaterally than the left one. The genioglossal tubercles are two small spines with a lower central spine, with the top of them 14.6 mm below the right I_1I_2 alveolar plane. There is neither an internal transverse torus nor a planum alveolare.

Comparative Dimensions and Angulation

The symphyseal height of the Tianyuan 1 mandible (Table 6-4) is relatively modest, falling below the means of all of the comparative samples and below the ranges of variation or close to the lower limits of the comparative ranges of variation (Table 6-2). However, plotting its symphyseal height against its superior length (Figure 6-5) places Tianyuan 1 at the bottom of the size range for Late Pleistocene humans but close to the middle of their proportional distribution. In this, despite the generally larger Middle Paleolithic versus Upper Paleolithic mandibles, there is little difference in relative symphyseal heights.

The size and robusticity of the symphysis are better quantified by using the full subperiosteal cross section of the symphysis, modeled as a solid beam given the combination of cortical bone and dense trabeculae through the symphysis (Table 6-5). Following Daegling (1993, 2001; see discussion in Dobson and Trinkaus, 2002), vertical bending rigidity should be scaled to dental arcade breadth, and labiolingual bending rigidity should be scaled to mandible length.

In the comparison of maximum to minimum second moments of area, which quantifies overall distribution of bone in the symphyseal cross section, Late Pleistocene mandibles are generally similar (Figure 6-6). The three ~40 ka cal BP

Table 6-4. Corpus heights and breadths of the Tianyuan 1 mandible, in millimeters

	Corpus height		*Corpus breadth (thickness)*	
	Right	*Left*	*Right*	*Left*
Symphysis (M-69)[1]	(27.5)		—	
Infradentale-pogonion[1]	(19.0)		—	
I_2-C_1	27.8	—	10.9	—
C_1-P_3	28.1	—	11.2	—
P_3-P_4	29.9	—	12.0	—
P_4-M_1 (MF) (M-69(1), M-69(3))[2]	29.7	(24.7) path	12.1	10.6 path
M_1-M_2 (M-69(2), M-69b)[3]	29.6	—	11.7	—
M_2-M_3	30.3	—	—	—

Note: Estimated values, due to antemortem tooth loss, are in parentheses. "path" indicates pathologically altered bone.

[1] Symphyseal height and the infradentale-pogonion height use the right $I_1 I_{1\text{-}2}$ interdental septum to define the alveolar plane (infradentale).

[2] The P_4-M_1 height is the same as the mental foramen (MF) height, hence the two Martin numbers for the same measurements. The left side values were altered by antemortem tooth loss.

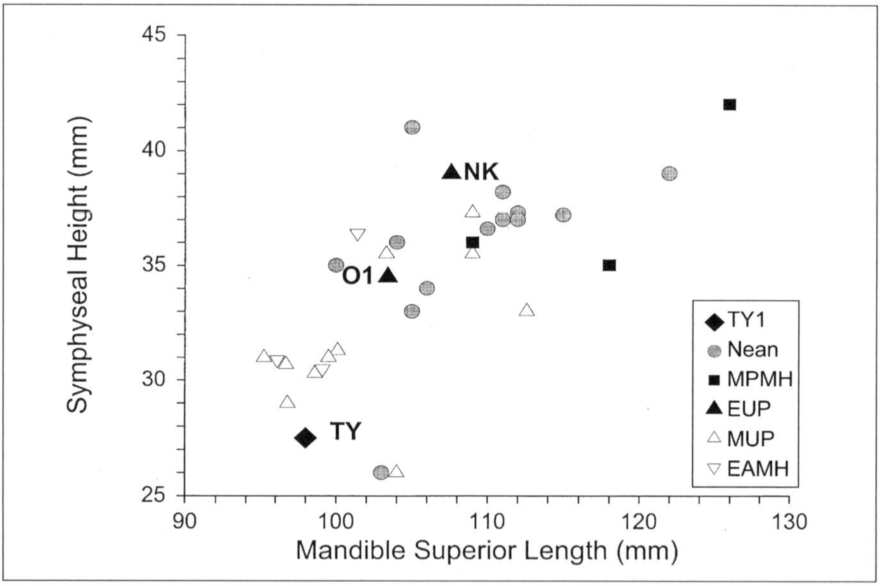

Figure 6-5. Bivariate plot of symphyseal height versus mandible superior length for Tianyuan 1 (*TY*) and Late Pleistocene comparative samples. *TY:* Tianyuan 1; *Nean:* Neandertals; *MPMH:* Middle Paleolithic modern humans; *EUP:* western Early Upper Paleolithic; *MUP:* western Eurasian Mid Upper Paleolithic; *EAMH:* eastern Asian modern humans; *NK:* Nazlet Khater 2; *O1:* Oase 1.

Table 6-5. Anterior symphyseal angle and symphyseal cross-sectional geometric parameters of the Tianyuan 1 mandible

Total cross-sectional area (mm^2)	212.7
Vertical second moment of area (I_x) (mm^4)	(8876)
Anteroposterior second moment of area (I_y) (mm^4)	(1589)
Maximum second moment of area (I_{max}) (mm^4)	8904
Minimum second moment of area (I_{min}) (mm^4)	1562
Orientation of I_{max} (theta)	(94°)
Anterior symphyseal angle (M-79(1b))	(96°)

Note: Modeled as a solid beam (see Dobson and Trinkaus, 2002). Estimated values are due to estimation of the alveolar plane given resorption of some of the alveoli. Symphyseal orientation is relative to the alveolar plane.

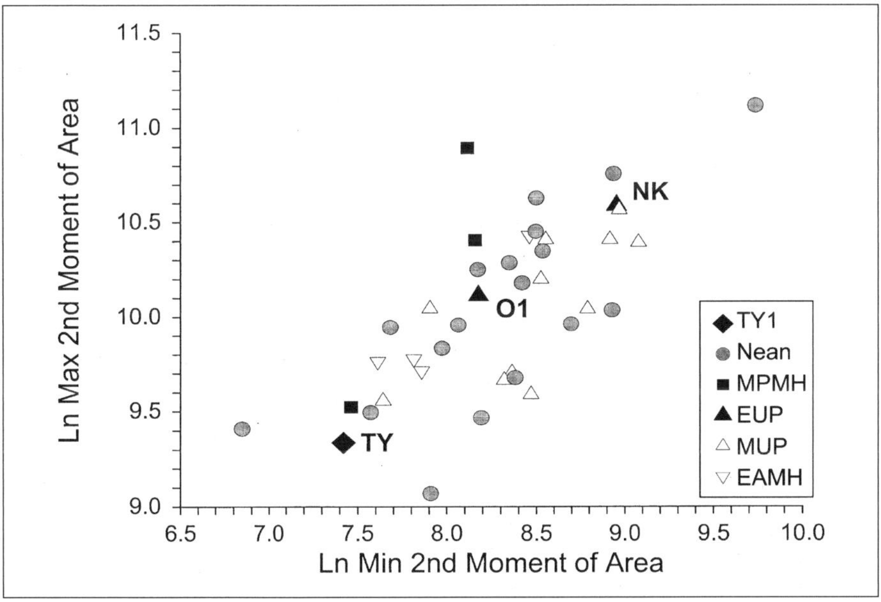

Figure 6-6. Bivariate plot of the symphyseal maximum second moment of area versus the minimum second moment of area for Tianyuan 1 (*TY*) and Late Pleistocene comparative samples. Values from modeling the symphysis as a solid beam. *TY:* Tianyuan 1; *Nean:* Neandertals; *MPMH:* Middle Paleolithic modern humans; *EUP:* western Early Upper Paleolithic; *MUP:* western Eurasian Mid Upper Paleolithic; *EAMH:* eastern Asian modern humans; *NK:* Nazlet Khater 2; *O1:* Oase 1.

mandibles—Tianyuan 1, Oase 1, and Nazlet Khater 2—all fall along the middle of the overall distribution, even though the first has among the smallest of the symphyses and the last has among the largest. The high outlier in size is the Kebara 2 Neandertal, and the high outlier in relative maximum second moment of area is the MPMH Skhul 4. There is no difference across the samples.

The scaling of the vertical (superoinferior) and horizontal (anteroposterior) second moments of area, oriented relative to the alveolar plane, provides little if any difference across the samples (Figure 6-7). As previously noted (Dobson and Trinkaus, 2002), these measures of symphyseal robusticity are independent of whether the mandible exhibits a "chin." One of the more robust symphyses, especially in a vertical direction, is that of Nazlet Khater 2, as previously noted (Crevecoeur and Trinkaus, 2004). Oase 1 falls in the middle of the overall distribution. And Tianyuan 1 is among the more gracile symphyses, in both directions and relative (as appropriate) to both mandibular breadth and length.

The orientation of the mandibular symphysis can be assessed using both the anterior symphyseal angle (infradentale-pogonion versus the alveolar plane) and the major axis through the cross section versus the alveolar plane (Table 6-5). The former, osteometric angle places Tianyuan 1 well above the Middle Paleolithic samples and the two EUP mandibles, but close to the middle of the MUP sample and that of the eastern Asian early modern human sample (Figure 6-8A). Despite some overlap in the ranges of variation and a couple of low outliers in the last two samples (Předmostí 3 and ZKD-UC 104), there is a shift from the Neandertals to the MPMH and EUP specimens to the later Upper Paleolithic samples. Tianyuan 1, despite its EUP age, falls with the last samples. The same pattern pertains to comparisons of the cross-sectional symphyseal angle (Figure 6-8B), with Tianyuan 1 falling among the later early modern humans and above the contemporaneous and earlier mandibles.

The Lateral Corpus

Morphology

The lateral corpus, from the symphyseal region to the beginning of the ramus, is gently convex inferiorly, both along its basilar margin and on the alveolar plane. At the same time, it remains relatively constant in height. This parallel nature of the superior and inferior corpus margins is reflected in corpus height measurements from C_1P_3 to M_2M_3, which vary by less than 2 mm (Table 6-4). This pattern of relatively parallel superior and inferior lateral corpus margins is common among both late archaic and early modern humans.

The one feature of note on the lateral surface of the corpus is the mental foramen. The right one is in the vertical middle of the corpus, 13.8 mm to the inferior margin (46.5% of the corpus height). It opens directly posteriorly and has a sharp anterior margin. The left one is a similar distance from the inferior margin (13.7 mm), and it opens posteriorly in a similar fashion. They are approximately equal distances from the symphyseal midline, such that the direct distances between the middle of the tuber symphyseos and each foramen are 30.0 mm on the right side

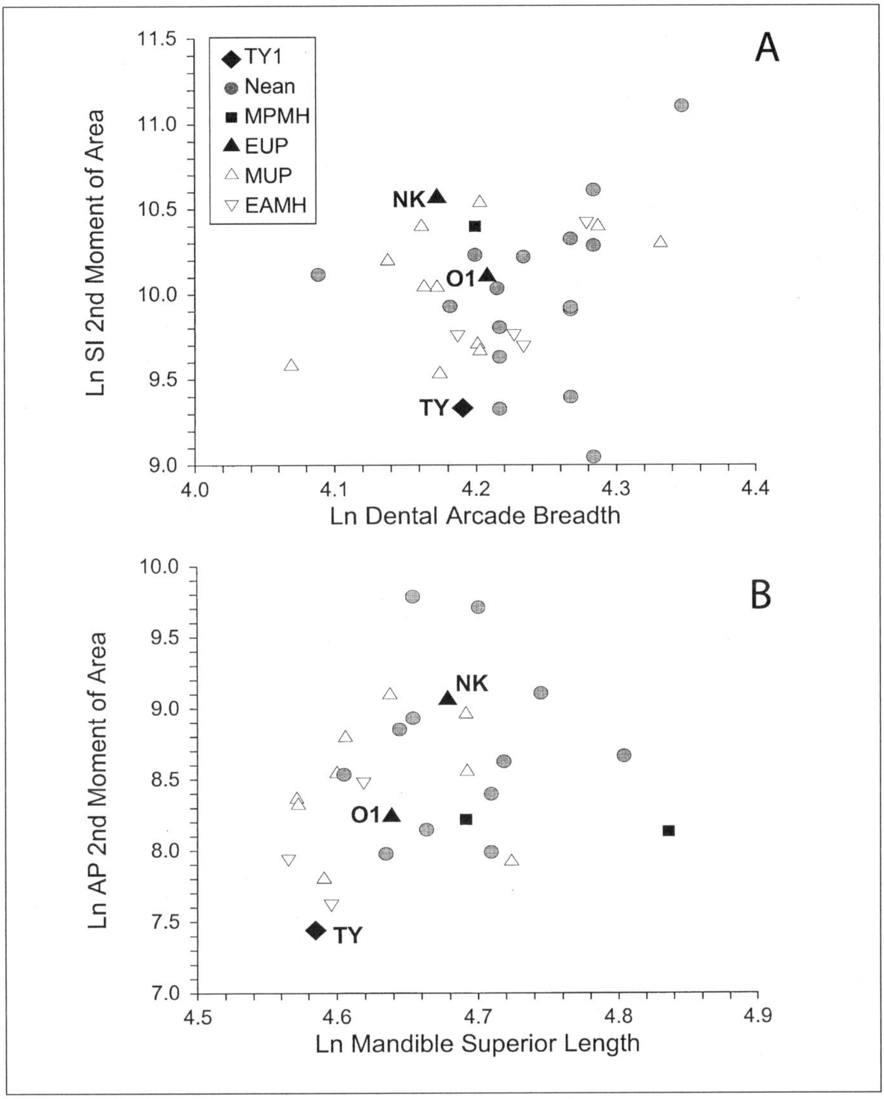

Figure 6-7. Bivariate plots for Tianyuan 1 (*TY*) and Late Pleistocene comparative samples of (*A*) symphyseal superoinferior second moment of area versus dental arcade breadth and (*B*) symphyseal anteroposterior second moment of area versus mandible superior length. Values from modeling the symphysis as a solid beam. *TY:* Tianyuan 1; *Nean:* Neandertals; *MPMH:* Middle Paleolithic modern humans; *EUP:* western Early Upper Paleolithic; *MUP:* western Eurasian Mid Upper Paleolithic; *EAMH:* eastern Asian modern humans; *NK:* Nazlet Khater 2; *O1:* Oase 1.

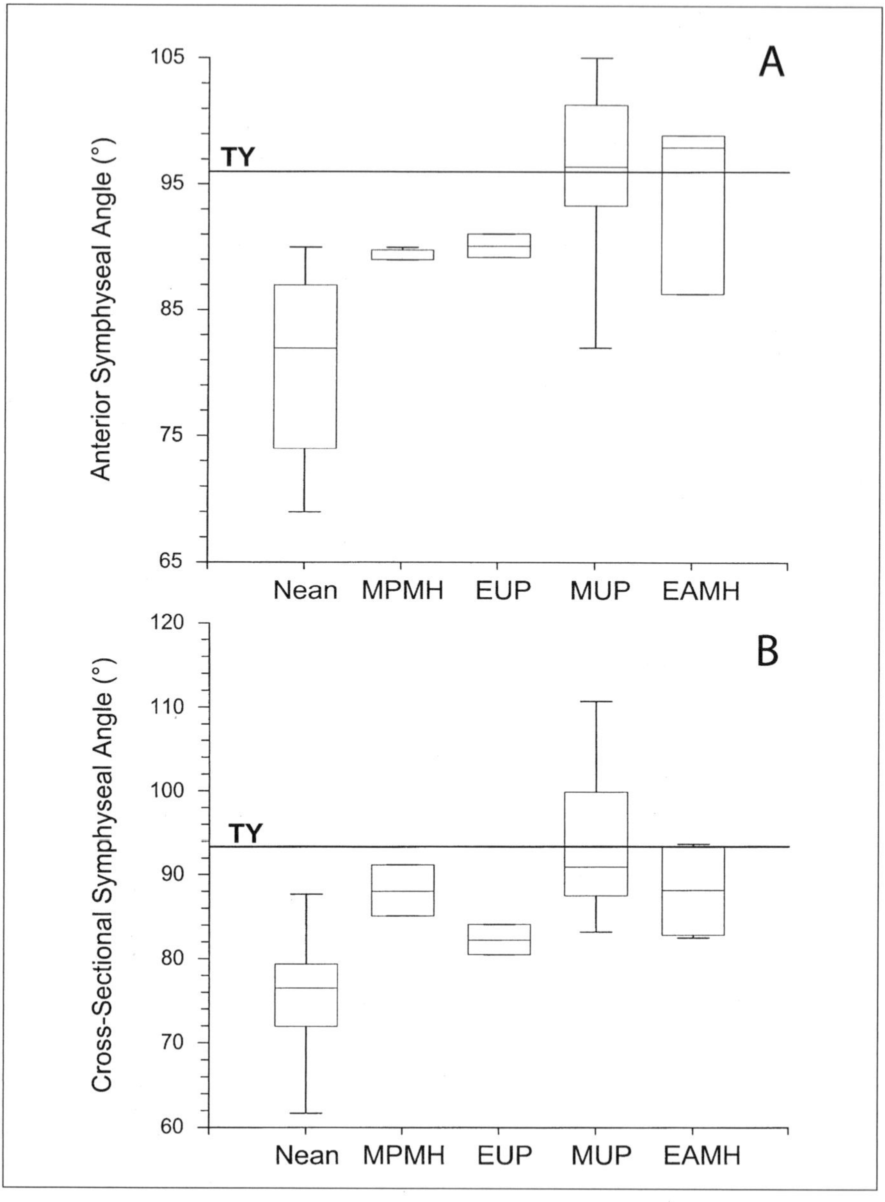

Figure 6-8. Box plots of (*A*) anterior symphyseal angle for Tianyuan 1 (*TY*) (~96°) and (*B*) cross-sectional symphyseal angle for Tianyuan 1 (~94°) versus distributions for comparative Late Pleistocene samples. *Nean:* Neandertals; *MPMH:* Middle Paleolithic modern humans; *EUP:* western Early Upper Paleolithic; *MUP:* western Eurasian Mid Upper Paleolithic; *EAMH:* eastern Asian modern humans.

and 28.5 mm on the left side. There are also two very small foramina, one below the right I_2 and one below the right C_1, that probably represent passages for additional fibers of the inferior alveolar nerve. Similar foramina are absent on the left side.

The right mental foramen is below the P_4M_1 interdental septum. It is possible that this relative position has been altered by interproximal wear and mesial dental drift, given the extensive wear on the teeth (Chapter 5), in which case placing it below the P_4 might approximate its original, young adult, position. However, if one were to correct it in this manner, one would have to do the same for a suite of other fossil mandibles (e.g., Banyoles 1, Brno 2, Caviglione 1, La Chapelle-aux-Saints 1, Dolní Věstonice 3 and 16, La Ferrassie 1, Guattari 2, KRM 41815, Minatogawa 1, Muierii 1, Nazlet Khater 2, Palomas 1, Pavlov 1, Shanidar 1 and 4, Sunghir 1, and ZKD-UC 101). Moving the mental foramen position one half tooth (midtooth to the next mesial interdental space or interdental space to the next mesial tooth) would shift the distributions of the positions of the mental foramina relative to the dentition mesially across all but the MPMH and the Aterian samples (Table 6-6).

Whether one uses the uncorrected or corrected mental foramen positions, it is apparent, as is well established (Trinkaus, 1993a), that the primary contrast in the western Old World Late Pleistocene is a shift, on average, from more distal foramina to more mesial ones between the Neandertals and early modern humans (Table 6-6). The one Aterian specimen has a relatively mesial foramen, and the South African MSA sample is variable. The eastern Asian early modern humans are, to some degree, intermediate between the Neandertals and the western early modern humans. In this context, the position of the Tianyuan 1 mental foramen most closely matches the position seen in the more recent eastern Asian man-

Table 6-6. Mental foramen position for Tianyuan 1 and comparative Late Pleistocene samples

	P_3 (%)	P_3/P_4 (%)	P_4 (%)	P_4/M_1 (%)	M_1 (%)	N
Tianyuan 1			(X)	X		
East Asia EMH			31.3 (56.3)	56.3 (31.3)	12.5 (12.5)	8
West MUP	10.7 (10.7)	0.0 (19.6)	69.6 (51.8)	12.5 (10.7)	7.1 (7.1)	28
West EUP		0.0 (20.0)	80.0 (80.0)	20.0 (0.0)		5
MPMH			57.1	28.6	14.3	7
Neandertals		0.0 (4.8)	11.3 (11.3)	40.3 (48.4)	48.4 (35.5)	31
South Africa MSA	0.0 (33.3)	33.3 (0.0)	16.7 (16.7)	33.3 (33.3)	16.7 (16.7)	3
Aterian			100.0			1

Note: EMH: early modern humans; *MUP:* Mid Upper Paleolithic; *EUP:* Early Upper Paleolithic; *MPMH:* Middle Paleolithic modern humans; *MSA*: Middle Stone Age. The uncorrected mental foramen positions are provided, followed in parentheses by the distributions after correction for mesial drift from pronounced interproximal wear. There is no correction for the MPMH and Aterian samples.

dibles, whether one uses the uncorrected or corrected position. Given that these data are indirect reflections of overall facial length, dental arcade length, and inferior alveolar nerve canal length (Trinkaus, 1993a, 2006a), it is hard to separate out which is more important in determining mental foramen position. However, the rather short mandible length, and hence facial length, of Tianyuan 1, suggests that its moderately more mesial foramen position is due in large part to its facial shortness.

More posteriorly the lateral eminence emerges below the M_2M_3 interdental septum and only becomes a distinct angle leading up to the anterior ramus below the M_3. This relatively posterior position of the lateral eminence has been considered a Neandertal characteristic (Rosas, 2001), and it is probably related in those humans to their midfacial prognathism and the associated retromolar spaces. The more distal position of the lateral eminence on the Tianyuan 1 mandible may be related to its worn, and hence more mesial, dentition and its small retromolar space. There is little else of note on the external surface. (The fenestration around the distal M_1 root is discussed in Chapter 12.)

Medially there is a rugose mylohyoid line from the P_4M_1 to the distal M_3. It is especially rugose below the M_1, thinning and becoming more of a crest below the M_3 (Table 6-7). As indicated by its heights below the alveolar plane, it rises gently posteriorly but remains well below the alveolar plane, even at the M_3 level. There is also a bilateral rugosity, from the superior digastric fossae rising laterally and slightly superiorly of the inferior mylohyoid line in the area of the M_1.

Comparative Dimensions

Across the Late Pleistocene samples, there is little change in lateral corpus heights at either the level of the mental foramen or at the M_1M_2 interdental septum. There are a few mandibles with pronounced heights, but the samples remain insignificantly different from each other (Table 6-2). In this context, the corpus heights of Tianyuan 1 (Table 6-4) are close to or slightly below the means of the other samples for all comparisons except the eastern Asian EMH M_1M_2 height (Table 6-2).

Table 6-7. Position and dimensions of the Tianyuan 1 right mylohyoid line, in millimeters

	Height to alveolar plane	*Rugosity thickness*
Mid M_1	16.1	6.5
Mid M_2	13.6	4.6
Mid M_3	(10.1)	3.7

Note: The height below the M_3 is estimated given pathological alteration of the M_3 alveolus.

At the same time, there is a clear reduction in corpus breadth, especially at the mental foramen, between the Middle Paleolithic samples and the Upper Paleolithic samples, independent of geography and of late archaic versus early modern human affinities. It is also apparently independent of molar crown (or root) breadths, given the absence of differences in molar breadths across these samples (see Table 5-5). Two of the thickest mandibles, Kebara 2 and Nazlet Khater 2, have unexceptional molar breadths, and the specimen with the broadest molars, Oase 1, has a corpus breadth in the middle of the Upper Paleolithic means. The corpus breadth at the mental foramen of Tianyuan 1 is similar to those of most of the Upper Paleolithic mandibles, falling on the east Asian mean and within one standard error of the MUP mean (Table 6-2). Its breadth at the M_1M_2 remains small, despite a general increase in the other samples, which is related to the relatively distal position of the lateral eminence in Tianyuan 1.

Detailed data on dental arcade dimensions, taken directly as oblique distances or estimated by doubling the distance from the midline to the arcade position, are provided in Table 6-8.

The Mandibular Ramus

Despite the loss of the condyle and condylar neck and of the coronoid process tip and anterior margin, the Tianyuan 1 ramus presents data on a series of discrete traits (Figure 6-9). These are traits that have been shown to change markedly in their distributions (but not presence/absence) between Neandertals and western Old World early modern humans (Stefan and Trinkaus, 1998a,b; Richards et al., 2003; Crevecoeur and Trinkaus, 2004; Soficaru et al., 2006; Trinkaus, 2007a). Some probably represent epigenetic traits in themselves, but others are reflections of the relative dimensions and/or positions of other aspects of the mandible or the facial

Table 6-8. Dental arcade measurements of the Tianyuan 1 mandible, in millimeters

	Direct oblique distance from ID	*Coronal plane arcade breadths*
I_2-C_1	8.9	—
C_1-P_3	16.6	(31.0)
P_3-P_4	21.5	(41.0)
P_4-M_1	27.8	(48.0)
M_1-M_2	37.8	(58.0)
M_2-M_3	46.8	(61.0)

Note: Direct distances are from infradentale (ID) to the buccal alveolus between each pair of teeth. Arcade breadths are from the mandibular midline, estimated from the symphyseal morphology, to the buccal alveolus between each pair of teeth, times 2. Since the infradentale position is estimated based on the right I_1-I_2 interdental septum, there is a small margin of error in the direct distances. Given possible angular error in determining the midline, the arcade breadth measurements have error, which should increase slightly distally.

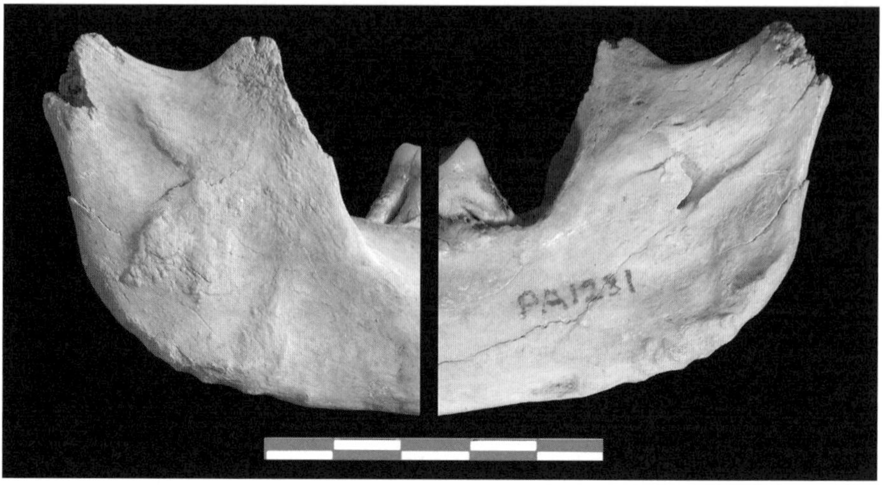

Figure 6-9. Lateral (*left*) and medial (*right*) views of the Tianyuan 1 mandibular ramus. Scale in centimeters.

skeleton overall (Franciscus and Trinkaus, 1995; Anderson, 2000). They can nonetheless be scored as discrete traits on incomplete and/or isolated mandibles, such as that of Tianyuan 1.

In strict *norma lateralis,* the anteroinferior margin of the ramus, where it crosses the alveolar plane, crosses the apical portion of the distal M_3 root. The M_3 is in an abnormal supererupted and tilted position, which has moved the crown mesially and the apical portions of the roots distally. As such, the impression is that there is not a retromolar space (a clear separation between the distal M_3 crown and the anterior ramus in *norma lateralis,* or in the parasagittal plane of the anterior ramus and M_3). However, if one repositions the crown of the M_3 against the distal M_2 crown, in its original occlusal position, there is a small retromolar space, 1 to 2 mm wide. At the same time, the extensive interproximal attrition of the whole dentition (Chapter 5) may have shortened the dental arcade by a millimeter or two.

As a result, given the current state of the Tianyuan 1 mandible and a reasonable positioning of the M_3 in its occlusal position, one would have to conclude that Tianyuan 1 possessed a small retromolar space (contra Shang et al., 2007). The effect of interproximal wear is secondary, especially since it is rarely (e.g., Nara et al., 1998) taken into account in scoring retromolar space presence/absence. However, given that this is based on M_3 repositioning, the "presence" of the space in Tianyuan 1 is placed in parentheses in Table 6-9.

Retromolar spaces are common among the Neandertals, and small ones occur in three of the MPMH mandibles (all from Skhul) and a minority of the Upper Pa-

Table 6-9. Discrete trait distributions for later Pleistocene human mandibles

	Retromolar space (% absent [N])	*Mandibular notch symmetry (% present [N])*	*Mandibular foramen (% open [N])*	*Medial pterygoid tubercle (% absent [N])*
Tianyuan 1	(Present)	Present	Open	Absent
East Asia EMH	78.6 (7)	100 (4)	80.0 (5)	50.0 (4)
West MUP	82.0 (25)	95.8 (24)	100 (21)	90.0 (10)
West EUP	100 (5)	66.7 (3)	83.3 (3)	100 (3)
MPMH	62.5 (8)	100 (6)	100 (7)	100 (8)
Neandertals	24.1 (29)	30.8 (13)	60.9 (23)	18.8 (16)
South Africa MSA	100 (2)	—	—	—
Aterian	100 (1)	100 (1)	100 (1)	—

Note: EMH: early modern humans; *MUP:* Mid Upper Paleolithic; *EUP:* Early Upper Paleolithic; *MPMH:* Middle Paleolithic modern humans; *MSA:* Middle Stone Age. The polarities are arranged so as to count the percentage of specimens with the derived Middle Paleolithic and Upper Paleolithic modern human pattern. Features that are asymmetrical or intermediate between categories are scored as 0.5 in each.

leolithic humans (Table 6-9). One is also present in Minatogawa 7, and ZKD-UC 101 has the anterior ramus on the distal edge of the M_3. The presence of a small retromolar space in Tianyuan 1 therefore places it among a minority of the early modern humans.

Among Neandertals, the mandibular notch tends to be asymmetrical with a high coronoid process and its lowest point just anterior of the condylar neck, even though several of them lack this configuration (Stefan and Trinkaus, 1998a). Among early modern humans, out of 37 rami, all but two mandibles (94.6%) have symmetrical notches with the lowest point in the middle of the notch; the two exceptions are the EUP Muierii 1 mandible and the late MUP Ohalo 2 mandible. The Tianyuan 1 mandibular notch lacks its anterior and posterior ends, but the middle third is present and preserves the lowest point of the curve. The lowest point of the Tianyuan 1 mandibular notch is in the middle, directly above the mandibular foramen, making it certain that its notch was essentially symmetrical. It therefore contrasts with the majority of the Neandertals and matches the overwhelming majority of early modern human mandibles. Earlier eastern Asian mandibular notches are rare and incomplete, but three partial ones from Zhoukoudian Locality 1 (G1, H1 and B1) all appear to have asymmetrical notches similar to those of the Neandertals (Weidenreich, 1936).

The mandibular foramen of the Tianyuan 1 mandible has a small and minimally projecting, fully intact lingula that borders an open notch 4.5 mm long and ~2.0 mm wide. The notch leads inferoanteriorly to an open mylohyoid groove 26.5 mm long with no bony bridging. It is therefore the open, or funnel-shaped (Kallay, 1970), form, and it lacks the closed, horizontal-oval or lingular bridging configuration (Kallay, 1970; Smith, 1978; Jidoi et al., 2000). The open foramen appears to be the ancestral human pattern (Trinkaus, 2006a), and it is the most common pattern found among recent humans (Yamano and Yamaguchi, 1976; Smith, 1978; Jidoi et al., 2000). It is also found in the majority of the Neandertals, even though 39% of them exhibit foraminal closure. Among early modern humans (N = 36), one specimen (ZKD-UC 104) has the horizontal-oval form, and one specimen (Oase 1) exhibits it unilaterally. Earlier eastern Asian archaic *Homo* mandibles (N = 4) from Chenjiawo and Zhoukoudian Locality 1 all have open foramina (Weidenreich, 1936; Woo, 1964).

The gonial angle of Tianyuan 1 is evenly rounded, with no notch or change in curvature at its anteroinferior end and only a slight change in its posterosuperior contour leading up to the condylar neck. The even curvature, or absence of a notch, at the anteroinferior gonial margin is the most common one among Late Pleistocene humans. A notch is similarly absent in the 6 available east Asian early modern human mandibles, all but 3 of 22 MUP mandibles, 2 of 3 EUP mandibles, 1 of 2 MPMH mandibles, but only 12 of 20 Neandertal specimens. Pronounced notches, with inferiorly directed gonial areas, are present only in Skhul 5 and Oase 1. Among the earlier Zhoukoudian Locality mandibles, 1 (G1) appears to have had a small notch, but it was absent from the H1 mandible (Weidenreich, 1936).

There is a modest concavity in the medial pterygoid area, with the gonial region curving medially. The tubercles for the medial pterygoid insertion are modest and located posteroinferiorly and inferiorly around the gonial region. There is no enlargement of the superior medial pterygoid tubercle, such as is commonly found among Neandertals and half of the four eastern Asian early modern human mandibles with the ramus preserved (Minatogawa 1 and 5) (Table 6-9).

The external masseteric area has a ridge rising from the anteroinferior end of the gonial curve, 8.6 mm wide at its base and extending anterosuperiorly ~20 mm from the edge. In the middle of the lateral ramus, there is a large rugose area, 18 mm high and 15.5 mm wide. It is a raised area but without indications of fiber orientations; it may represent a broad enthesopathy for the deep portion of masseter.

Summary

The Tianyuan 1 mandible presents a suite of distinctively modern human features that align it particularly with contemporaneous and more recent Upper Paleolithic early modern human mandibles. There is a clear projecting mentum osseum, but one lacking prominent lateral tubercles. This combines with the symphyseal cross section to produce anterior and cross-sectional symphyseal angles above those of Middle Paleolithic mandibles and even those of western Old World Early Upper Paleolithic modern humans. Its mental foramen position is intermediate for these comparative samples, but its lateral corpus breadths are modest, similar to those of almost all Upper Paleolithic mandibles. Tianyuan 1 probably had a small retromolar space, a feature not unknown among early modern humans, and the remainder of its ramal discrete traits align it with early modern humans even though individual traits are found among the Neandertals. All of this occurs in the context of a rather small mandible, again similar to those of more recent Upper Paleolithic humans from both eastern and western Eurasia.

Chapter 7

Body Size and Body Proportions of Tianyuan 1

THE ASSESSMENT of various aspects of the Tianyuan 1 skeletal remains and of more general aspects of its paleobiology requires an estimation of its body size and body proportions. Given that at least one long bone from each of the major limb segments is sufficiently preserved to provide a bone length, and that there are several lower limb articulations preserved (Chapters 3, 9, and 10), it is possible to provide some assessment of these parameters. However, the almost complete absence of the pelvis and of the axial skeleton (all except for portions of the sternum and one cervical vertebra; Chapter 8) means that it is not possible to assess body core transverse diameters or trunk length.

Limb Segment and Articular Dimensions

The estimated lengths of the left humerus, the left radius, the right femur, and the right tibia are provided in Table 7-1, along with the standard errors of the estimates (SE_{est}) (see notes to tables 9-2, 9-7, 10-1, and 10-4 for estimation details). The SE_{est} are all ≤1.4% of the estimated lengths; they therefore provide reliable estimates of the original lengths of the bones. When they are compared to the lengths of other Late Pleistocene humans (Table 7-1; Figures 7-1 and 7-2), it is apparent that the Tianyuan 1 limb bones are relatively long compared to the Neandertal sample and especially to the eastern Asian early modern humans. The latter sample includes the very diminutive Minatogawa sample (Baba and Endo, 1982), as well as a single, purportedly female, femur length from ZKD-UC (Weidenreich, 1941). Except for the radius, the Tianyuan 1 long bone lengths are moderately short compared to most of the MPMH humans, but they fall very close to the middle of the combined distribution for the western EUP and MUP early modern humans.

A contrasting pattern is evident in the estimated femoral head diameter of Tianyuan 1 of 53.8 mm (± 1.7 mm). It is at the upper limits of the Late Pleistocene

Table 7-1. Major long bone lengths, talar length and femoral head diameter for Tianyuan 1 and Late Pleistocene comparative samples, in millimeters

	Humerus maximum length	*Radius maximum length*[1]	*Femur bicondylar length*[2,3]	*Tibia maximum length*	*Talar length*	*Femoral head diameter*[4]
Tianyuan 1	331.4 ± 3.6	251.8 ± 3.4	460.4 ± 1.8	385.2 ± 5.2	56.6	53.8 ± 1.7
East Asia EMH	276.3 ± 11.3 (4)	209.2 ± 7.7 (4)	391.4 ± 32.5 (4)	301.0, 305.5, 322.0	43.0, 47.0	41.3 ± 4.4 (6)[5]
West MUP	334.7 ± 24.2 (25)	259.9 ± 18.6 (22)	460.8 ± 36.3 (31)	390.7 ± 27.1 (22)	52.0 ± 3.3 (15)	46.9 ± 3.7 (26)
West EUP	315.5, 373.5	240.5	483.3	—	57.4	40.1, 44.0, 50.7
MPMH	355.1 ± 25.2 (4)	249.0 ± 19.4 (5)	481.1 ± 23.0 (5)	365.0, 433.0, 439.0	52.8 ± 3.8 (7)	47.0 ± 1.8 (4)
Neandertals	311.5 ± 13.8 (11)	233.8 ± 16.0 (13)	437.5 ± 25.7 (13)	343.5 ± 25.7 (10)	52.0 ± 3.2 (11)	50.3 ± 3.2 (9)
K-W P-value	<0.001	<0.001	0.007	<0.001	0.135	0.013

Note: EMH: early modern humans; *MUP:* Mid Upper Paleolithic; *EUP:* Early Upper Paleolithic; *MPMH:* Middle Paleolithic modern humans. Mean ± standard deviation (N) are provided for samples >3. For the estimated Tianyuan 1 long bone lengths, mean ± SE_{est} are provided; for estimation details, see Tables 9-2, 9-8, 10-1, and 10-4. Kruskal-Wallis (K-W) P-values are provided across the comparative samples; all but talar length are significant at the $P < 0.05$ level after a sequentially reductive multiple comparison correction.

[1] Seven of the fossil radial maximum lengths were predicted from the skeleton's ulnar maximum length, based on a least squares regression from the pooled Late Pleistocene fossil remains (RadMaxLen = 0.964 x UlnMaxLen—11.02, $r^2 = 0.973$, N = 25).

[2] The ZKD-UC femoral maximum length of 438 mm provided by Weidenreich (1941) was converted to a bicondylar length of 435 mm using a least squares regression from a recent human sample (FemBclLen = 1.01 x FemMaxLen—8.38, $r^2 = 0.998$, N = 40).

[3] The femora of Nazlet Khater 2 are abnormally short relative to his humeri and apparently to his trunk dimensions (Crevecoeur, 2008). As a result, his average femoral bicondylar length of 407.0 mm is not included in the western EUP sample (the single individual is the estimate for Mladeč 27 [Trinkaus, Smith, et al., 2006]). However, in the comparison of femoral head diameter to length (Figure 7-7), these two values are included.

[4] Seven of the femoral sagittal head diameters were predicted from the skeleton's acetabular height, based on a least squares regression from a recent human sample (FemHdDia = 0.988 x AcetHt—8.39, $r^2 = 0.896$, N = 39).

[5] Two of the femoral head diameters for the eastern Asian early modern human sample are for the now lost ZKD-UC specimens, 105 and 117. Weidenreich (1941) did not provide head diameters for these femora, but his plates 21, 22, and 24 provide scaled radiographs or photos of the original femora. Measurement of the plates and correction for his indicated scaling provide vertical head diameters of 48.9 and 46.8 mm for the ZKD-UC 105 "male" femur (average: 47.9 mm) and 42.8 mm for the ZKD-UC 117 "female" femur. These have been rounded off to 48 and 43 mm.

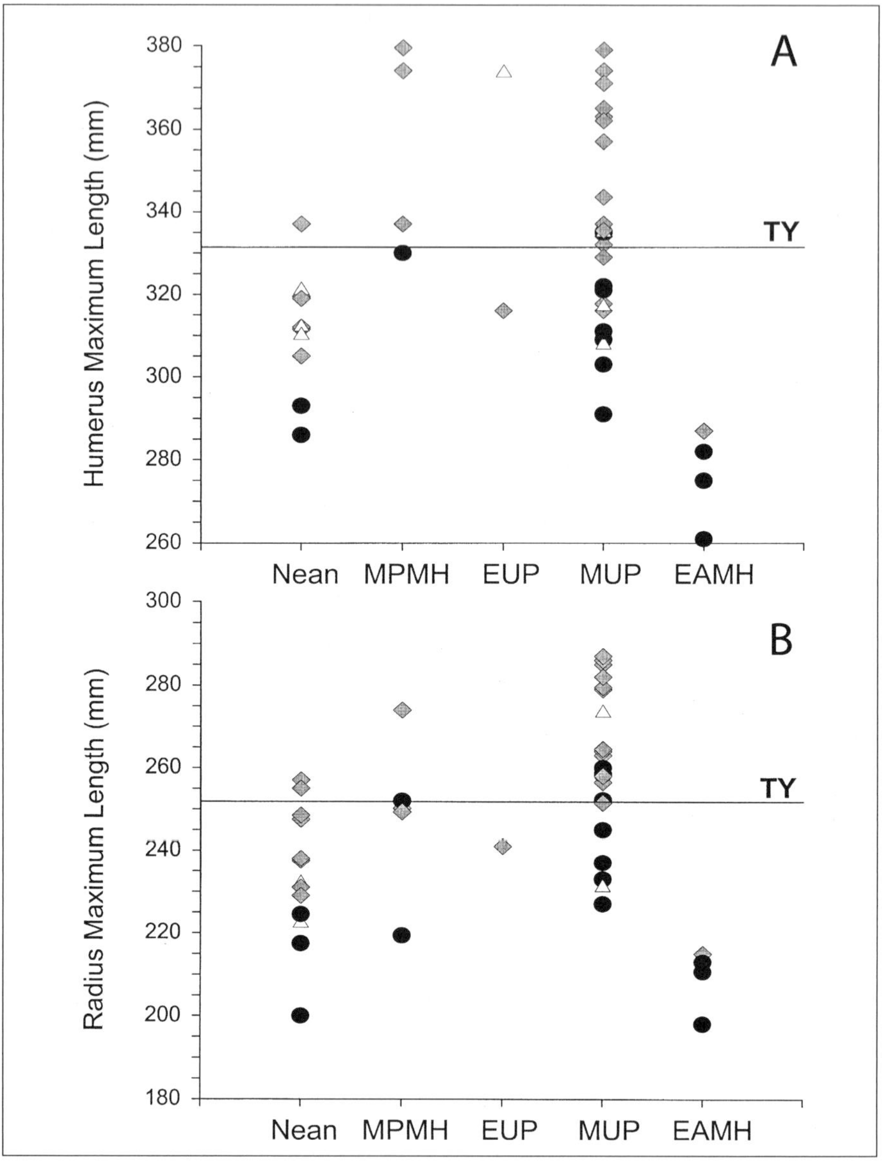

Figure 7-1. Plots of (*A*) humerus maximum length and (*B*) radius maximum length (B) for Tianyuan 1 (*TY*) compared to Late Pleistocene samples. *Nean:* Neandertals; *MPMH:* Middle Paleolithic modern humans; *EUP:* western Early Upper Paleolithic; *MUP:* western Eurasian Mid Upper Paleolithic; *EAMH:* eastern Asian modern humans. *Gray diamonds:* males; *black circles:* females; *open triangles:* sex uncertain.

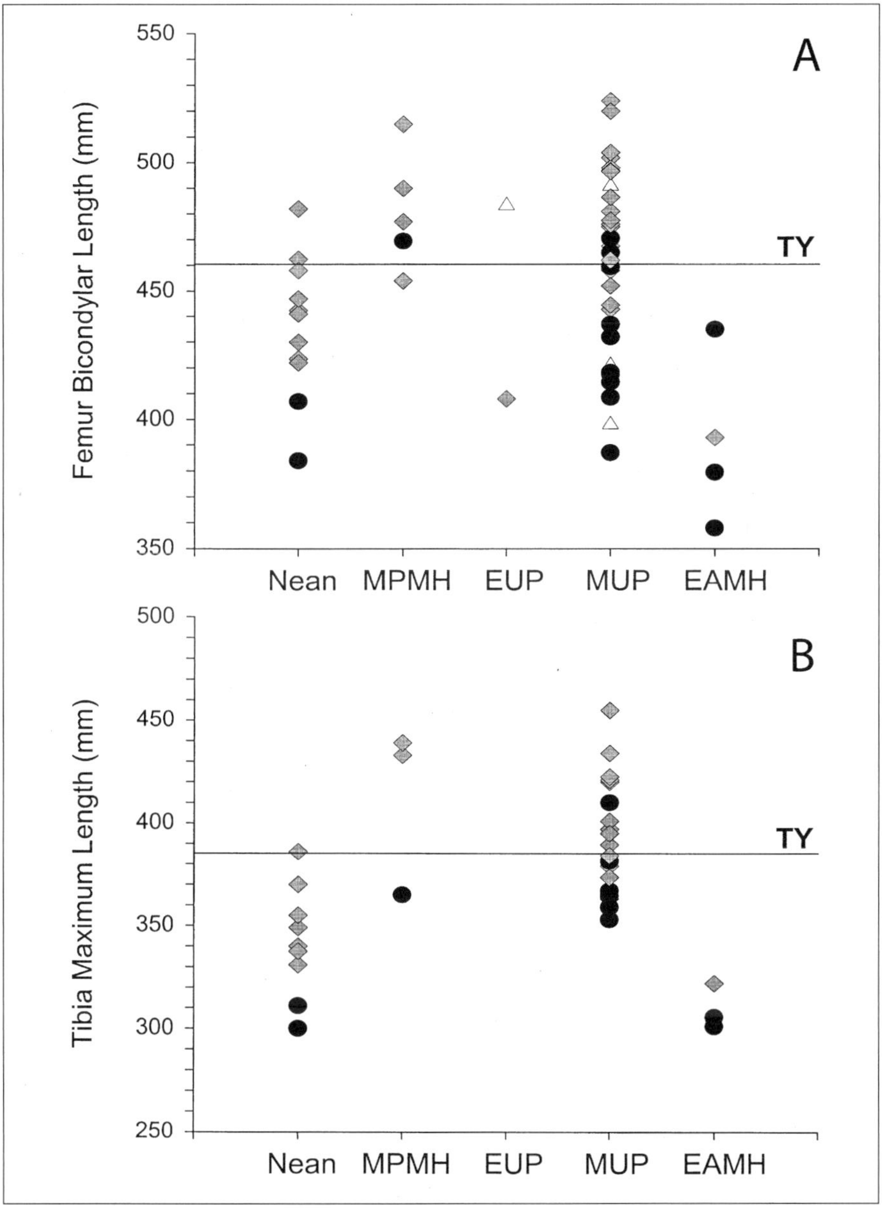

Figure 7-2. Plots of (*A*) femur bicondylar length and (*B*) tibial maximum length for Tianyuan 1 (*TY*) compared to Late Pleistocene samples. *Nean:* Neandertals; *MPMH:* Middle Paleolithic modern humans; *EUP:* western Early Upper Paleolithic; *MUP:* western Eurasian Mid Upper Paleolithic; *EAMH:* eastern Asian modern humans. *Gray diamonds:* males; *black circles:* females; *open triangles:* sex uncertain.

ranges of variation. It is exceeded only by the western European La Ferrassie 1 and Spy 2 Neandertals, and it is approached (≥52 mm) only by two MUP specimens (Barma Grande 2 and Grotte-des-Enfants 4) and by two other western European Neandertals (La Chapelle-aux-Saints 1 and Feldhofer 1) (Figure 7-3A). In eastern Asia, however, the Minatogawa femora, especially of the three females, have very small femoral heads, and estimated values for the ZKD-UC femora from Weidenreich (1941; Table 7-1) are only moderately larger than the same-sex Minatogawa values.

Relatively large (≥50 mm) femoral head diameters are present among earlier Middle Pleistocene humans. The lower limit of the estimated Tianyuan 1 range (minus 2 standard errors at 50.4 mm) is close to the value of 50.2 mm estimated from the acetabular height of the late Middle Pleistocene female Jinniushan 1 skeleton (Rosenberg et al., 2006). Three European Middle Pleistocene specimens have similarly large femoral head diameters, all estimated from their acetabular heights (Arago 44: 52.1 mm; AT-SH pelvis I: 51.4 mm; AT-SH 835/2501: 50.6 mm), and two African, probably Middle Pleistocene femora (Berg Aukas 1 and Broken Hill 907) have large head diameters (57.6 and 52.5 mm respectively) (Grine et al., 1995; Trinkaus, 2009).

It should be emphasized that the estimated femoral head diameter is based on the anteroposterior diameter of the right lateral condyle, from the posterior condyle to the anterior patellar surface margin (see Table 10-1). The value employed for the lateral condylar anteroposterior diameter (~75 mm) is a rounded off value from the preserved anteroposterior dimension of the lateral condyle (74.6 mm), given that the preserved bone may not have quite reached the full anterior extent of the lateral patellar margin (Chapter 3). As a result, the value of 75 mm for the condylar dimension is a minimum value, and therefore the estimated femoral head diameter of 53.8 ± 1.7 mm may slightly underestimate the original dimensions of the Tianyuan 1 hip articulation.

The talar length of Tianyuan 1 follows a similar pattern (Figure 7-3B). This is partially to be expected, since the talus is largely an articular cog in the universal joint across the talocrural and subtalar articulations. The length (flexor hallucis longus sulcus to distal head) of 56.6 mm is matched or exceeded among Late Pleistocene humans only by Cro-Magnon 4337, La Ferrassie 1, Mladeč 30, and Skhul 6. Neither of the Minatogawa tali come close to its size.

Given that the femoral head diameters of Tianyuan 1 are estimated from its distal femoral lateral articular depth (see Table 10-1) and that talar length is not strictly an articular measurement, one could question whether these provide an accurate reflection of the lower limb baseline joint reaction forces of Tianyuan 1 and hence of its body mass. If one compares its distal femoral depth measurement (~75 mm), hence avoiding the least squares estimation of its femoral head diameter, with those of the few other Late Pleistocene femora that preserve this

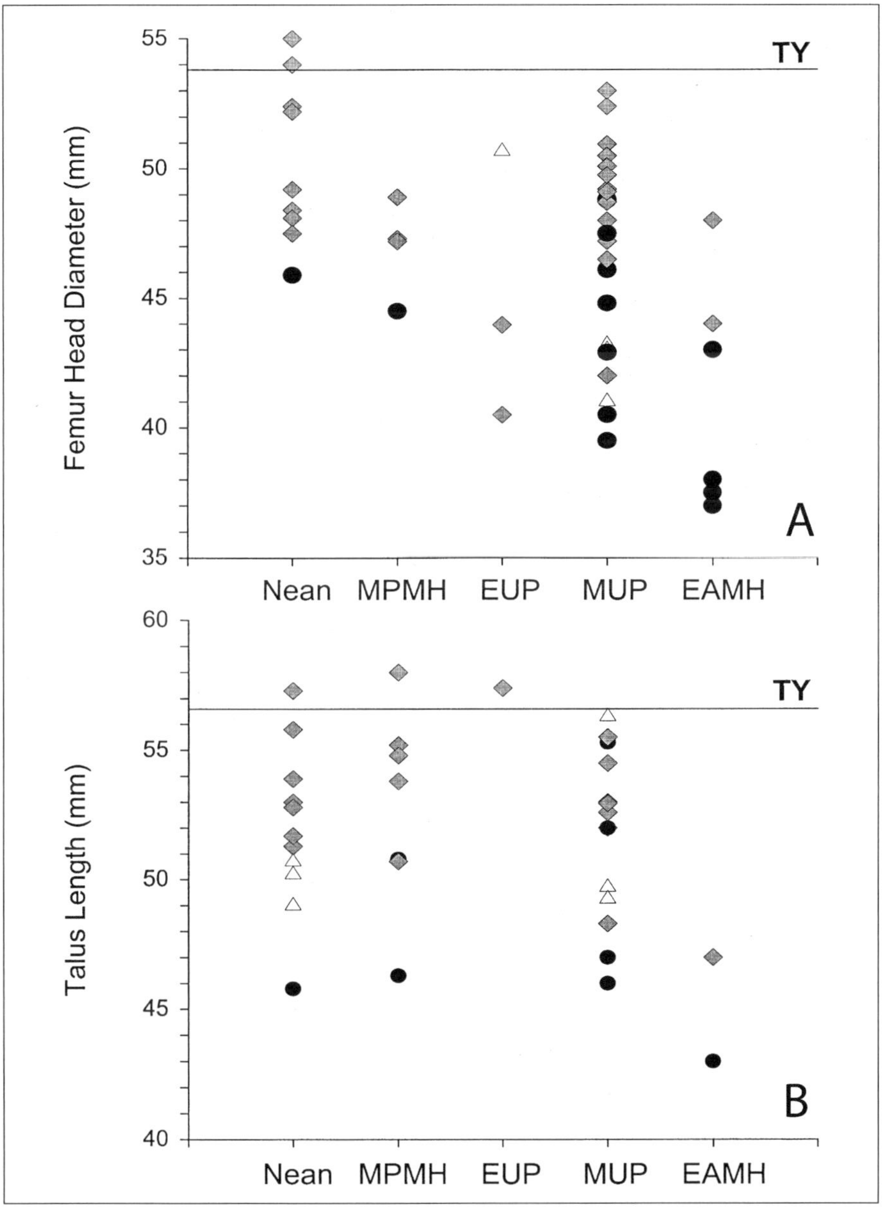

Figure 7-3. Plots of (*A*) femur head diameter and (*B*) talus length for Tianyuan 1 (*TY*) compared to Late Pleistocene samples. *Nean:* Neandertals; *MPMH:* Middle Paleolithic modern humans; *EUP:* western Early Upper Paleolithic; *MUP:* western Eurasian Mid Upper Paleolithic; *EAMH:* eastern Asian modern humans. *Gray diamonds:* males; *black circles:* females; *open triangles:* sex uncertain.

taphonomically fragile region, it remains at the top of the range of variation. Of the six Late Pleistocene specimens with data, only one (Feldhofer 1: 73.2 mm) approaches the Tianyuan 1 value, although the large Dolní Věstonice 14 (69.3 mm), Skhul 4 (68.1 mm), and Spy 2 (69.1 mm) femora are close. Moreover, as detailed in Chapter 10, using the body mass estimate (see Figures 10-9 and 10-14) derived from its femoral head diameter provides an appropriate scaling for the femoral and especially tibial diaphyseal robusticity of Tianyuan 1, a level of robusticity well within expected Late Pleistocene levels. The large femoral head diameters and talar length of Tianyuan 1 are therefore reliable in indicating a large body mass for this individual.

Body Proportions

The primary aspect of body proportions that can be directly assessed with the Tianyuan 1 remains is limb segment proportions, namely brachial (radius-to-humerus) and crural (tibia-to-femur) indices. These indices correlate strongly with mean annual temperature among recent humans (Trinkaus, 1981), such that colder climate native populations tend to have shorter distal segments. This pattern holds in the Late Pleistocene with respect to crural indices, such that the long-term resident higher latitude Neandertals tend to have lower indices, and the more equatorial MPMH humans have, or appear to have in less complete individuals, higher indices (Trinkaus, 1981; Holliday, 2000; Figure 7-4A). The relatively high indices of the European MUP remains have consequently been used as an indication of substantial, relatively recent, lower latitude ancestry among them (Trinkaus, 1981; Holliday, 1997a), possibly combined with selection for locomotor efficiency (Weaver and Steudel-Numbers, 2007; but see Holliday and Falsetti, 1995). As a result, the different Late Pleistocene samples are highly significantly different from each other (Kruskal-Wallis $P < 0.0001$). The pattern is less obvious among Late Pleistocene brachial indices (Trinkaus, 1981; Figure 7-4B), in which there is considerable scatter and little sample differentiation (Kruskal-Wallis $P = 0.207$). At least with respect to crural indices, this pattern appears to have considerable antiquity among Early and Middle Pleistocene *Homo*, whether reflected in comparisons of limb segment lengths (Ruff and Walker, 1993; Carretero et al., 2004) or appropriate biomechanical scaling of lower limb diaphyses (Trinkaus, Stringer, et al., 1999).

In these comparisons, the Tianyuan 1 brachial index of 76.0 is in the middle of the Late Pleistocene distributions. Its crural index of 83.7, however, is completely separate from the Neandertals, below the one MPMH specimen (Skhul 4), in the middle of the European MUP sample, and at the top of the small eastern Asian early modern human (Minatogawa) sample. On a modern human global scale (Trinkaus, 1981; Holliday, 1997a), its crural index is among subtropical to warm temperate populations, well above the means of modern arctic populations. Given

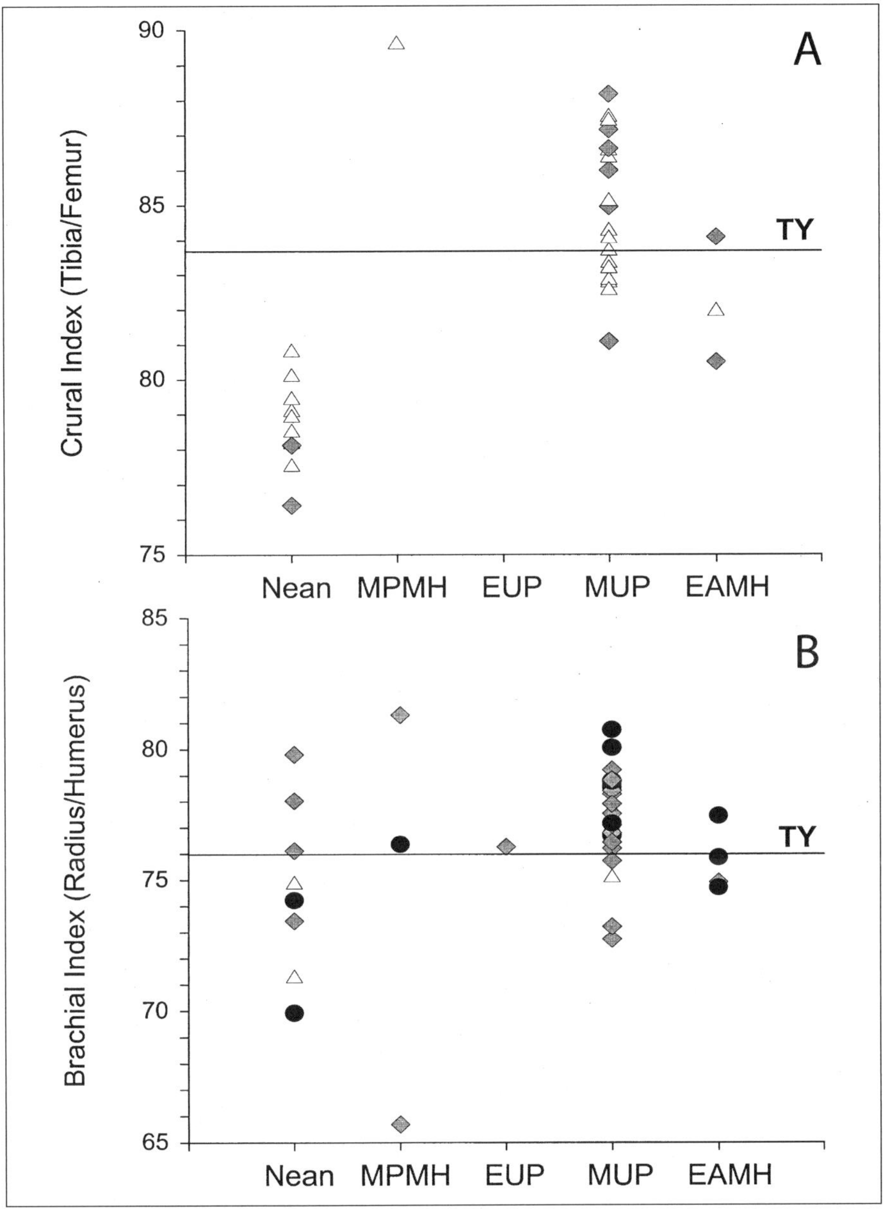

Figure 7-4. Plots of Tianyuan 1 (*TY*) and Late Pleistocene comparative samples for (*A*) crural indices [(tibia maximum length/femur bicondylar length) x 100] and (*B*) brachial indices [(radius maximum length/humerus maximum length) x 100]. *Nean:* Neandertals; *MPMH:* Middle Paleolithic modern humans; *EUP:* western Early Upper Paleolithic; *MUP:* western Eurasian Mid Upper Paleolithic; *EAMH:* eastern Asian modern humans. The EUP brachial index specimen is Nazlet Khater 2; there are no sufficiently preserved EUP specimens providing a crural index. *Gray diamonds:* males; *black circles:* females; *open triangles:* sex uncertain.

the position of Tianyuandong ~40° N latitude and the associated temperate faunal remains (Chapter 2), this latitudinally intermediate position of its crural index corresponds reasonably well with its apparent interpleniglacial climatic zone.

In contrast, the pattern of large articular dimensions and average long bone lengths for an early modern human suggests that Tianyuan 1 had a large body mass relative to its stature. Long bone lengths, especially of the lower limb, are strongly correlated with stature within populations (Trotter and Gleser, 1952; Feldesman et al., 1990), and femoral head diameter scales with body mass, at least during late development (Ruff et al., 1991, 1993; Grine et al., 1995; Auerbach and Ruff, 2004). Without going through the intermediate steps of estimating stature and body mass from these respective values, with their additional errors of estimation, a plot of femoral head diameter versus femoral bicondylar length (Figure 7-5A) confirms this impression. The Tianyuan 1 average estimated values for these two dimensions places it completely outside of the Eurasian early modern human (Middle and Upper Paleolithic) ranges of variation and among the European Neandertals with the relatively largest femoral heads. Increasing femoral length and decreasing femoral head diameter, each by one standard error (462.2 and 52.1 mm, respectively) would still keep it separate from the early modern humans and among the Neandertals. Placing each measure 2 standard errors from its mean estimate (464.0 and 50.4 mm respectively), an unlikely combination ($P < 0.01$), would only place it on the upper margin of the western Eurasian MUP plus Minatogawa 1 distribution and still very far from the three MPMH individuals.

Femur head diameter does scale positively allometrically with femur length, to approximately the 1.385 power (Ruff et al., 1993). Raising femur bicondylar length to that power changes little in the distribution, keeping Tianyuan 1 with the male European Neandertals, separate from the Eurasian Upper Paleolithic samples, and totally separate from the three MPMH individuals.

It could be argued that the comparison of femoral head diameter only to femur length does not fully reflect body mass to stature proportions, given between sample variation in crural indices (Figure 7-4A). To correct for this, femoral head diameter was also plotted against the sum of femoral bicondylar length and tibial maximum length (Figure 7-5B). Comparative sample sizes are reduced slightly, but the three sufficiently complete European Neandertals (La Chapelle-aux-Saints 1, La Ferrassie 1, and Spy 2) are still separate from the MUP distribution; two southwest Asian Neandertals (Amud 1 and Shanidar 5) plus the La Ferrassie 2 female fall in the upper portion of the MUP distribution; and the one sufficiently complete MPMH (Skhul 4) is at the bottom of the Late Pleistocene distribution. Tianyuan 1 is securely among the male European Neandertals in size and proportion, despite its higher crural index.

Weight-bearing articular dimensions may not be the best measure of absolute body mass if comparisons are made across diverse populations since subchondral

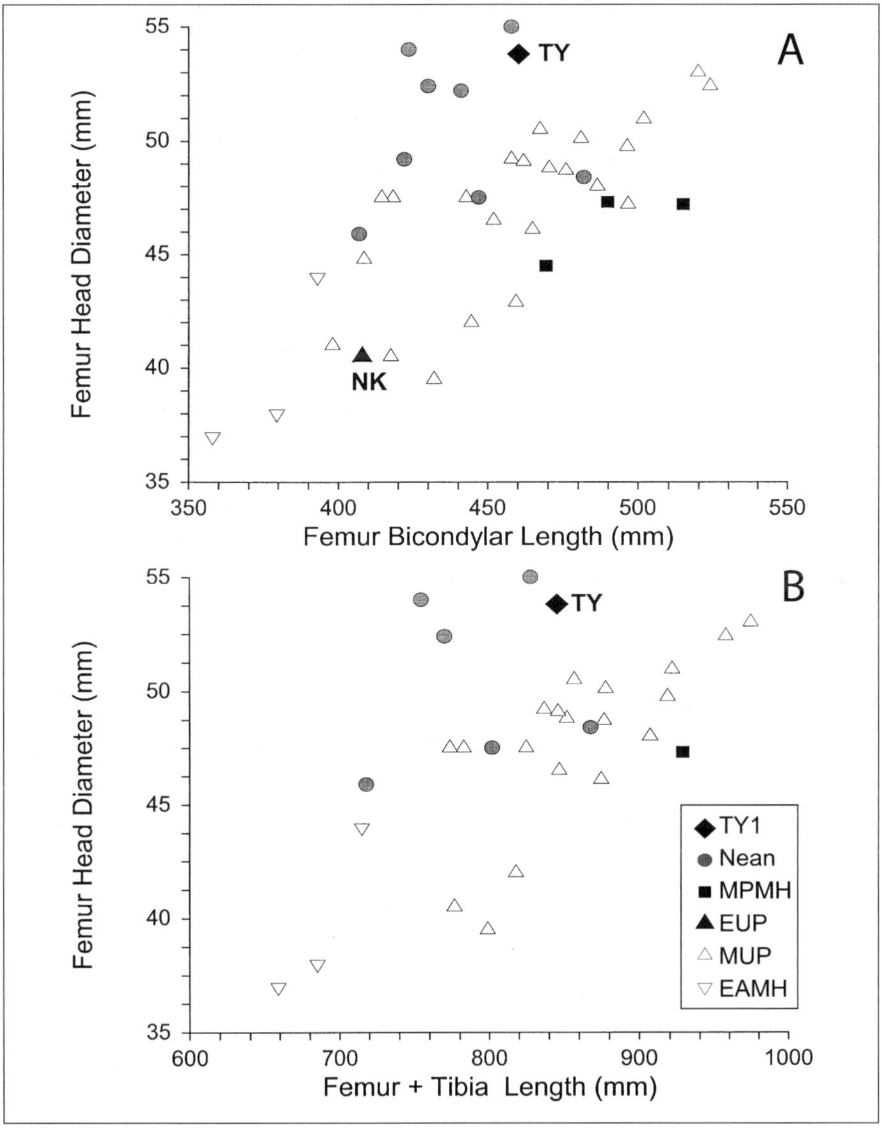

Figure 7-5. Bivariate scatter plots of Tianyuan 1 (*TY*) an.d comparative Late Pleistocene samples for (*A*) femoral head diameter versus femoral bicondylar length and (*B*) femoral head diameter versus femoral bicondylar plus tibial maximum lengths. *TY*: Tianyuan 1; *Nean:* Neandertals; *MPMH:* Middle Paleolithic modern humans; *EUP:* western Early Upper Paleolithic; *MUP:* western Eurasian Mid Upper Paleolithic; *EAMH:* eastern Asian modern humans; *NK:* Nazlet Khater 2.

bone development is responding to habitual joint reaction forces, which are a combination of body mass and forces (muscular contraction and body momentum) from activity levels. Higher activity populations may therefore exhibit greater articular dimensions for a given body mass. This issue should not be a concern here, since all of these Late Pleistocene humans (including Tianyuan 1) appear to have had similarly elevated activity levels as is reflected in their lower limb diaphyseal hypertrophy (Ruff et al., 1993; Trinkaus, 2006d, 2009; see Chapter 10).

Consequently, in terms of recent human ecogeographical patterns of body proportions, Tianyuan 1 presents an anomalous situation. On the one hand, its limb segment proportions are ambiguous (brachial index) or suggest midlatitude interglacial conformity (crural index). On the other hand, its lower limb articular dimensions indicate a high body mass to stature ratio, a relationship associated with cold climates (Holliday, 1997b). Did Tianyuan 1 have a very broad body core, possibly reflected in a large bi-iliac breadth, similar to one evident in the earlier archaic human Jinniushan 1 skeleton (Rosenberg et al., 2006), as well as the Neandertals and several earlier *Homo* individuals (Arsuaga et al., 1999; Simpson et al., 2008)? Might its moderately high crural index reflect relatively recent ancestry from a warmer climate, or simply the interpleniglacial climate of China in the vicinity of Zhoukoudian?

Stature and Body Mass Estimation

The skeletal bases for assessing stature (bone lengths) and body mass (femoral head diameter) involve some estimation, and using them to then estimate these body parameters compounds the potential error of the estimates. However, there are comparisons for which a stature value is useful, and the proper scaling of at least lower limb (and probably upper limb) robusticity (especially diaphyseal) requires a baseline input of body mass (Ruff et al., 1993; Trinkaus and Ruff, 2000; Ruff, 2000). Therefore, these values are estimated for Tianyuan 1.

Stature

Stature estimates are dependent upon the selection of the correct reference sample, the one that most closely approximates the body proportions of the specimen in question (Holliday and Ruff, 1997). Given this, and the absence of any indication of trunk length for Tianyuan 1, the crural index provides the best indication of an appropriate sample. It could fit well with the Trotter and Gleser (1952) Euroamerican sample body proportions or those of their Afroamerican sample. It is also appropriate to estimate it from a relevant sample from the same region of the Old World, namely northern China, for which the male regression formulae from Stevenson (1929) are employed. Table 7-2 provides stature estimates for each of the long bone lengths of Tianyuan 1 for each of these reference samples.

Table 7-2. Estimated stature in centimeters for Tianyuan 1 from its four long bones

	Euroamerican formulae[1]	*Afroamerican formulae*[1]	*Northern Chinese formulae*[2]
Humerus (maximum)	170.2	169.3	173.5
Radius (maximum)	172.9	174.3	172.9
Femur (maximum)	171.8	168.5	173.4
Femur (bicondylar)	171.4	168.6	—
Tibia (maximum)	175.7	173.2	—
Tibia (lateral total)	176.8	173.9	172.5
Average stature[3]	172.7	171.4	173.1
Coefficient of variation[4]	1.49	1.62	0.27

Note: Length estimates were made using the Trotter and Gleser (1952) male regression formulae for Euroamericans and Afroamericans and Stevenson's (1929) (male) formulae for northern Chinese. SE_{est}s for the regression formulae vary from 1.9 to 5.0 cm.

[1] To correct for hanging cadaver stature, 2.5 cm were subtracted from each result.

[2] To correct for supine cadaver stature, 1.2 cm were subtracted from each result.

[3] For the Euroamerican and Afroamerican results, the 2 femoral estimates were averaged and the 2 tibial estimates were averaged prior to calculating the overall average.

[4] Standard deviation of the four estimates (using the average femoral and tibial estimates for the Euroamerican and Afroamerican estimations) divided by the mean value.

The resultant stature estimates (Table 7-2) provide a range from ~169 cm to ~177 cm, with averages for all three sets of formulae ~172 cm. Since these values are based on the long bone lengths compared in Table 7-1, it is not surprising that the Tianyuan 1 estimates are close to the means for the MPMH and European earlier Upper Paleolithic samples (169.9 ± 8.8, N = 9 and 168.4 ± 8.2, N = 39 respectively) and above those for Neandertals (165.2 ± 5.9, N = 20) and the small Minatogawa sample (151.8 ± 3.8, N = 4).

To assess consistency across the different long bones, coefficients of variation (CV) were calculated for the results from each reference sample; the northern Chinese estimates have a markedly smaller CV (0.27) than the other two sets of formulae (1.49 and 1.62). The greater consistency in the Chinese-based estimates suggests that the overall limb segment proportions of Tianyuan 1 more closely match those of the Chinese sample. It is not clear why this might be the case, since there is no reason to expect consistency in such limb segment proportions across 40,000 years in eastern Asia, especially given the changes evident in western Eurasia during the same time period (Trinkaus, 1981; Holliday, 1995).

Body Mass

Body mass estimates, given the absence of a pelvis and hence bi-iliac breadth, require the use of a mechanically based estimate, which assumes that the lower limb articular dimensions (in this case the femoral head) reflect baseline joint reaction forces at the time of full articular development, or late adolescence. In active

Pleistocene hunter-gatherers, it may moderately exaggerate the perceived body mass, given elevated joint reaction forces from activity throughout development. Nonetheless, the estimation formula that appears (Auerbach and Ruff, 2004) to best estimate body mass for large-bodied Pleistocene humans is the one provided by Grine et al. (1995) (BM = 2.268 × FHD—36.5, $r^2 = 0.846$, SE_{est}: 4.3), computed here using its estimated vertical femoral head diameter of 53.6 mm. This provides a mean estimate of 85.1 kg (±2σ range: 76.5–93.7 kg). If one takes into account the ±2σ range for the femoral head estimation, in addition to the ±2σ range for the body mass estimation, it provides a very large estimation range of 68.8–101.4 kg, which is statistically honest but probably too large a range. Alternative formulae provided by McHenry (1992) and Ruff et al. (1997) for males provide mean estimates of 80.1 kg and 92.0 kg respectively. The value of 85.1 kg is used in limb bone scaling here, with the understanding that the range of estimates might affect the resultant interpretations.

It should be emphasized that this body mass estimate is used to scale several aspects of the limb bones of Tianyuan 1, especially in the weight-bearing lower limbs but also for the humeri (Chapters 9 and 10). If its body mass estimate is too high, the more likely direction of error given its very high value relative to those of other Late Pleistocene humans, this will have a conservative effect in almost all of the comparisons, since it will make the Tianyuan 1 limbs appear more gracile. The exception to this regards the second pedal proximal phalanx robusticity, but in that case the bone is so gracile that an adjustment would have little effect on the implications (Chapter 10).

Interestingly, if a body mass index (BMI) is calculated for Tianyuan 1 using the mean lean mass estimate and the Chinese-based stature estimate, the result is 28.4 (body mass/stature^2, in kg and m). For modern urban European and North American populations, a BMI >25 is considered "overweight," and a BMI >30 is considered "obese." Although such a comparison is not appropriate for a Pleistocene foraging population, the estimated lean BMI of Tianyuan 1 reinforces the impression that the individual had a high body mass for its stature.

Summary

The Tianyuan 1 skeletal remains indicate a relatively tall individual with respect to the Neandertals and the other available eastern Asian early modern humans but a near average stature compared to western Eurasian early modern humans. This is combined with unusually large lower limb (hence weight-bearing) articulations, at the top of the Late Pleistocene range of variation. Whether reflected in articular diameter to bone length comparisons or in estimated lean body mass to stature comparisons, it is apparent that Tianyuan 1 had a large body core. In this respect, it is close to the European Neandertals. However, a similar pattern is also evident in the late Middle Pleistocene eastern Asian archaic human from Jinniushan, it may be generally characteristic of Middle Pleistocene archaic humans, and it is echoed in early Holocene human remains from the New World (Auerbach, 2007). Those more recent Native Americans show variation in crural indices in the context of their wide trunks, paralleling in part the moderately high crural index of Tianyuan 1 combined with its absolutely and relatively large weight-bearing articulations.

Chapter 8

The Tianyuan 1 Axial Remains

DESPITE THE relative completeness of the Tianyuan 1 limbs, the axial skeleton is poorly represented. It consists of an essentially complete second cervical vertebra (C2, or axis), a partial manubrium, and two elements of the mesosternum. In addition to minor degenerative osteoarthritis on the axis (Chapter 12), the axis is of note primarily for its asymmetry. The sternal pieces provide some indication of their size.

Second Cervical Vertebra

The C2, or axis, of Tianyuan 1 presents few details of note for a human second cervical vertebra (Figure 8-1). The caudal body has a fully attached annular ring, which remains prominent dorsally and ventrally. The cranial facets are largely flat laterally, and then dorsoventrally convex closer to the dens. The caudal facets are essentially flat. The dorsal tip of the spinous process is bifurcated caudally, but it is single cranially with a prominent midline crest running along the cranial side of the process.

The overall height of the Tianyuan 1 axis (maximum, or ventral height) of 36.7 mm (Table 8-1) is unexceptional for a Late Pleistocene human second cervical vertebra (Table 8-2), or for a recent human one (Hasebe, 1912; Lanier, 1939; Stewart, 1962a). There is considerable variation in C2 heights; curiously two of the tallest individuals in the comparative sample [Qafzeh 3 and Skhul 5 (McCown and Keith, 1939; Vandermeersch, 1981)] have rather short axes (Table 8-2).

The Dens

The dens appears slightly waisted adjacent to the cranial intervertebral facets, but the "waisting" is produced by the lateral expansion of the facet for the an-

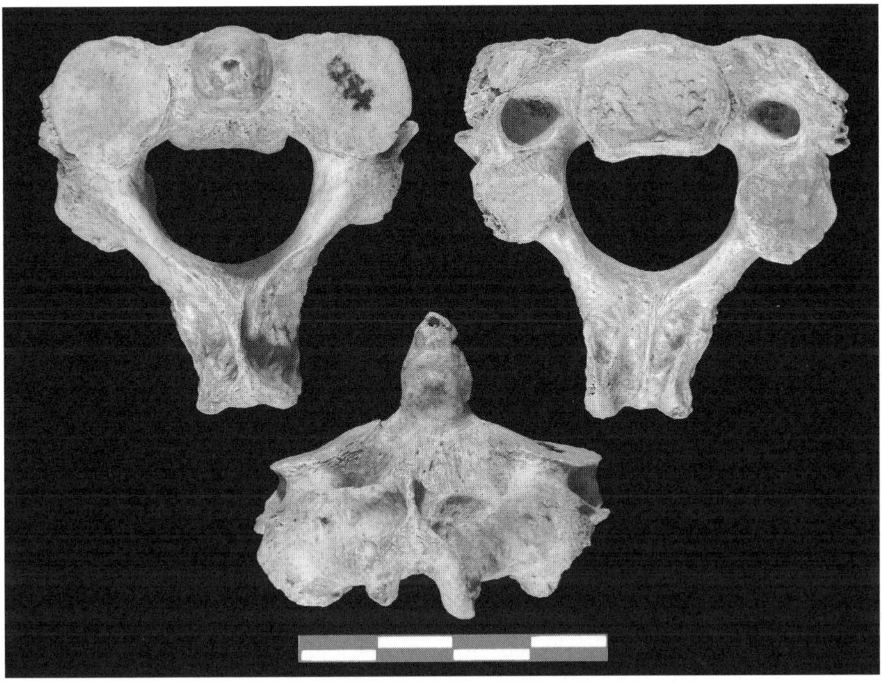

Figure 8-1. Tianyuan 1 2nd cervical vertebra (C2: axis). *Upper left:* cranial view; *upper right:* caudal view; *bottom:* dorsal view. The vertebra is oriented with respect to the cranial articular facets. Scale in centimeters.

terior arch of the atlas. The dorsal side of the dens shows little lateral expansion. The groove for the transverse ligament of the atlas on the dorsal dens is weakly expressed. The main aspect of note is the dens's relatively small height, or craniocaudal distance from the medial ends of the cranial facets to the dens tip (not including the osteophyte on the dens tip). Its height of 13.2 mm (Table 8-1) is among the smaller of the few sufficiently complete Late Pleistocene axes for whom data are available. It is matched by a third of the other Late Pleistocene axes providing data, including those from the small Minatogawa 1 and 2 skeletons, the small Dolní Věstonice 3, the (possibly pathological) Dolní Věstonice 15 C2, the otherwise small Qafzeh 3 and Skhul 5 axes, and the early Neandertal Krapina 103 axis (Table 8-2). But if its dens height is compared to its total ventral height, or maximum height (Figure 8-2), the proportions for Tianyuan 1 place it below all of the other Late Pleistocene specimens and a recent human eastern Asian sample (Hasebe, 1912). There is little overall difference across the comparative samples, either Late Pleistocene or recent. It is unclear what the significance of this relatively, but not absolutely, small dens height might be.

Table 8-1. Osteometric dimensions of the Tianyuan 1 second cervical vertebra (axis), in millimeters except for the foraminal areas.

	Right	*Left*
Dorsoventral diameter	52.3	
Cranial external articular diameter	48.2	
Cranial internal articular diameter	14.7	
Caudal external articular diameter	47.8[1]	
Caudal internal articular diameter	22.2	
Canal dorsoventral diameter (M-10)	16.3	
Canal transverse diameter (M-11)	22.7	
Lamina height[2]	11.4	13.7
Spinous process length—maximum	20.2	
Spinous process length—midline[3]	19.7	
Spinous process tip height[4]	13.9	11.7
Total height to dens tip—dorsal[5]	32.0	
Total height to dens tip—ventral[5]	36.7	
Body ventral height (M-1)[6]	23.5	
Body dorsal height (M-2)[6]	18.8	
Dens height (M-1a)[5]	13.2	
Body caudal dorsoventral diameter (M-5)	15.9	
Body caudal transverse diameter (M-8)	20.7	
Foramina transversaria—cranial maximum diameter[7]	6.6	4.6
Foramina transversaria—cranial minimum diameter[7]	5.1	3.8
Foramina transversaria—cranial area (mm^2)[7]	26.4	13.7
Foramina transversaria—caudal maximum diameter[7]	8.1	6.9
Foramina transversaria—caudal minimum diameter[7]	6.9	4.8
Foramina transversaria—caudal area (mm^2)[7]	43.9	26.0

[1] Two times the distance from the midline to the right side.
[2] Craniocaudal height taken at the middle of each lamina.
[3] From the craniodorsal margin of the spinal canal to the sulcus between the bifid dorsal tubercles.
[4] Craniocaudal heights of the two dorsal tubercles.
[5] Craniocaudal distances taken from the dorsal or ventral midlines of the caudal intervertebral disk surfaces. Heights including the dens are taken avoiding the small enthesopathy on the cranial tip of the dens.
[6] Craniocaudal heights from the caudal intervertebral disk surface to the transverse plane tangential to the medial margins of the cranial articular facets.
[7] Since the foramina approximate ellipses, the "minimum" diameters are taken perpendicular to the maximum diameters. For the cranial openings, the maximum diameters are approximately dorsoventral. For the caudal openings, the maximum diameters are approximately mediolateral. Areas are computed using the ellipse formula.

The Spinous Process

The spinous process presents a prominent crest cranially, which rises above the cranial margins of the laminae and reaches the level of the medial margins of the cranial facets, or the level of the base of the dens. Caudally the process is distinctly bifid, with completely separated right and left caudal margins and dorsocaudal tubercles. In cranial view, the spinous process does not appear to be bifid, since it

Table 8-2. Comparative dimensions of the Tianyuan 1 and Late Pleistocene dens, in millimeters

	Total (ventral) height	*Dens height*
Tianyuan 1	36.7	13.2
East Asia EMH	34.0, 35.0	14.0, 14.0
West MUP	37.8 ± 2.8 (5)	17.6 ± 3.5 (9)
MPMH	30.0, 32.0	11.0, 14.0
Neandertals[1]	35.3 ± 3.1 (4)	15.6 ± 1.8 (4)
Recent East Asian[2]	36.0 ± 2.3 (30)	17.3 ± 1.8 (30)
K-W P-value	0.098	0.225

Note: EMH: early modern humans; *MUP:* Mid Upper Paleolithic; *MPMH:* Middle Paleolithic modern humans. Kruskal-Wallis (K-W) P-values are provided for comparisons across the Late Pleistocene comparative samples.

[1] The Neandertal sample includes the Regourdou 1 and Shanidar 2 axes, plus two from the earlier Neandertals from Krapina (103 and 104). Other known Neandertal axes (La Chapelle-aux-Saints 1, La Ferrassie 1, La Quina 5) are too fragmentary to provide metric data.

[2] Data from Hasebe (1912).

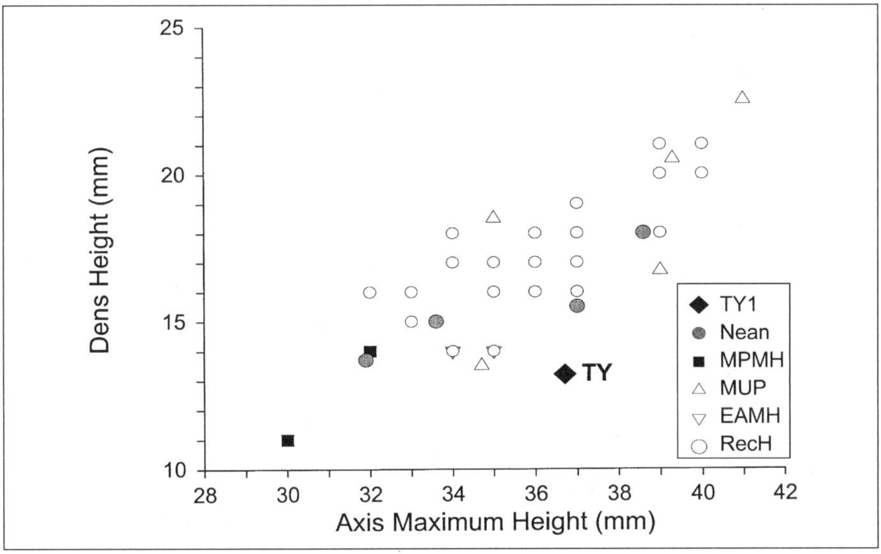

Figure 8-2. Bivariate plot of the dens height (height from the transverse line connecting the craniomedial cranial facets to the cranial dens tip) versus the total (ventral) axis craniocaudal height. *TY:* Tianyuan 1; *Nean:* Neandertals; *MPMH:* Middle Paleolithic modern humans; *MUP:* western Eurasian Mid Upper Paleolithic; *EAMH:* eastern Asian modern humans; *RecH:* recent eastern Asians (N = 30; Hasebe, 1912). Note that the two EAMH specimens overlap two recent human values.

only has a slight concavity between the right and left tubercles. The considerable data available on cervical spinous process bifidity since Cunningham (1886; e.g., Shore, 1933; Lanier, 1939) provide frequencies of C2 bifidity from 18% to 100% with an average of 69%, but it is unclear to what extent the variation is real or due to variance in definitions of C2 bifidity (see Shore, 1933). The degree of bifidity of the Tianyuan 1 axis is well within the range of variation of these recent human samples. Few Late Pleistocene axes preserve the dorsal spinous process, but of those that do, the Tianyuan 1 spinous process resembles those of La Ferrassie 1, Krapina 104, Předmostí 10, and Dolní Věstonice 13 and 15, but contrasts with the clearly bifid Dolní Věstonice 3 C2.

Asymmetry

The Tianyuan 1 axis exhibits considerable asymmetry. Some of this asymmetry is reflected in bilateral measurements, such as the larger left lamina height and larger right spinous process tip height. But much of it is reflected in other, harder to quantify aspects. Using the sagittal plane through the dens and the vertebral body for reference, the articular facets are higher on the right side, especially with respect to the caudal facets. The right lamina, however, is distinctly more caudal than the left one. Both of the laminae are concave on their cranial surfaces, but the right one dips down below the level of the cranial articular facet to a greater extent than the left one does. The spinous process deviates to the right, which is evident in the midline crests both cranially and caudally. The right spinous process tip is both taller and projects more caudally than the left one. Ventrally, the caudal body margin extends farther caudally on the right side.

In addition, the right foramen transversarium is distinctly larger than the left one. Assuming that the foraminal cross sections approximate an ellipse and computing the area of the foramen using the maximum and minimum diameters at their cranial and caudal ends, an index of asymmetry provides a cranial index for Tianyuan 1 of 63.3 and a caudal one of 51.2. Comparative Late Pleistocene data are unavailable, but two samples of recent humans (Figure 8-3) indicates that the Tianyuan 1 caudal index is toward the top of these recent human samples in asymmetry. Only three (7.5%, N = 40) have higher levels of asymmetry than Tianyuan 1. The recent human samples are, nonetheless, almost evenly divided between those with the right or left side larger; 46.2% (N = 39, one symmetrical) have the right foramen larger. It is unclear what significance this foraminal asymmetry might have since the vertebral arteries that pass through them join intracranially into the basilar artery.

At the same time, however, the vertebral canal is highly symmetrical. The canal itself is symmetrical, and it is centered on the sagittal plane defined by the body and the dens.

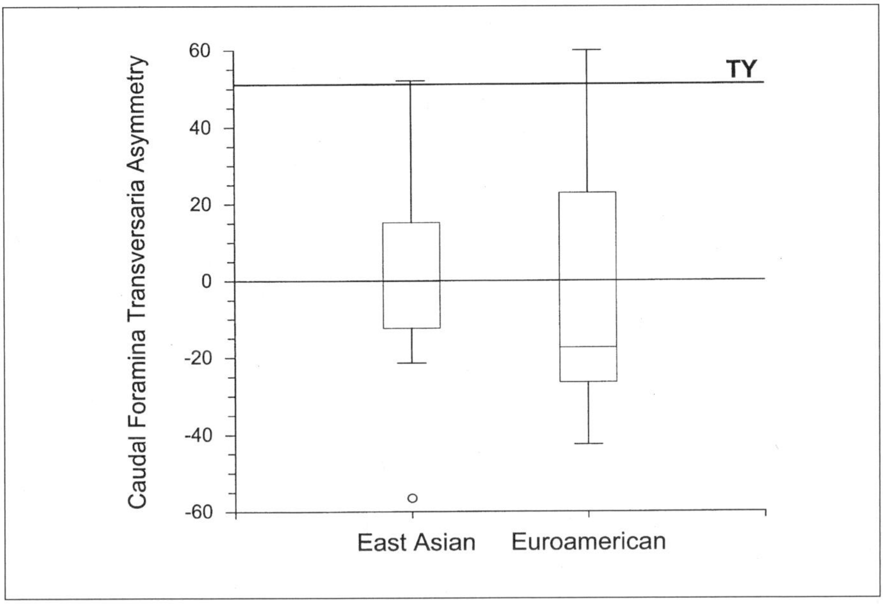

Figure 8-3. Box plot of the Tianyuan 1 caudal foramina transversaria asymmetry versus distributions for samples of recent East Asians (N = 23) and Euroamericans (N = 17). Positive values indicate a larger right foramen; the line at zero indicates symmetry.

The Manubrium and Mesosternum

The symphysis between the manubrium and the first sternebra was completely open (Figure 8-4). The two bones are separate, and there is no evidence that they were even partially fused. This is the usual condition, since sternomanubrial fusion is relatively rare, occurring in only 5.7% of individuals up to age 30 and 10.3% between 30 and 60 in one sample (Trotter, 1934), and in 5.2% of males up to age 40 in another sample (McKern and Stewart, 1957), both of recent North Americans. There are nonetheless several small bony spurs along the dorsal margin of the cranial first sternebra, which may indicate the beginning of minor ossification of the cartilaginous connection between the two sternal elements.

The Manubrium

The caudal manubrial piece is strongly transversely convex ventrally and transversely concave dorsally. The sternomanubrial surface slopes caudoventrally to craniodorsally, suggesting a clear angle between the manubrium and the sternal body. Articulation of the two bones (ignoring the effects of the intervening cartilage) provides an angle of ~170°, which is similar to those recorded on living in-

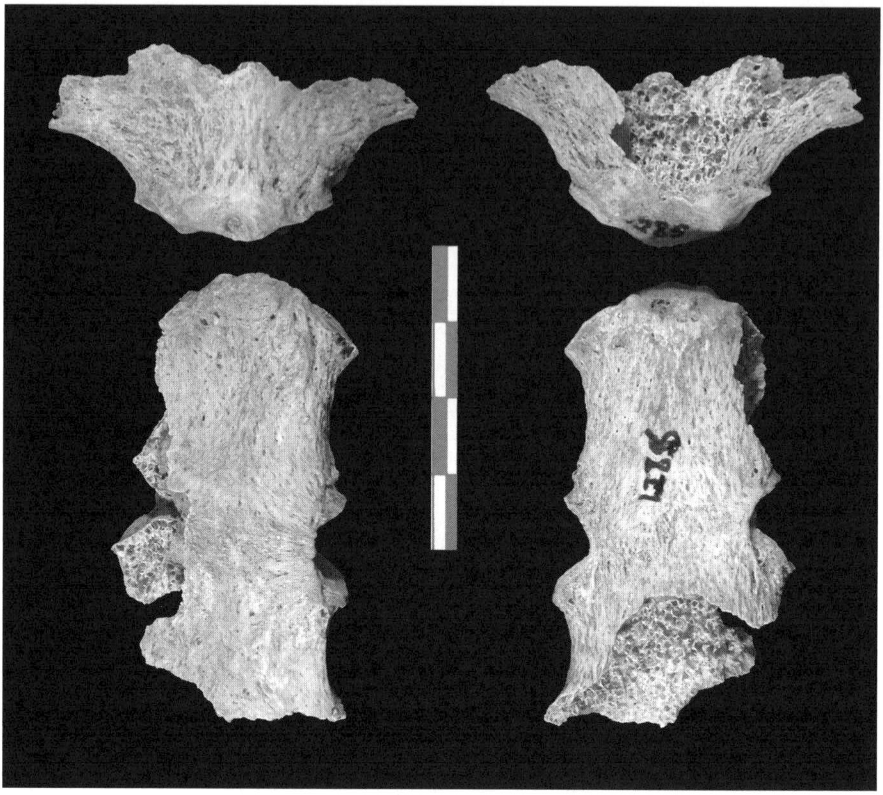

Figure 8-4. Ventral (*left*) and dorsal (*right*) views of the Tianyuan 1 manubrium and cranial 2 segments of the sternal body (first and second sternebrae). The sternomanubrial synchondrosis is completely unfused, whereas the first and second sternebrae are completely fused with only a slight separation at the bases of the 3rd costal cartilage attachments. Scale in centimeters.

dividuals [relaxed respiration: 165.7° ± 7.1, N = 61 (Constantinesco, 1974)]. The second rib facets are displaced dorsally relative to the sternomanubrial cartilage surface. Their broadest depths are caudal, and their greatest heights are dorsal. The inferolateral margins of the manubrium are concave as they rise superolaterally toward the first rib articulations, rather than being relatively straight.

The Mesosternum

Since the sternal body consists of most of the first sternebra with damage to the right side and cranial portions of the second sternebra, observations are limited to those portions. The two retained sternebrae are fully fused to each other between the third rib facets, even though the fusion line is evident dorsally and ventrally

Table 8-3. Osteometric dimensions of the Tianyuan I sternum

	Right	*Left*
Manubrium: sternomanubrial facet max. dorsoventral diameter	14.1	
Manubrium: sternomanubrial facet artic. dorsoventral diameter	12.9	
Manubrium: sternomanubrial facet transverse diameter	12.8	
Manubrial 2nd rib facet dorsoventral diameter	8.8	9.7
Manubrial 2nd rib facet craniocaudal diameter	7.8	7.3
Sternebra: sternomanubrial facet max dorsoventral diameter	15.2	
Sternebra: sternomanubrial facet artic. dorsoventral diameter	11.1	
Sternebra: sternomanubrial facet dorsal transverse diameter	(18.0)	
Sternebra: sternomanubrial facet ventral transverse diameter	11.5	
Sternebral 2nd rib facet dorsoventral diameter	—	8.6
Sternebral 2nd facet craniocaudal diameter	—	10.4
1st sternebra craniocaudal length—ventral	33.8	
1st sternebra craniocaudal length—dorsal	31.6	
Manubrial facet to mid 3rd rib notch	32.2	29.5
Intercostal 2–3 distance	—	19.7

and there is a pit from the fusion at the bottom of the left third costal cartilage attachment. Complete fusion of the first and second sternebral segments to each other normally occurs by early in the third decade It may occur later in the third decade, but it is rare, and considered anomalous, for fusion not to have happened at older ages (McKern and Stewart, 1957). This supports an older age-at-death for Tianyuan 1 (see Chapter 11).

The Tianyuan 1 1st sternebra has a length (Table 8-3) that falls well within the variation of the few sternebrae available for Late Pleistocene humans. Its maximum length of 33.8 mm is above those of Dolní Věstonice 13 (27.8 mm) and 15 (30.8 mm), similar to those of Kebara 2 (32.1 mm) and Shanidar 4 (32.9 mm), and below the high values of 38.0 mm and 38.5 mm for Regourdou 1 and Nazlet Khater 2, respectively (Holliday, 2006; Crevecoeur, 2008). It is well within the ranges of variation of two samples of recent humans (Euroamericans: 32.2 ± 3.7; Amerindians: 32.4 ± 4.8) (Franciscus, 1989). If the Tianyuan 1 sternebral length is normalized for body size by its humeral maximum length, the resultant index of 10.2 is above that of Dolní Věstonice 13 (8.3), below those of Nazlet Khater 2 (12.2) and Regourdou 1 (12.3), and close to those of Dolní Věstonice 15 (10.1), Kebara 2 (10.0), and Shanidar 4 (10.8).

There is little more of note on the sternum. The sternebrae appear to be relatively narrow, but damage precludes proper assessment of their breadths. There is no evidence of a sternal perforation, at least on the two preserved cranial segments. The notches for the third costal cartilages, as well as the facets for the second ones adjacent to the manubrium, are relatively deep, with prominent margins. The prominence of the margins may be more a reflection of age-at-death (Chapter 11) than a primary indicator of sternal morphology.

Chapter 9

The Tianyuan 1 Upper Limb Remains

THE UPPER limb skeletons of Tianyuan 1 are reasonably well represented. The fossil preserves portions of both scapulae, both humeri, an ulna, both radii, three carpal bones, three proximal manual phalanges, a middle manual phalanx, and a distal manual phalanx (see Figure 3-2). In addition, it is possible to estimate reliably the original lengths of the left humerus and radius. But only the hand bones are largely complete. The preserved arm articulations consist of only one scapular glenoid fossa and one distal humerus; therefore, most of the comparisons are diaphyseal.

Given evidence for marked asymmetry, especially in the humeral diaphyses, of the Tianyuan 1 upper limb and those of most other Late Pleistocene partial skeletons (Trinkaus et al., 1994; Churchill and Formicola, 1997; Trinkaus, 2006b), with the apparent exception of the Minatogawa ones (Baba and Endo, 1982), most of the comparisons for the shoulder and arm bones are made only to bones of the same side.

Scapulae

The scapulae provide information on both lateral spines and acromion processes and for the left glenoid fossa, coracoid process, infraglenoid tubercle, and cranial axillary border. Given the incomplete natures of these scapulae, there are limited osteometrics for them (Table 9-1), and several areas of interest are not in evidence.

Spines and Acromion Processes

The scapular spines (Figure 9-1) are relatively smooth with little rugosity for the attachments of deltoideus and trapezius. Near the middle of each spine is the

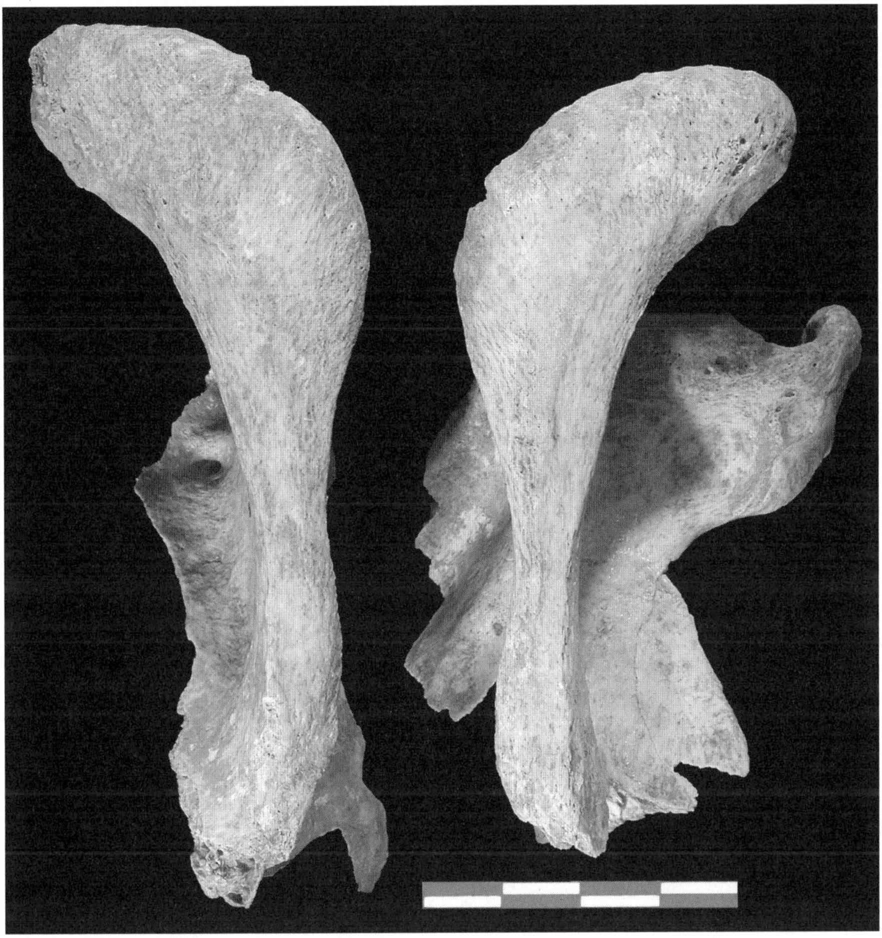

Figure 9-1. *Left to right:* Dorsal view of the Tianyuan 1 right and left scapular spines and acromion processes. Scale in centimeters.

caudal expansion over the infraspinatus surface, which is rounded on both spines but more projecting on the right side. It forms more of an inferiorly directed lip of bone on the right side, whereas it is mostly a thickening of the spine on the left side. This difference, as well as the general asymmetry reflected in the minimum spine thicknesses, is evident in the larger right side thickness values (Table 9-1). These values provide asymmetry indices of 13.4 and 22.2, respectively. The acromion processes are similar in terms of their breadths, but the right one projects ~5 mm farther than the left one. Both of the acromion processes nonetheless have normal facets for the distal clavicle, moderate levels of rugosity dorsally, and a series of fine pits on their ventral surfaces.

Table 9-1. Osteometric dimensions of the Tianyuan 1 scapulae, in millimeters

	Right	*Left*
Glenoid maximum height (M-12)	—	39.2
Glenoid maximum breadth (M-13)	—	29.7
Glenoid articular height[1]	—	35.5
Glenoid articular breadth[1]	—	27.8
Coracoid thickness	—	7.9
Coracoid length[2]	—	43.9
Spine mid maximum thickness[3]	12.7	11.1
Spine lateral minimum thickness[4]	9.0	7.2
Acromial base maximum thickness (M-9)	26.9	27.0
Acromial projection[5]	58.1	53.0

[1] Diameters of the subchondral bone surface, not including the glenoid labrum attachment area (Churchill and Trinkaus, 1990).
[2] Length from the angle to the tip of the coracoid process.
[3] The maximum craniocaudal thickness of the dorsal spine, approximately in its mediolateral middle.
[4] The minimum craniocaudal thickness of the dorsal spine, between the maximum thickness and the expansion for the acromion.
[5] Direct distance from the spine base on the dorsal margin of the glenoid fossa to the tip of the acromion process.

Coracoid Process

Tianyuan 1 is one of the few Late Pleistocene specimens with a complete coracoid process (Figure 9-2). At its proximal, or dorsomedial, end it projects slightly over the supraspinatus surface, forming a hollow between itself and the lateral end of the supraspinatus surface. It is moderately rugose on its dorsocranial surface, but it is not possible to discern distinct attachments for the muscles and ligaments that attach to it, including the biceps brachii short head origin and the coracoacromial arch attachment.

Glenoid Fossa

The right glenoid fossa is completely absent, but the left one is intact and has sustained only trivial damage to its inferior margin (Figure 9-2). The left glenoid fossa is strongly pyriform. Its caudal two-thirds closely approximate a circle, or a portion of a sphere given its concavity. The dorsal margin continues as an evenly convex curve up to the supraglenoid tubercle. The ventral margin becomes slightly concave as it approaches the inferior root of the coracoid process, and then convex up to the supraglenoid tubercle. While the inferior two-thirds are strongly concave, the superior portion is only slightly so.

The glenoid margin has a slight lipping, or ossification into the glenoid labrum, along the entire ventral margin. Dorsally it forms a rounded right angle onto the

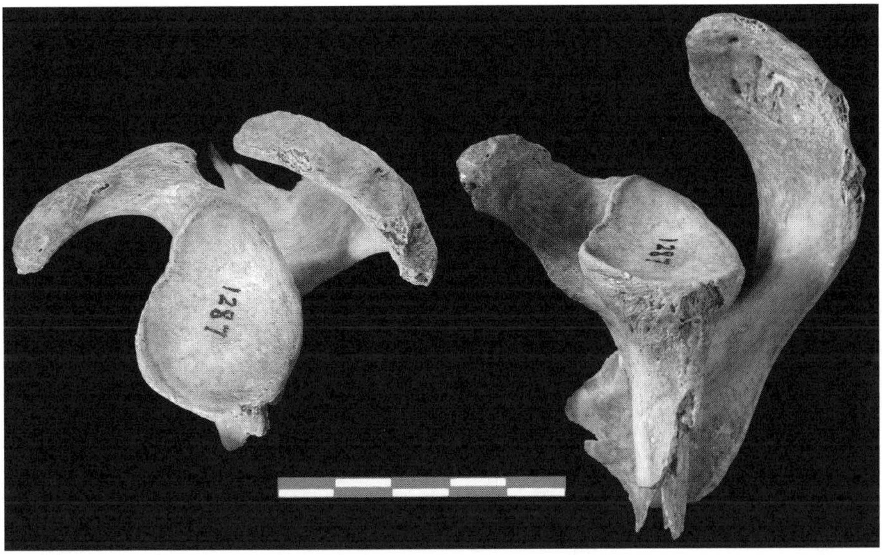

Figure 9-2. *Left to right:* Lateral view of the Tianyuan 1 left scapular glenoid fossa and coracoid process; caudolateral view of the left infraglenoid tubercle and axillary border. Scale in centimeters.

adjacent bone without any of the same lipping. There is no evidence of subchondral bone degeneration.

It is well established that there is an average increase, on the right side, in western Eurasia in relative glenoid fossa breadth between Middle Paleolithic late archaic humans (and probably Middle Paleolithic modern humans (MPMH)) and Mid Upper Paleolithic (MUP) humans (Churchill and Trinkaus, 1990; Trinkaus, 2006b). The one right glenoid fossa of an Early Upper Paleolithic (EUP) human, from Muierii 1, also exhibits the narrow glenoid fossa of earlier humans (Soficaru et al., 2006; Trinkaus, 2008a). This shift has been attributed to changes in shoulder loading patterns related to projectile use (Churchill and Trinkaus, 1990; Churchill and Rhodes, 2009; Trinkaus, 2008a). However, the left glenoid fossae of Late Pleistocene humans appear to show little change through time (Trinkaus, 2006b). A plot of left glenoid breadth versus its height (Figure 9-3) reinforces this pattern. There is only a hint of shift from the Neandertals to the MUP humans. The very approximate values (McCown and Keith, 1939; Stewart, 1962b) for the MPMH Skhul 5 scapula suggest a relatively narrow glenoid fossa. The same applies to the very small Minatogawa 3 scapula. Nazlet Khater 2 falls in the middle of the distribution, and Tianyuan 1 is along the upper margin of the MUP human distribution. Despite the lack of clear separation between the Middle and Upper Paleolithic humans

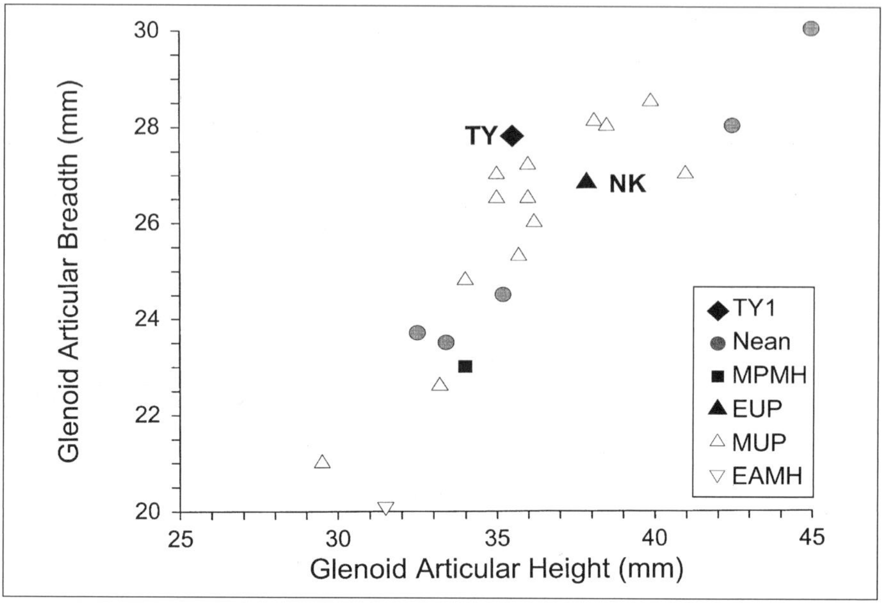

Figure 9-3. Bivariate plot of left glenoid breadth versus glenoid height. *TY:* Tianyuan 1; *Nean:* Neandertals; *MPMH:* Middle Paleolithic modern humans; *EUP:* western Early Upper Paleolithic; *MUP:* western Eurasian Mid Upper Paleolithic; *EAMH:* eastern Asian modern humans; *NK:* Nazlet Khater 2.

in left glenoid fossa proportions, Tianyuan 1 is nonetheless notable for its rather broad glenoid fossa.

Supraglenoid and Infraglenoid Tubercles

The supraglenoid tubercle, for the long head of biceps brachii, is barely discernible from the glenoid labrum and glenohumeral capsular attachments. There is no elevation or distinct rugosity for its origin.

The infraglenoid tubercle for the long head of triceps brachii, however, is prominent and moderately thick, but only minimally rugose (Figure 9-2). It is distinctly dorsally located below the glenoid fossa, such that it grades smoothly onto the infraspinatus surface but has a narrow sulcus between it and the craniolateral corner of the subscapularis surface.

Axillary Border

The axillary borders of both Tianyuan 1 scapulae are largely absent, the right one completely (Chapter 3). On the left side (Figure 9-2), at the level of the infraglenoid tubercle and distally for 12 mm more ventrally, there is the craniolateral end of a distinct ventral bar. It is a raised, rounded, cortically thick, buttress ~7 mm wide. Along its lateral side there is a distinct sulcus 4–6 mm wide, between the bar and the inferior end of the infraglenoid tubercle. If this configuration continued distally, then Tianyuan 1 would have had the ventral sulcus pattern to its axillary border. If, however, the ventral bar decreased in prominence distally, the crest descending from the infraglenoid tubercle rotated to the midaxillary border, and a modest dorsal bar emerged, then it might conform to the bisulcate pattern. Given that scapulae with a dorsal sulcus normally have the infraglenoid tubercle deviating or displaced ventrally and the emergence of a dorsal bar by the distal end of the infraglenoid tubercle, it can be concluded that Tianyuan 1 did not have the dorsal sulcus pattern.

Given these considerations, it is most likely that the Tianyuan 1 axillary border conformed to the ventral sulcus pattern of Eickstedt (1925), or at the most, had a morphology intermediate between the ventral sulcus and bisulcate patterns. Axillary borders intermediate between the ventral sulcus and bisulcate patterns are present among MPMH (Skhul 4: McCown and Keith, 1939), early Neandertals (Krapina 127: Trinkaus, 2006c), MUP humans (Dolní Věstonice 3: Trinkaus and Jelínek, 1997), and eastern Asian early modern humans (Minatogawa 3: Baba and Endo, 1982). Late Pleistocene scapulae with the full ventral sulcus pattern are otherwise known only from 3 MUP individuals (11.5%, N = 26). It is only with early *Homo* and then with recent humans that the ventral sulcus pattern is the dominant form of the axillary border (Carretero et al., 1997; Trinkaus, 2008b). The two other EUP individuals with scapulae, Muierii 1 (Trinkaus, 2008b) and Nazlet Khater 2 (Crevecoeur, 2008) have the bisulcate pattern (as do the early MUP scapulae from Cro-Magnon and Paviland), the intermediate form most common for both MPMH and Eurasian MUP humans. Scapulae are unknown from eastern Eurasia prior to the time of Tianyuan 1.

If Tianyuan 1 did indeed have the ventral sulcus pattern, as seems likely from its preserved cranial axillary border, then it is relatively unusual among early modern humans in exhibiting this ancestral, Early Pleistocene, human pattern. It would also be the oldest Late Pleistocene human scapula to exhibit a distinctly ventral sulcus pattern, as opposed to one intermediate between the ventral sulcus and bisulcate patterns.

Humeri

The Tianyuan 1 humeri consist of a complete left diaphysis, a right diaphysis from the surgical neck to the mid-distal diaphysis, and most of the left distal epiphysis (Figure 9-4). Using landmarks on the left greater tubercle and its distal epiphysis, it is possible to estimate reliably the original lengths of the left humerus

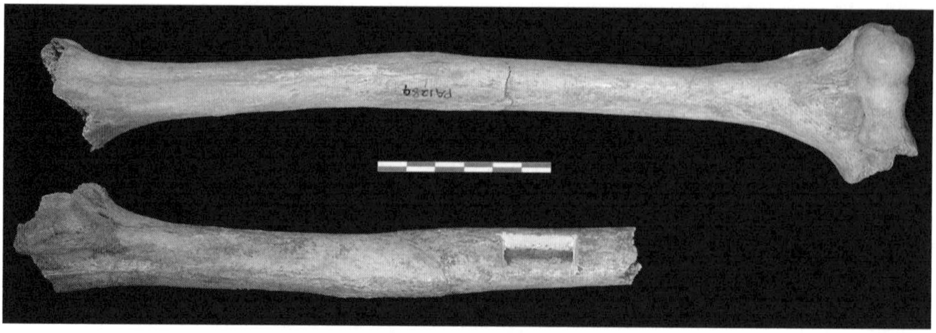

Figure 9-4. Anterior view of the Tianyuan 1 right (*below*) and left (*above*) humeri. Proximal is on the left. Scale in centimeters.

(Table 9-2); it is assumed that the right humerus had the same lengths, given minimal asymmetry in long bone lengths.

Proximal Epiphysis

The right humerus preserves most of the lesser tubercle and the adjacent intertubercular (bicipital) sulcus. The former has a distinct proximodistal ridge on it for the insertion of subscapularis, resulting in a peak toward the distal side of the tubercle. The sulcus is moderately broad and has a lateral insertion line extending proximally from the pectoralis major tuberosity; it is otherwise unremarkable. The left humerus only preserves the posterolateral end of the greater tubercle, with the insertion area for teres minor, as well as much of the intertubercular sulcus anteriorly. The teres minor insertion area is prominent but not rugose. The sulcus is similar to the right one.

Diaphyses

Diaphyseal Muscular Impressions

The humeral diaphyses are generally smooth with moderate muscle markings (Figure 9-5). The right deltoid tuberosity is broad, moderately rugose, and rises little from the contour of the midlateral diaphysis. More anteriorly it blends with the distal end of the muscle line extending distally from the pectoralis major insertion. It then fades out abruptly at midshaft. The left one is similarly broad and diffuse, extending laterally from the anterior crest. The maximum breadths of the tuberosities (Table 9-2) are located proximally. This configuration of the deltoid tuberosity, broad but with little relief, is characteristic of late archaic humans and contrasts with the more pronounced development of the tuberosity among many MUP humans (Churchill and Rhodes, 2009). Nazlet Khater 2 shares the

Table 9-2. Osteometric dimensions of the Tianyuan 1 humeri, in millimeters except as noted

	Right	*Left*
Humerus maximum length (M-1)	—	(331.4)[1]
Humerus articular length (M-2)	—	(328.9)[1]
Humerus biomechanical length	—	(327.4)[1]
Midshaft maximum diameter (M-5)	23.6	20.0
Midshaft minimum diameter (M-6)	17.7	15.1
Midshaft circumference (M-7a)	68.0	59.0
Distal minimum circumference (M-7)	(64.0)[2]	55.5
Pectoralis major tuberosity breadth[3]	6.6	6.1
Teres major tuberosity breadth[3]	9.5	10.0
Deltoid tuberosity breadth[4]	(20.0)	(18.0)
Distal articular breadth (M-12a)	—	(45.0)
Trochlear breadth (M-11)	—	22.2
Trochlear maximum anteroposterior diameter (M-13)	—	25.0
Trochlear minimum anteroposterior diameter (S-2)[5]	—	16.3
Olecranon fossa breadth (M-14)	—	28.4
Medial pillar thickness (S-12)	—	7.6
Cubital angle (M-16)	—	85°
Septal aperture	—	absent

Note: Estimated values are in parentheses.

[1] The articular length (to the distal capitulum) of the left humerus was estimated from its posterodistal greater tuberosity to the preserved capitulum (291.0 mm) using a least squares regression based on recent human humeri [ArtLen = (TY1Len × 1.099) + 9.1); r^2 = 0.965; N = 34; estimated length = 328.9 ± 3.6 mm; SE_{est} = 1.10%]. Based on the proximodistal distances from the capitulum to the medial (+2.5 mm) and lateral (−1.5 mm) trochlear margins, the maximum (to the medial trochlea) and biomechanical (to the lateral trochlea) lengths were calculated from the estimated articular length.

[2] The right distal minimum circumference is indicated as though it were estimated since it was taken at the distal break of the diaphysis. The distal break is close to, and may be at, the actual minimum circumference.

[3] The maximum breadths of the tuberosities for the insertions of the pectoralis major and teres major muscles on the proximal diaphysis.

[4] The maximum breadths of the deltoid tuberosities on the mid-diaphyses. The parentheses indicating estimation are due to the indistinct natures of the boundaries of the tuberosities on the bones.

[5] S-## indicates the number of the equivalent measurement in Senut (1981).

pronounced deltoid tuberosity with the MUP humans (Crevecoeur, 2008), but the Mladeč 24 humerus has a smooth one similar to those on the Tianyuan 1 humeri (Trinkaus, Smith, et al., 2006).

The right and left pectoralis major tuberosities are moderately rugose along clear anterior crests. Their anteromedial sulcal borders are distinct, but their lateral borders consist of only a cessation of the insertion rugosity. There are no concavities or raised nubbins of bone on them. Their breadths of 6.6 (right) and 6.1 mm (left) are close to the available values for both Middle and Upper Paleolithic associated early modern humans, but substantially lower than those of the Neandertals (Table 9-3). Since there is little correlation among Late Pleistocene humans between pectoralis major tuberosity breadth and humeral length (right:

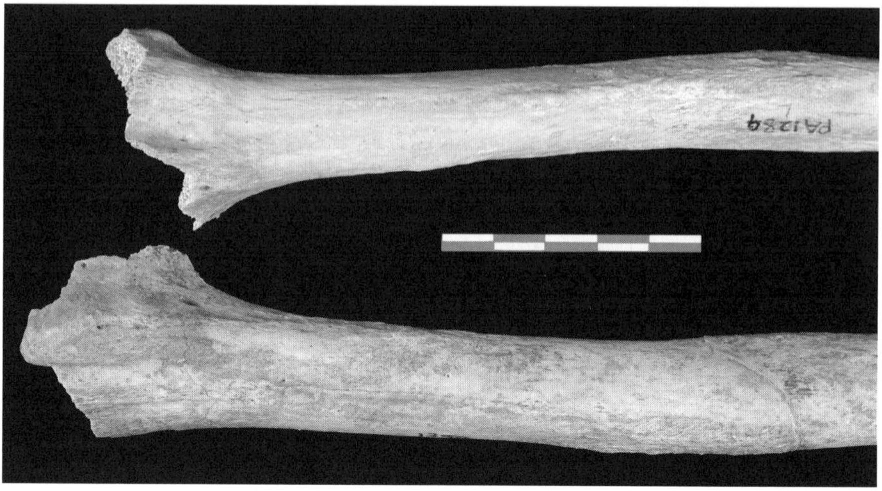

Figure 9-5. Anterior view of the Tianyuan 1 humeral proximal epiphyses and diaphyses. Proximal is on the left. Scale in centimeters.

Table 9-3. Pectoralis major tuberosity breadths in millimeters for Tianyuan 1 and comparative samples

	Pectoralis major breadth—right	*Pectoralis major breadth—left*
Tianyuan 1	6.6	6.1
East Asia EMH	6.4 ± 2.1 (4)	4.3, 7.1
West MUP	6.1 ± 1.9 (12)	6.1 ± 1.7 (14)
West EUP	4.7	—
MPMH	6.7, 8.6	—
Neandertals	10.1 ± 1.8 (13)	9.8 ± 1.5 (7)
K-W P-value	<0.001	0.003

Note: EMH: early modern humans; *MUP:* Mid Upper Paleolithic; *EUP:* Early Upper Paleolithic; *MPMH:* Middle Paleolithic modern humans. Kruskal-Wallis (K-W) P-values across the comparative samples are provided.

r = –0.046, N = 31; left: r = 0.202, N = 23), these values are compared using their absolute values.

The lateral heads of the triceps brachii muscles have created crests, running distally from the greater tuberosities down the midposterior shafts. The right one is prominent, ~30 mm long and up to 3 mm wide. The left one is less prominent and ≤3 mm wide but slightly longer (~36 mm).

Diaphyseal Robusticity

Most of the considerations of humeral diaphyses involve assessment of the quantity and distribution of diaphyseal cortical bone, for which both linear osteometrics (Table 9-2) and cross-sectional geometric parameters (Table 9-4) are provided.

The percent cortical areas of Tianyuan 1, or cortical area versus total subperiosteal area (Table 9-5), are moderately elevated for Tianyuan 1 (Figure 9-6). There is considerable individual variation but little between sample variation in percent cortical area across the Late Pleistocene samples. At the 35% section, only three right and three left humeri (8.6%, N = 35 each) have greater percent cortical areas than Tianyuan 1, although at midshaft six humeri on each side (right: 18.8%, N = 32 and left: 20.7%, N = 29) have higher values for this index. Since percent cortical areas reflect the individual's patterns of endosteal resorption and subperiosteal deposition through life (Ruff and Hayes, 1983b; Ruff et al., 1994), it is only an approximate measure of diaphyseal robusticity (Ruff, 2006).

Table 9-4. Cross-sectional geometric parameters of the Tianyuan 1 humeri

	Right	*Left*
35%		
Total area (mm^2)	357.2	247.0
Cortical area (mm^2)	300.4	208.9
Medullary area (mm^2)	56.8	38.1
Anteroposterior second moment of area (I_x) (mm^4)	(10,613)[1]	5,138
Mediolateral second moment of area (I_x) (mm^4)	(9,403)	4,482
Maximum second moment of area (I_{max}) (mm^4)	11,312	5,230
Minimum second moment of area (I_{min}) (mm^4)	8,704	4,389
Polar moment of area (mm^4)	20,016	9,620
Orientation of I_{max} (theta)	(109°)	121°
50%		
Total area (mm^2)	380.1	283.9
Cortical area (mm^2)	307.2	219.1
Medullary area (mm^2)	72.9	64.8
Anteroposterior second moment of area (I_x) (mm^4)	(9,564)	6,018
Mediolateral second moment of area (I_x) (mm^4)	(12,974)	6,691
Maximum second moment of area (I_{max}) (mm^4)	13,511	7,747
Minimum second moment of area (I_{min}) (mm^4)	9,027	4,961
Polar moment of area (mm^4)	22,538	12,709
Orientation of I_{max} (theta)	(160°)	142°
65%		
Total area (mm^2)	387.2	283.6
Cortical area (mm^2)	281.7	214.2
Medullary area (mm^2)	105.5	69.4
Anteroposterior second moment of area (I_x) (mm^4)	(11,911)	6,925
Mediolateral second moment of area (I_x) (mm^4)	(10,309)	5,406
Maximum second moment of area (I_{max}) (mm^4)	11,912	7,077
Minimum second moment of area (I_{min}) (mm^4)	10,308	5,254
Polar moment of area (mm^4)	22,220	12,331
Orientation of I_{max} (theta)	(92°)	107°

[1] The right I_x, I_y, and theta are in parentheses since the anatomical planes of the right humerus have been approximated using the more complete left humerus.

Table 9-5. Percent cortical areas (%CA) for Tianyuan 1 and comparative sample humeri

	Right 35% %CA	*Left 35% %CA*	*Right 50% %CA*	*Left 50% %CA*
Tianyuan 1	85.2	84.6	80.8	77.2
East Asia EMH	76.5 ± 5.6 (4)	75.7 ± 5.1 (4)	74.7 ± 7.7 (4)	74.1 ± 7.3 (4)
West MUP	76.6 ± 7.5 (15)	75.2 ± 6.1 (19)	68.6 ± 9.1 (12)	67.6 ± 7.4 (17)
West EUP	76.6, 81.7	82.9	66.6, 81.0	83.0
MPMH	80.1 ± 6.2 (5)	79.9 ± 10.8 (4)	76.2 ± 7.4 (4)	61.6, 92.1
Neandertals	79.6 ± 5.0 (9)	81.2 ± 6.6 (7)	75.3 ± 6.2 (10)	76.4 ± 2.7 (5)
K-W P-value	0.632	0.250	0.136	0.062

Note: EMH: early modern humans; *MUP:* Mid Upper Paleolithic; *EUP:* Early Upper Paleolithic; *MPMH:* Middle Paleolithic modern humans. Includes comparisons at the mid-distal (35%) and midshaft (50%) locations. Kruskal-Wallis (K-W) P-values across the comparative samples are provided.

Diaphyseal robusticity can be measured osteometrically by comparing the distal minimum circumference to humeral length (Bräuer, 1988) and the mid-distal polar second moment of area to humeral length (Trinkaus, 2006b). However, it has also been argued, based on recent human comparisons and resultant consistency across samples of humeral diaphyseal robusticity (Ruff et al., 1993; Ruff, 2000; Shackelford, 2007), that body mass (times bone length) should be incorporated into the scaling of humeral properties. Incorporating estimated body mass, through its multiplication with bone biomechanical length to scale second moments of area, is the appropriate method for weight-bearing limbs (see Polk et al., 2000; Ruff, 2000). Yet, human humeri are not normally weight bearing, so it is unclear whether such considerations apply. At the same time, incorporating weight estimates into the scaling of humeral cross-sectional parameters is conservative, since it tends to decrease the perceived differences in robusticity between samples, especially given the contrasts in body proportions across Late Pleistocene human samples (Holliday, 1997b; 2000; Chapter 7). Therefore, even though the traditional osteometric data and index of robusticity are provided (Tables 9-2 and 9-6), the principal comparisons of the Tianyuan 1 humeral diaphyseal robusticity to other samples involve the mid-distal (35%) polar second moments of the area versus estimated body mass times humeral biomechanical length (Figure 9-7).

The comparisons of humeral robusticity indices (Table 9-6) provide little difference across the samples for the right humerus, but a significant difference for the left ones. In the right humerus, the Tianyuan 1 value is in the middle of the ranges of variation, remaining within one standard deviation of each of the means. In the left humerus, the Tianyuan 1 value is relatively low, being below all of the Neandertal values and above only those of Skhul 5 and 25% of the MUP humeri (N = 20).

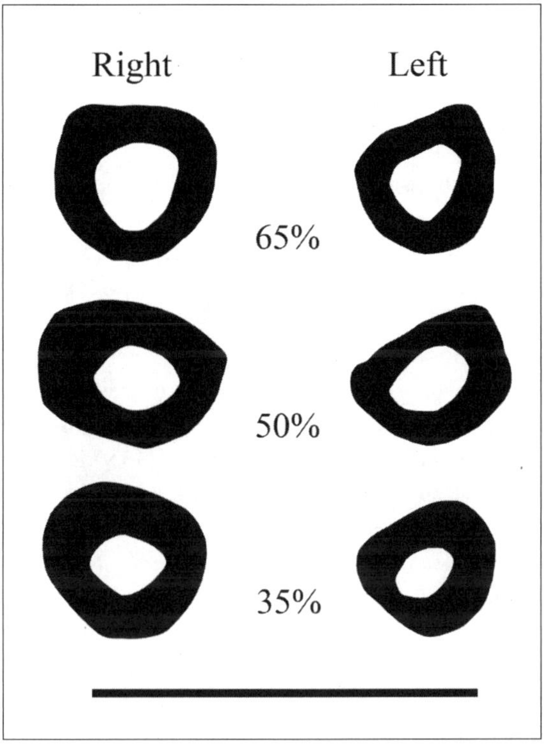

Figure 9-6. Diaphyseal cross sections of the right and left humeri. Sections are in distal view, anterior is above, medial is to the inside, and 0% is distal. Scale bar equals 50 mm.

Table 9-6. Robusticity indices and midshaft indices for Tianyuan 1 and comparative Late Pleistocene samples

	Right humeral robusticity index[1]	*Left humeral robusticity index*[1]	*Right humeral midshaft index*[2]	*Left humeral midshaft index*[2]
Tianyuan 1	19.4	16.8	130.9	139.5
East Asia EMH	19.0 ± 1.1 (4)	18.7 ± 1.2 (4)	137.3 ± 8.8 (4)	133.5 ± 6.5 (4)
West MUP	18.9 ± 1.1 (16)	17.8 ± 1.2 (20)	126.0 ± 8.1 (18)	122.2 ± 9.6 (21)
West EUP	15.8, 20.5	21.1	116.9, 141.3	114.1
MPMH	18.5 ± 1.2 (4)	15.8, 17.5	131.8 ± 15.2 (7)	117.1, 123.5, 127.4
Neandertals	20.8 ± 1.8 (11)	20.0± 1.9 (8)	135.9 ± 8.3 (12)	137.2 ± 7.3 (8)
K-W P-value	0.999	0.014	0.251	0.002

Note: EMH: early modern humans; *MUP:* Mid Upper Paleolithic; *EUP:* Early Upper Paleolithic; *MPMH:* Middle Paleolithic modern humans. Kruskal-Wallis (K-W) P-values are across the comparative samples.

[1] (distal minimum circumference/maximum length) × 100.

[2] (maximum diameter/minimum diameter) × 100.

Comparisons of the mid-distal (35%) polar second moments of area to humerus length times body mass estimations provides little proportional separation of the comparative samples, even though the Neandertals cluster toward the upper end of the abscissa. The Tianyuan 1 right humerus remains in the middle of the distribution, whereas Mladeč 24 is very gracile and Nazlet Khater 2, with its low estimated body mass based on its femoral head diameter, appears very robust (Figure 9-7A). With respect to the left humerus, Tianyuan 1 is rather gracile, matched by the La Ferrassie 1 Neandertal (Figure 9-7B). Nazlet Khater 2 remains very robust. In these plots the very large Barma Grande 2 MUP individual, with its exceptionally robust right humerus and high level of upper limb asymmetry (Churchill and Formicola, 1997), has not been included.

This difference in right versus left robusticity assessments is evident as well in comparisons of mid-distal cortical area and polar moment of area asymmetry (Figures 9-6 and 9-8). In the cortical area comparison, Tianyuan 1 (35.9) is at the tops of the Late Pleistocene ranges of variation. Only Barma Grande 2, with its exceptional (51.0) and probably pathological upper limb asymmetry (Churchill and Formicola, 1997), is noticeably higher. Two other specimens, the Spy 2 Neandertal (37.3) and the MUP Dolní Věstonice 14 (35.4), have similar levels of asymmetry. In the polar moment of area comparison, Tianyuan 1 (70.2) is only exceeded by the exceptional Barma Grande 2 (92.0), although Dolní Věstonice 14 (66.0), La Quina 5 (~68.5) and Spy 2 (66.0) are close. The MPMH sample and Nazlet Khater 2 (21.1) have relatively low asymmetry, and the Minatogawa sample is exceptional for a Late Pleistocene sample in its virtual absence of humeral asymmetry. Also, the majority of these specimens, including Tianyuan 1, have larger right humeral cortical areas and greater right polar moments of area (84.6%, N = 26 and 88.0%, N = 25 respectively), suggesting right-handedness for these individuals.

Diaphyseal Shape

Variation in humeral diaphyseal shape consists principally in the degree to which the humeral midshaft is subcircular as opposed to relatively anterolaterally to posteromedially elongated. Quantitative comparisons involve an index of maximum and minimum diameters plus distributions of maximum and minimum second moments of area. In the right humerus, the midshaft index provides only a modest separation of the samples, mostly with the MUP sample having rounder diaphyses than the other samples (Table 9-6). There is more contrast in the left humeral indices, with the Neandertals and eastern Asian early modern humans having more ovoid humeri on average. This pattern is largely maintained in the comparisons of the midshaft second moments of area (Figure 9-9), except that the MPMH sample clusters more with the Upper Paleolithic early modern humans in both humeri. In these comparisons, the Tianyuan 1 humeri and the Mladeč 24

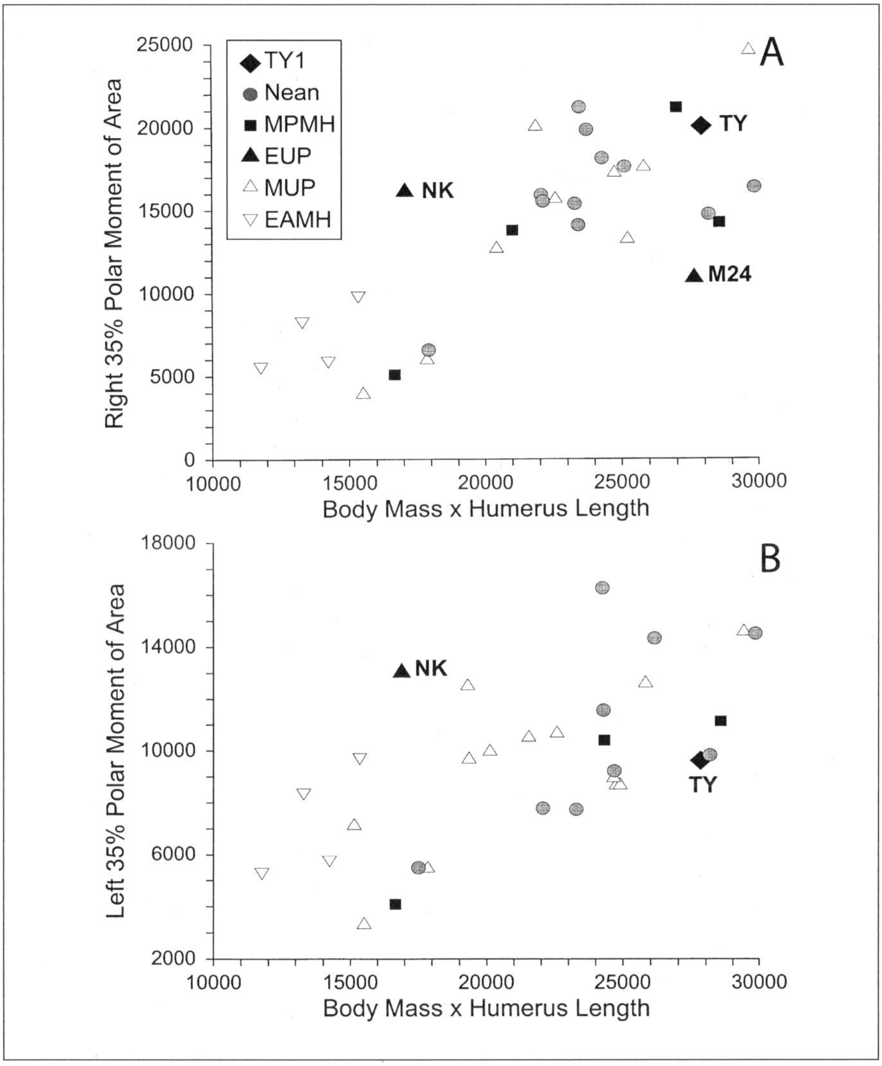

Figure 9-7. Bivariate plots of (*A*) right and (*B*) left mid-distal polar second moments of area versus humeral lengths times estimated body masses. *TY:* Tianyuan 1; *Nean:* Neandertals; *MPMH:* Middle Paleolithic modern humans; *EUP:* western Early Upper Paleolithic; *MUP:* western Eurasian Mid Upper Paleolithic; *EAMH:* eastern Asian modern humans; *NK:* Nazlet Khater 2; *M24:* Mladeč 24.

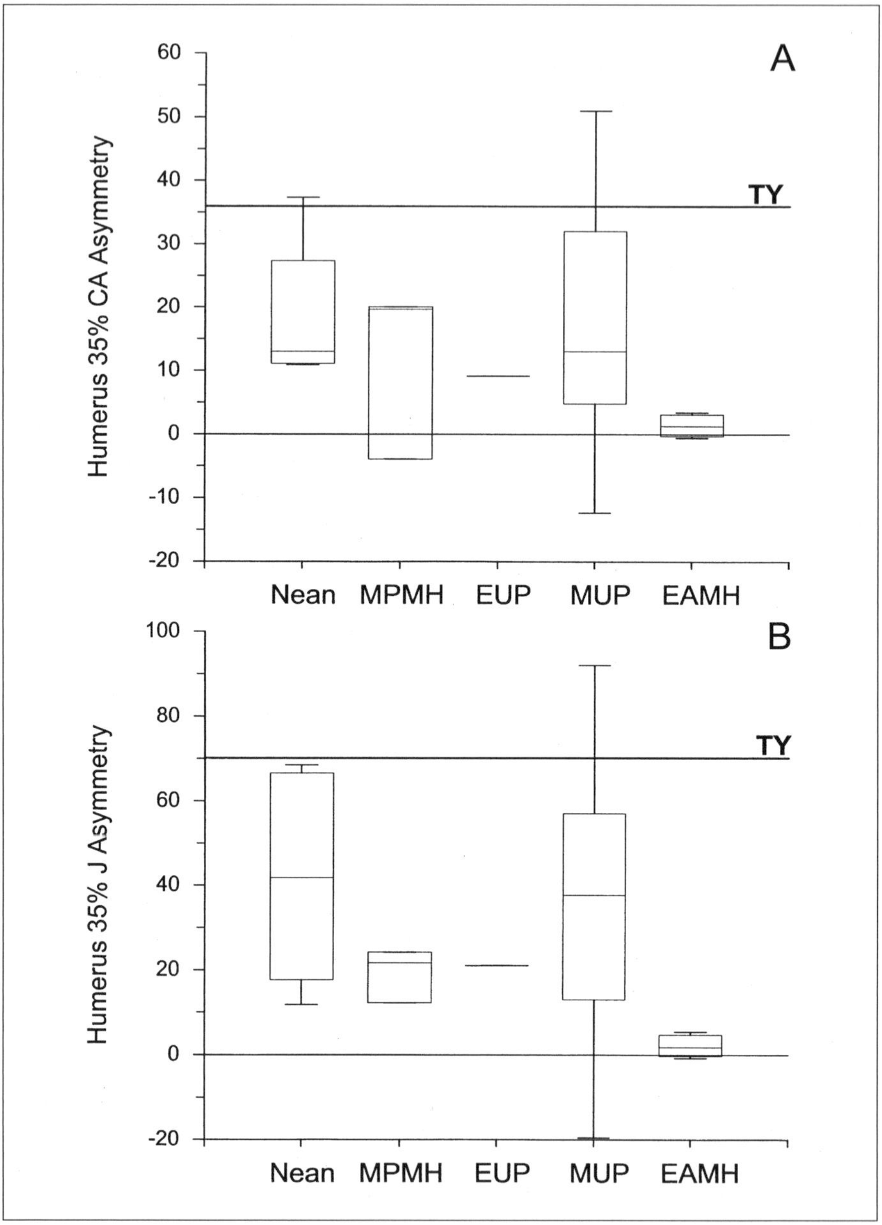

Figure 9-8. Box plots of (*A*) asymmetry in mid-distal humeral cortical area (CA) and (*B*) polar second moments of area (J). *TY:* Tianyuan 1 values of 35.9 and 70.2. Positive values indicate a larger right side; the line at 0 indicates symmetry. *Nean:* Neandertals; *MPMH:* Middle Paleolithic modern humans; *EUP:* western Early Upper Paleolithic; *MUP:* western Eurasian Mid Upper Paleolithic; *EAMH:* eastern Asian modern humans. EUP specimen: Nazlet Khater 2.

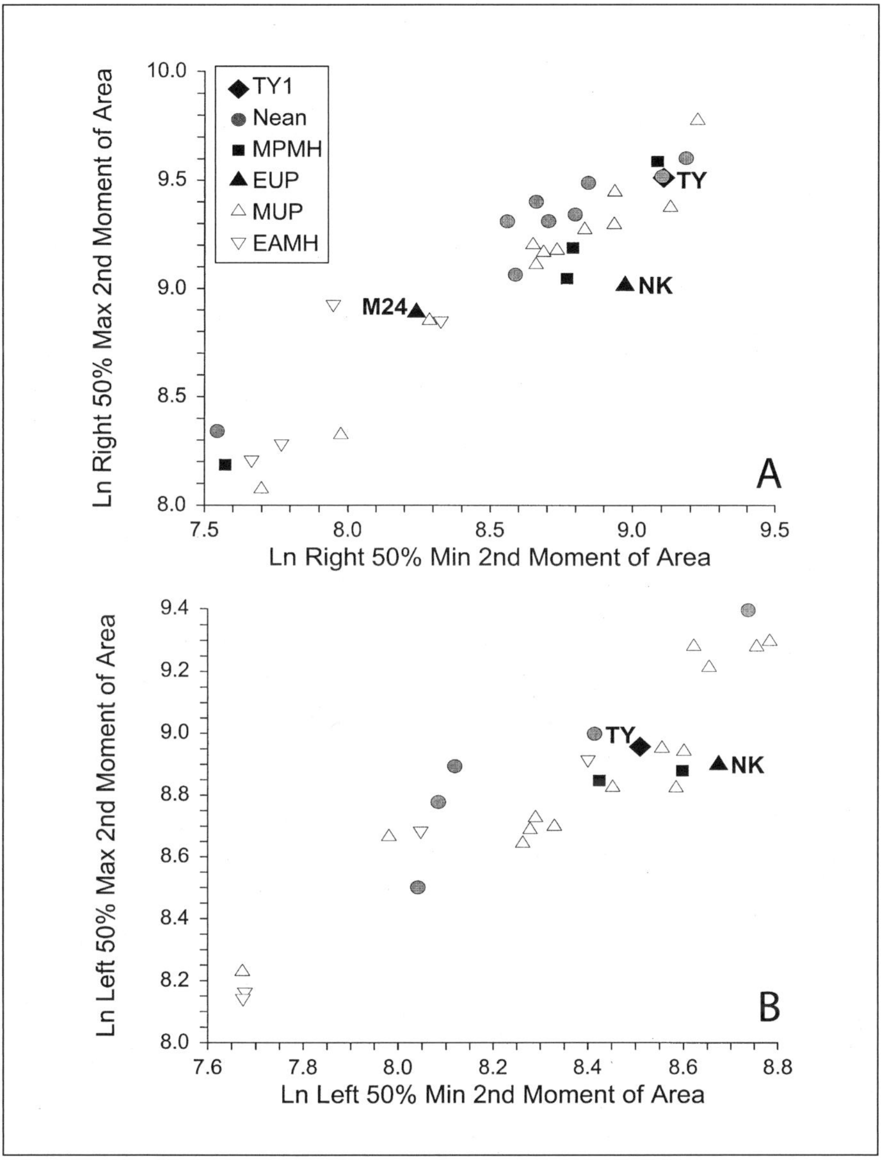

Figure 9-9. Bivariate plots of (*A*) right and (*B*) left humeral midshaft maximum to minimum second moments of area. *TY:* Tianyuan 1; *Nean:* Neandertals; *MPMH:* Middle Paleolithic modern humans; *EUP:* western Early Upper Paleolithic; *MUP:* western Eurasian Mid Upper Paleolithic; *EAMH:* eastern Asian modern humans; *NK:* Nazlet Khater 2; *M24:* Mladeč 24.

right humerus are in the middles of the overall distributions, but the Nazlet Khater 2 humeri are quite round.

Distal Epiphysis

The Tianyuan 1 left humerus retains most of the cubital articular surfaces, the olecranon and coronoid fossae, and the medial pillar (Figure 9-10). The epicondyles are absent, as are the lateral pillar and the lateral supracondylar crest. The articular surfaces are unexceptional, with a moderately projecting medial trochlear flange and a poorly delineated lateral one. The coronoid fossa is discrete, relatively deep, and separated from the radial head fossa by a sharp crest. The olecranon fossa is deep and slightly irregular on its floor. It grades smoothly onto both the posterior distal diaphysis and the proximoposterior trochlea.

The Tianyuan 1 humerus does not exhibit a septal aperture. Among Late Pleistocene humans, pooling together right and left observations given little evidence for asymmetry (only in Nazlet Khater 2), they are relatively common among the Neandertals (44.4%, N = 18) but less so among early modern humans. One of four MPMH exhibits one. Mladeč 24 lacks one but Nazlet Khater 2 has it unilaterally.

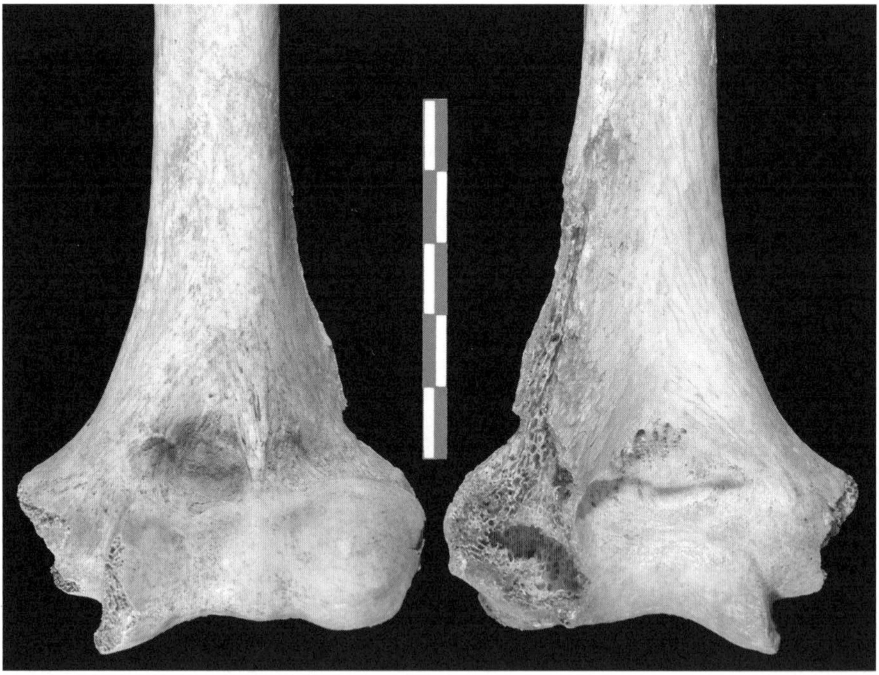

Figure 9-10. Anterior (*left*) and posterior (*right*) views of the Tianyuan 1 left distal humerus. Scale in centimeters.

None of the nineteen MUP humans has them, and neither of the Minatogawa individuals has them.

Neandertals have been described as having large olecranon fossae, but relative to other Late Pleistocene humeri their fossa breadths are unexceptional, either as raw measurements or compared to other distal epiphyseal measures (Hambücken, 1993). The Tianyuan 1 olecranon fossa, with a breadth of 28.4 mm, is similar to most other Late Pleistocene humeri. At the same time, the medial pillar thickness of Tianyuan 1 of 7.6 mm is modest. Neandertals tend to have relatively thin medial pillars compared to recent humans (Carretero et al., 1997, 2009; Yokley and Churchill, 2006; Trinkaus et al., 2007), and similarly thin medial pillars are variably found among earlier *Homo* humeri (Carretero et al., 2009). The Neandertal to modern human contrast is largely supported by a plot of the left medial pillar thickness against olecranon breadth (Figure 9-11). The primary overlap between the Neandertals and early modern humans concerns the low values for Dolní Věstonice 13 and 15. Tianyuan 1 falls among the Neandertals, although it is also close to several MUP humeri. It is proportionately distant from Skhul 4 and Nazlet Khater 2.

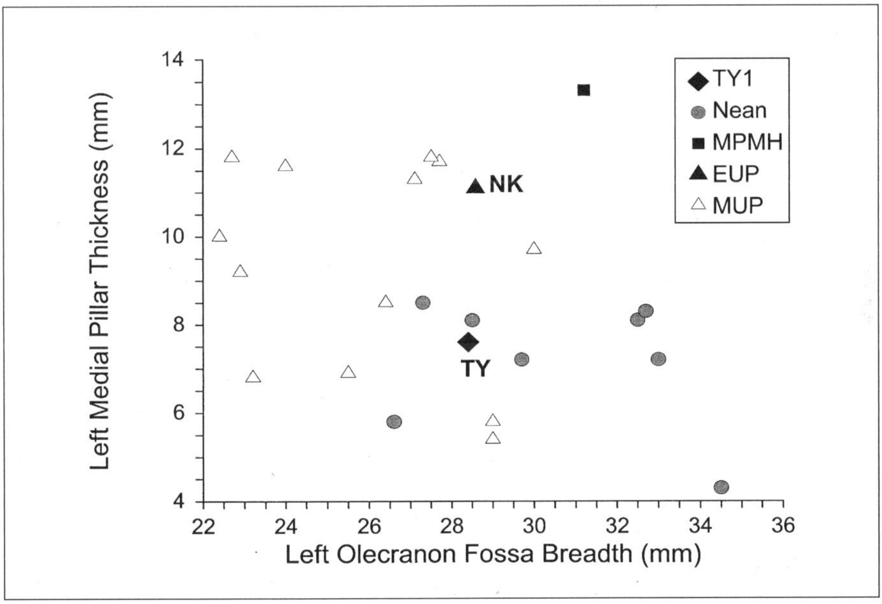

Figure 9-11. Bivariate plot of left medial pillar thickness versus olecranon breadth. *TY:* Tianyuan 1; *Nean:* Neandertals; *MPMH:* Middle Paleolithic modern humans; *EUP:* western Early Upper Paleolithic; *MUP:* western Eurasian Mid Upper Paleolithic; *NK:* Nazlet Khater 2.

Right Ulna

Consisting of a partial diaphysis from just distal of the brachialis tuberosity to the proximal end of the pronator quadratus tuberosity, the Tianyuan 1 right ulna presents little of note (Table 9-7; Figure 9-12). There is a distinct proximal posterior convexity to the shaft. The interosseus crest is strongly projecting, even though its full extent is probably not evident, given damage to the proximal crest. There is the proximal end of a distinct pronator quadratus tuberosity or crest, minimum measurements for which are provided in Table 9-6. This crest is likely to have been prominent, a feature observed in both late archaic and early modern humans.

Radii

The left radius is complete except for its articular surfaces, both proximally and distally, and the right radius preserves only a small midproximal diaphyseal section (Table 9-8; Figures 9-13 and 9-14). The two bones are similar in the preserved

Table 9-7. Osteometric dimensions of the right ulnar diaphysis of Tianyuan 1, in millimeters

Midshaft anteroposterior diameter	14.2
Midshaft mediolateral diameter	15.6
Pronator quadratus maximum diameter[1]	(13.5)
Pronator quadratus minimum diameter[1]	(10.5)

Note: Estimated values are in parentheses.

[1] Maximum and minimum diameters at the maximum projection of the pronator quadratus crest on the distal medial ulna. Values are indicated as estimated, since the distal break of the bone is probably not at the full development of the crest.

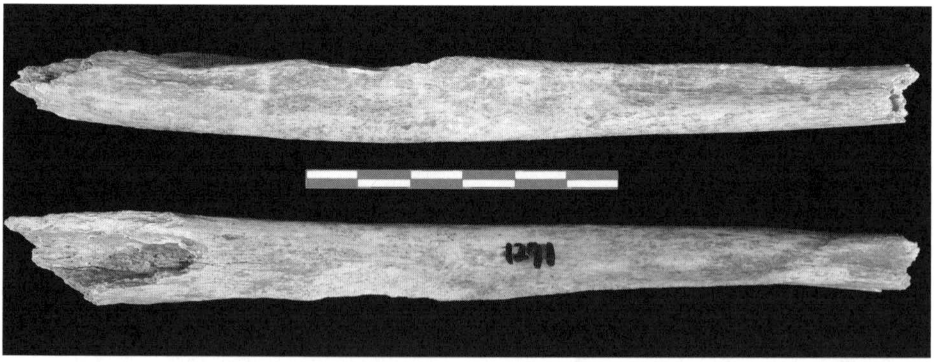

Figure 9-12. Anterior (*bottom*) and lateral (*top*) views of the Tianyuan 1 right ulna. Proximal is on the left. Scale in centimeters.

Table 9-8. Osteometric dimensions of the Tianyuan 1 radii, in millimeters except as noted

	Right	Left
Radius maximum length (M-1)	—	(251.8)[1]
Radius articular length (M-2)	—	(240.9)[2]
Lateral curvature chord[3]	—	189.0
Lateral curvature subtense (M-6)	—	5.0
Proximal anteroposterior diameter	—	12.1
Proximal mediolateral diameter	—	11.4
Crest anteroposterior diameter (M-5)	11.1	10.9
Crest mediolateral diameter (M-4)	(14.2)[4]	15.5
Midshaft anteroposterior diameter (M-5a)	—	11.1
Midshaft mediolateral diameter (M-4a)	—	13.1
Minimum circumference (M-3)	—	37.0[5]
Neck anteroposterior diameter (M-5(2))	—	12.7
Neck mediolateral diameter (M-4(2))	—	11.3
Neck maximum diameter	—	13.1
Neck minimum diameter	—	11.2
Tuberosity position[6]	—	2
Tuberosity length (S-1)[7]	—	23.7
Tuberosity breadth (S-4)	—	10.8
Tuberosity projection (S-8)	—	15.7
Neck-shaft angle (M-7)	—	(7°)

Note: Estimated values are in parentheses.

1 The maximum length of the left radius was estimated from the distance between the distal medial head and the proximal ulnar head facet (224.0 mm) using a linear regression based on a sample of recent human radii [MaxLen = ((TY1Len × 1.060) + 14.4); N = 34; r^2 = 0.974; estimated length = 251.8 ± 3.4 mm; SE_{est} = 1.36%].

2 The articular length of the left radius was estimated from the distance between the distal medial head and the proximal ulnar head facet (224.0 mm) using a linear regression based on a sample of recent human radii [ArtLen = ((TY1Len × 1.028) + 10.5); N = 3; r^2 = 0.990; estimated length = 240.9 ± 2.9 mm; SE_{est} = 1.22%].

3 The lateral curvature chord is from the lateral surface opposite the middle of the radial tuberosity to the lateral concavity just proximal of the distal diaphysis.

4 The right mediolateral diameter may be slightly underestimated, given that the distal break may not be at the full original extent of the crest.

5 The minimum circumference of the left radius is close to midshaft.

6 Radial tuberosity position relative to the interosseous crest following Trinkaus and Churchill (1988).

7 S-## indicates the number of the equivalent measurement in Senut (1981).

portions, with the right one having a slightly flatter lateral diaphyseal surface opposite the interosseus crest. Morphological assessments of the Tianyuan 1 radial morphology rely on the more complete left bone.

The neck is ovoid in cross section, with the maximum diameter anterolateral to posteromedial. The neck appears moderately constricted relative to the head, based on the preserved flare for the head epiphysis. The radial tuberosity is oriented anteromedially, such that a line from the interosseus crest goes through its posterior third; this is type 2 of Trinkaus and Churchill (1988). A similar antero-

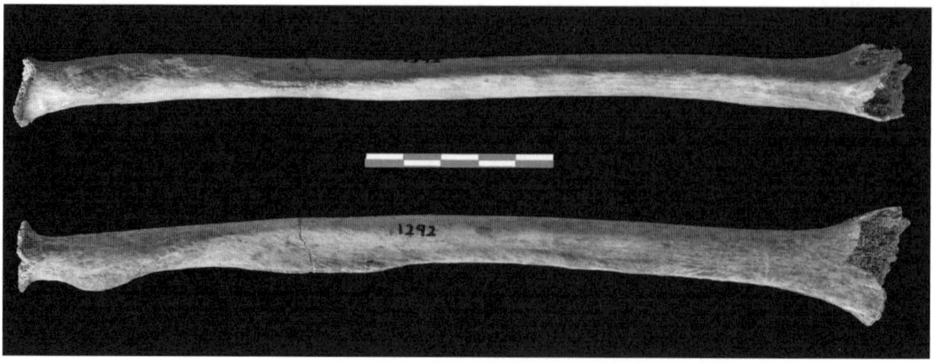

Figure 9-13. Medial (*top*) and anterior (*bottom*) views of the Tianyuan 1 left radius. Proximal is on the left. Scale in centimeters.

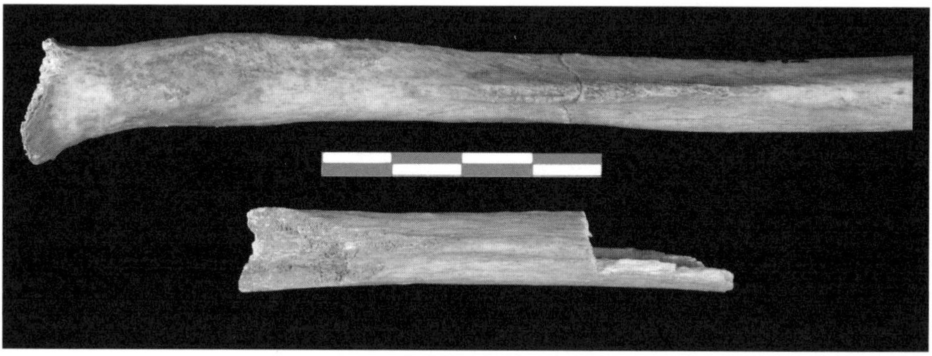

Figure 9-14. Proximal medial view of the Tianyuan 1 right (*bottom*) and left (*top*) proximal radii. The proximodistal position of the right radius relative to the left one is approximate. Scale in centimeters.

medial position for the radial tuberosity is the dominant one for Late Pleistocene modern humans (Table 9-9), as well as most recent humans (Trinkaus and Churchill, 1988; Hambücken, 1993; Churchill, 1994). It is the more common arrangement for Neandertal left radii, but if the right and left Neandertal radii are combined to provide an average score for each individual preserving a radius, then the more medial position 3 is the most common among them. Of the eleven Late Pleistocene individuals providing tuberosity position data for both radii, one (Regourdou 1) is asymmetrical, with the right side having the more dorsal tuberosity position. The late Middle Pleistocene Omo-Kibish 1 also has asymmetry of the tuberosity position, but its more dorsal tuberosity is left.

The left radius is relatively straight mediolaterally with a slight posterior convexity near midshaft and a similar anterior convexity in the distal shaft. The lateral curvature of the diaphysis is of particular interest, given its Late Pleistocene human variation (Vandermeersch and Trinkaus, 1995) and biomechanical implications for the mechanical efficiency of pronator teres (Galtés et al., 2008). The absolute curvature subtense for Tianyuan 1 is modest, even though it is within the ranges of variation of the comparative samples. Its curvature index is intermediate between the higher Neandertal values, the intermediate and highly variable (right radius only) MPMH indices, and the lower Upper Paleolithic ones (Table 9-10). At the same time, its neck-shaft angle, which should reflect to some extent proximal diaphyseal lateral deviation and hence lateral shaft curvature, is low compared to

Table 9-9. Radial tuberosity position for Tianyuan 1 and comparative samples

	Position 1 (%)	*Position 2 (%)*	*Position 3 (%)*	*N*
Tianyuan 1		2		
East Asia EMH	0.0 [0.0]	100 [100]	0.0 [0.0]	1 [3]
West MUP	15.6 [17.9]	78.1 [75.0]	6.3 [7.1]	16 [28]
West EUP	0.0 [0.0]	100 [100]	0.0 [0.0]	1 [2]
MPMH	50 [20.0]	0.0 [70.0]	50.0 [10.0]	2 [5]
Neandertals	7.1 [3.8]	64.3 [30.8]	28.6 [65.4]	7 [13]

Note: EMH: early modern humans; *MUP:* Mid Upper Paleolithic; *EUP:* Early Upper Paleolithic; *MPMH:* Middle Paleolithic modern humans. Frequencies for the left radii are followed, in brackets, with frequencies for the pooled right and left radii. Position 1: interosseous crest line posterior of tuberosity; position 2: crest line in posterior third of tuberosity; position 3: crest line in middle third of tuberosity (Trinkaus and Churchill, 1988).

Table 9-10. Comparative osteometrics for radial lateral diaphyseal curvature and neck-shaft angle

	Lateral curvature subtense	*Lateral curvature index*	*Neck-shaft angle*
Tianyuan 1	5.0	2.6	(7°)
East Asia EMH	—	1.8	11°, 13°
	—	[1.9, 2.0]	[9°, 13°, 13°]
West MUP	3.3 ± 1.1 (6)	1.8 ± 0.5 (6)	12.2° ± 2.8° (13)
	[4.4 ± 1.47 (14)]	[2.3 ± 0.7 (14)]	[11.7° ± 3.1° (16)]
West EUP	—	—	17° [16°]
MPMH	—	—	11°, 13°
	[4.3 ± 3.8 (4)]	[2.4 ± 2.2 (4)]	[10°, 12°, 14°]
Neandertals	7.3 ± 3.2 (6)	4.5 ± 1.8 (6)	14.0° ± 1.3° (6)
	[8.0 ± 2.7 (10)]	[4.9 ± 1.6 (10)]	[13.7° ± 2.2° (11)]
K-W P-value	0.153	0.006	0.003
	[0.079]	[0.005]	[<0.001]

Note: EMH: early modern humans; *MUP:* Mid Upper Paleolithic; *EUP:* Early Upper Paleolithic; *MPMH:* Middle Paleolithic modern humans. Mean and standard deviation (N) for the left radii are followed, in brackets, with data for the pooled right and left radii. Right and left values, when available for an individual, were averaged prior to the computation of the pooled summary statistics. Kruskal-Wallis (K-W) P-values across the comparative samples are provided.

all of the other samples. Even though its radial head is missing, the neck is complete and the angle should be within a degree of the 7° presented here.

The diaphyseal cross section is tear-drop shaped, similar to archaic human radii and many recent human ones, without longitudinal sulci other than those immediately adjacent to the interosseus crest. As with the right ulna, the interosseus crests are prominent and sharp. The more complete left crest deviates slightly dorsally along its medial edge. The interosseus crest then remains rugose as it extends distally to the anteromedial corner of the distal epiphysis. The crest is nonprojecting from midshaft to the distal epiphysis, such that the minimum circumference is close to midshaft.

The diaphyseal diameters at the maximum development of the interosseus crest separate out the late archaic and early modern human Middle Paleolithic samples for the right radius (Trinkaus and Biglari, 2006) and the left one (Figure 9-15), but the earlier Upper Paleolithic values span the ranges of these earlier samples. The Tianyuan 1 left radius, as well as the Nazlet Khater 2 one, fall in the middle of the overall Late Pleistocene distribution.

The distal articulations are missing, but the preserved portion of the metaphyseal region indicates that the dorsal extensor tendon sulci were well marked.

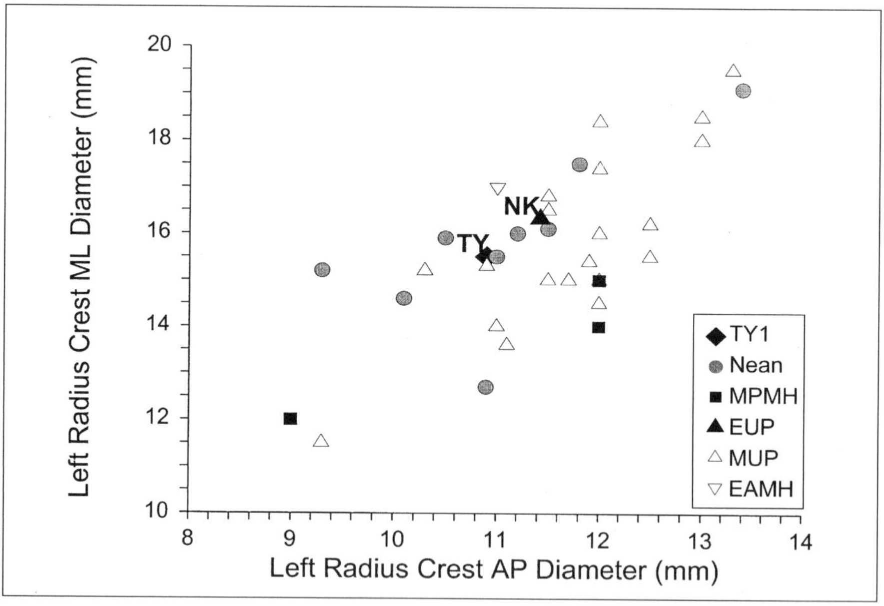

Figure 9-15. Bivariate plot of radial interosseus crest maximum mediolateral diameter versus the anteroposterior diameter at the same proximodistal position. *TY:* Tianyuan 1; *Nean:* Neandertals; *MPMH:* Middle Paleolithic modern humans; *EUP:* western Early Upper Paleolithic; *MUP:* western Eurasian Mid Upper Paleolithic; *EAMH:* eastern Asian modern humans; *NK:* Nazlet Khater 2.

Hand Remains

The Tianyuan 1 upper limb remains preserve three carpals (the right lunate and capitate and the left hamate), three proximal phalanges, a middle phalanx, and a distal phalanx. The identifications of the proximal phalanges to digit and side are morphologically evident (Chapter 3). The middle and distal phalanges derive from the same digit, based on matching osteoarthritic degenerations, and they may derive from the same digit as the left proximal phalanx 2. Other than the distal interphalangeal joint, the bones are free of lesions.

Lunate Bone

The complete right lunate bone is unremarkable for a recent human or for a Late Pleistocene archaic or modern human (Table 9-11). The dorsal and palmar surfaces are relatively smooth, with little rugosity for the various interarticular ligaments. The capitate and hamate facets are distinctly separated from each other, with a low and rounded ridge between them. Each one is radioulnarly concave, and the angle between their tangents (determined by their radial and ulnar borders) is 35°. The triquetral facet is then angled 40° from the hamate facet. The articular subchondral bone is smooth, but the individual facets have raised dorsal and palmar borders.

Capitate Bone

The complete right capitate bone (Table 9-12; Figure 9-16) has smooth articular facets with raised edges, especially on the hamate facet and the dorsal and palmar metacarpal 3 facet. The nonarticular areas are relatively smooth, including the palmar aspect. The dorsal nonarticular bone is concave between the lunate and metacarpal 3 facets, especially adjacent to the distodorsoradial raised corner for the metacarpal 3 styloid process.

Table 9-11. Osteometric dimensions of the Tianyuan 1 right lunate bone, in millimeters

Maximum breadth (M-2)	(18.0)
Maximum depth (M-3)	18.3
Radius articular height (M-10)	15.7
Radius articular breadth (M-11)	16.6
Distal articular breadth (M-7)	11.3
Capitate articular breadth (M-7a)	7.8
Hamate articular breadth (M-7b)	4.7
Triquetral articular height (M-8)	10.5
Triquetral articular breadth (M-9)	8.5

Note: Estimated value is in parentheses.

Table 9-12. Osteometrics of the Tianyuan 1 right capitate bone, in millimeters except as noted

Articular length (M-4)	22.7
Maximum breadth (M-2)	15.1
Maximum height (M-3)	20.2
Proximal maximum articular height (M-6)	14.7
Proximal maximum articular breadth (M-5)	9.9
Scaphoid articular length	17.6
Scaphoid articular height	14.5
Hamate articular length (M-9)	22.0
Hamate articular height (M-10)[1]	8.3
Metacarpal 2 articular height (M-7)	(14.4)
Metacarpal 2 articular breadth (M-8)	5.6
Metacarpal 2 articular depth[2]	1.7
Metacarpal 3 articular height	16.6
Metacarpal 3 articular breadth	10.9
Metacarpal 2/3 articular angle[3]	54°

Note: Estimated value is in parentheses.

[1] Maximum proximal height of the facet.

[2] Maximum depth from the articular height chord (Niewoehner et al., 1997).

[3] Angle in the coronal plane between the middles of the metacarpal 2 and 3 articular facets (Niewoehner et al., 1997).

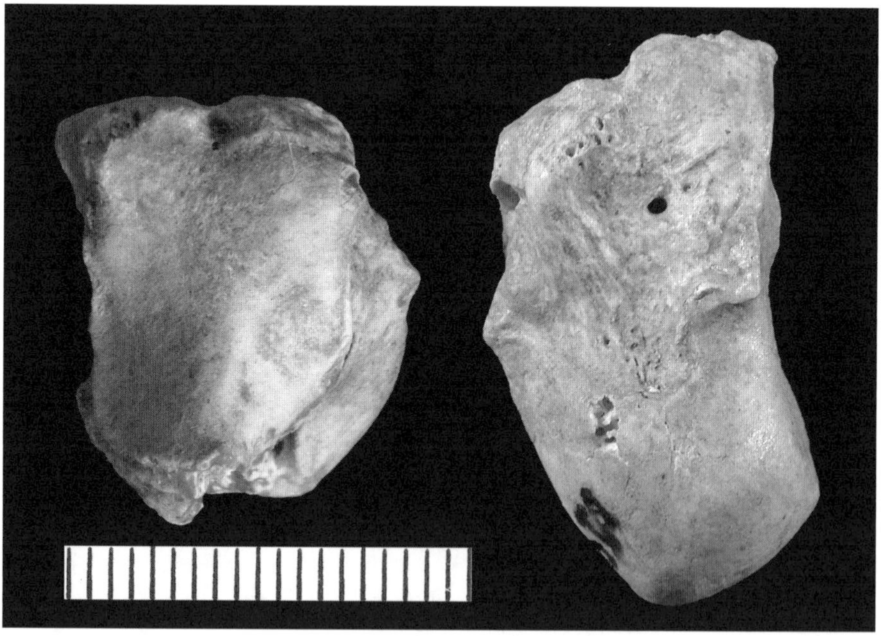

Figure 9-16. Distal (*left*) and dorsal (*right*) views of the Tianyuan 1 right capitate bone. Scale in millimeters.

The metacarpal facets exhibit two of the derived characteristics of recent humans, ones which are present in late archaic humans but occur among some of them in less pronounced forms (Niewoehner et al., 1997; Niewoehner, 2001). The dorsal margin of the metacarpal 3 facet slopes distoulnarly to proximoradially, with a distinct rounding of the distodorsoradial corner of the facet. This is for a substantial styloid process on the third metacarpal base. Variation in the size of the metacarpal 3 styloid process, which cannot be directly observed on Tianyuan 1, is related to overall metacarpal length among Late Pleistocene humans, independent of late archaic or early modern human affinities (Trinkaus et al., 2010). Given the moderately long limbs of Tianyuan 1, especially in relation to those of the Neandertals, the degree of capitate distodorsoradial rounding may reflect principally its upper limb length.

There is a torsion to the metacarpal 3 facet, such that the palmar portion of it faces distoulnarly (73° relative to the proximal hamate facet), whereas the dorsal portion of it faces slightly distoradially (92° relative to the proximal hamate facet), resulting in a 19° torsion to the facet.

In addition, there is a distinct, dorsopalmarly and radioulnarly concave facet for the base of the metacarpal 2. The form of the facet matches those of recent humans, but its orientation is toward the more sagittal end of recent human variation. The middle of the metacarpal 2 facet is oriented 54° relative to the middle of the metacarpal 3 facet. This value is moderately high for recent humans (Euroamericans: 46° ± 7°, N = 41; Amerindians: 39° ± 9°, N = 53) and early modern human variation (Qafzeh 9: 41°; Nazlet Khater 2: 55°; Dolní Věstonice 15 and 16: 40° and 25°), but it is close to the mean of a variable Neandertal sample (59°± 12°, 44°—75°, N = 6).

Both of these capitate-metacarpal articular facet configurations indicate full possession of the recent human morphology for the effective resistance of oblique joint reaction forces through the mid-carpometacarpal region (Marzke, 1983; Marzke and Marzke, 1987). The slightly more sagittal orientation of the capitate-metacarpal 2 facet, on both Tianyuan 1 and Nazlet Khater 2, suggests the more sagittal orientation of this articulation seen in the Neandertals (Niewoehner et al., 1997) but remains within recent human ranges of variation.

Hamate Bone

The left hamate bone (Table 9-13; Figure 9-17) echoes the pattern of smooth subchondral bone, slightly raised articular margins, and smooth nonarticular bone present on the other Tianyuan 1 carpal bones. The metacarpal 4 and 5 facets are divided by a raised vertical ridge, and the angle between them in the radioulnar plane is ~40°. The metacarpal 4 facet is dorsopalmarly and radioulnarly concave. The metacarpal 5 facet is distinctly concave dorsopalmarly and but straight to

Table 9-13. Osteometric dimensions of the Tianyuan 1 left hamate bone, in millimeters

Articular length (M-1)	20.2
Maximum breadth (M-2)	16.4
Maximum height (M-3)	23.0
Body height	13.8
Capitate articular length	21.7
Capitate articular height[1]	9.6
Triquetral articular length	19.3
Triquetral articular height[1]	11.4
Metacarpal articular height	10.9
Metacarpal articular breadth	16.4
Metacarpal 4 articular breadth	7.6
Metacarpal 5 articular breadth	9.6
Hamulus proximodistal length (M-14)	12.9
Hamulus radioulnar thickness (M-15)	5.1
Hamulus dorsopalmar projection (M-5)	9.2

[1] Maximum proximal dorsopalmar dimension of the articular facet.

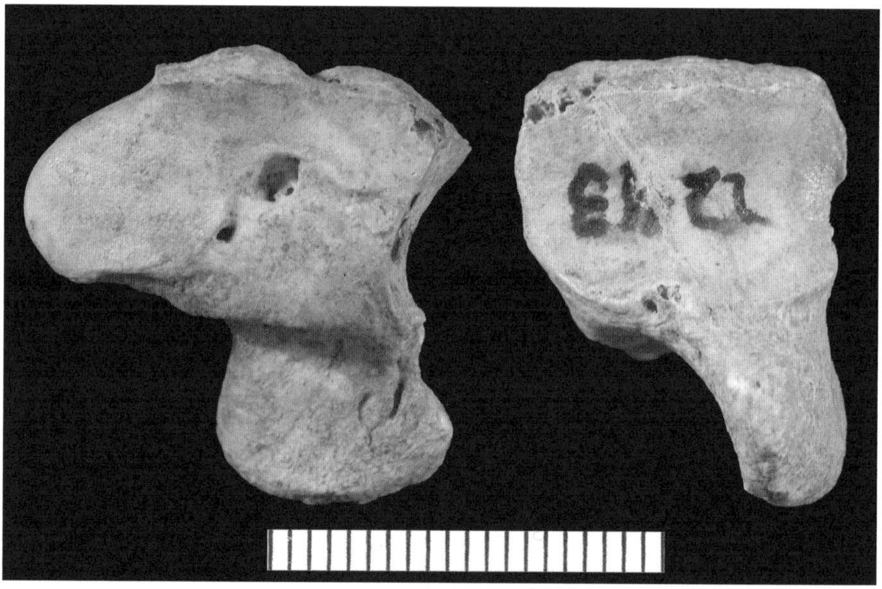

Figure 9-17. Radial (*left*) and distal (*right*) views of the Tianyuan 1 left hamate bone. Scale in millimeters.

trivially convex radioulnarly. It is therefore probable that the base of the Tianyuan 1 metacarpal 5 had the sellar-shaped facet of early and recent modern humans rather than the more condyloid facet of the Neandertals.

There is reduction in the size and projection of the hamulus between late archaic and early modern humans (Trinkaus, 1983a; Villemeur, 1994). The palmar projection of the Tianyuan 1 hamulus (from the palmar body to the palmar hamu-

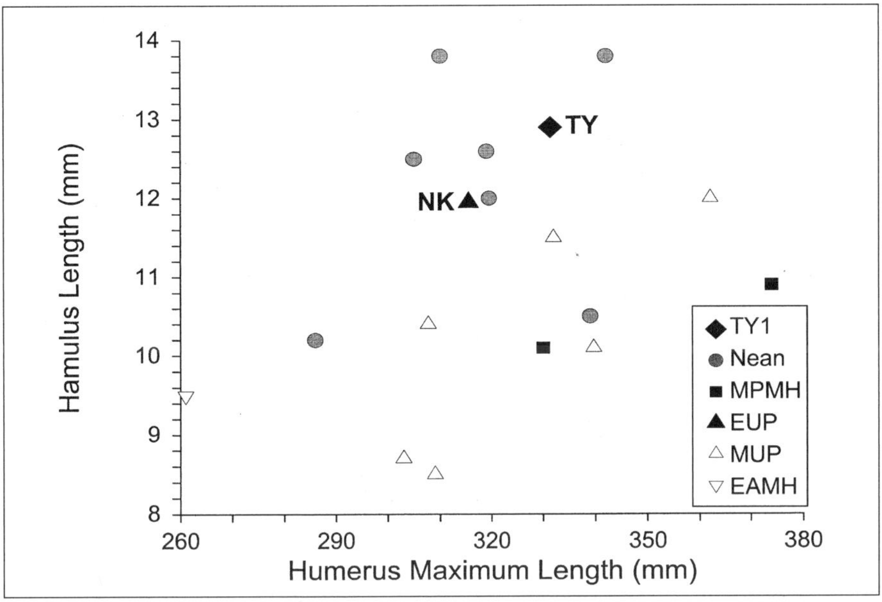

Figure 9-18. Bivariate plot of hamulus proximodistal length versus humerus maximum length. *TY:* Tianyuan 1; *Nean:* Neandertals; *MPMH:* Middle Paleolithic modern humans; *EUP:* western Early Upper Paleolithic; *MUP:* western Eurasian Mid Upper Paleolithic; *EAMH:* eastern Asian modern humans; *NK:* Nazlet Khater 2.

lus) is moderate for an early modern human, since its value of 9.3 mm is exceeded among Late Pleistocene early modern humans only by Qafzeh 8 and 9 (10.9 and 9.4 mm) and Sunghir 1 (10.9 mm). However, it is below those of all of the Neandertals (11.9 ± 1.4 mm, 10.1–13.9 mm, N = 8). It proximodistal length, however, is elevated. Its absolute measurement of 12.9 mm exceeds those of all other Late Pleistocene early modern humans (8.5–12 mm, N = 13) and falls among the higher Neandertal values (12.1 ± 1.3 mm, 10.2–13.8 mm, N = 8). If it is scaled to body size, using humeral maximum length (Figure 9-18), the Tianyuan 1 hamulus length falls with the Neandertals and distinct from the MPMH and the western Eurasian Upper Paleolithic modern humans. However, Nazlet Khater 2 is proportionately similar to Tianyuan 1 in this feature, as is the very small Minatogawa 2.

Proximal Manual Phalanges

The preserved proximal manual phalanges, from the left digits 2 and 3 and from the right digit 5, present few distinctive attributes (Table 9-14; Figure 9-19). The sides for them have been determined by the radial 1st and 2nd dorsal interosseus tubercles on the second and third phalanges, and by the abductor digiti minimi tubercle on the 5th one. The flexor sheath crests are variably marked. On the

Table 9-14. Osteometric dimensions of the Tianyuan 1 manual proximal phalanges, in millimeters except as noted

	2 left	*3 left*	*5 right*
Maximum length	41.7	44.2	34.5
Articular length	39.5	41.9	31.6
Midshaft height	5.8	6.5	5.8
Midshaft breadth	9.2	8.9	8.0
Proximal maximum height	12.1	12.0	10.4
Proximal maximum breadth	16.0	—	14.8
Proximal articular height	10.7	10.5	9.9
Proximal articular breadth	13.7	—	12.5
Distal articular height	8.3	8.0	6.7
Distal articular breadth	12.3	12.0	9.8
Head horizontal angle	3° radial	7° radial	8° ulnar

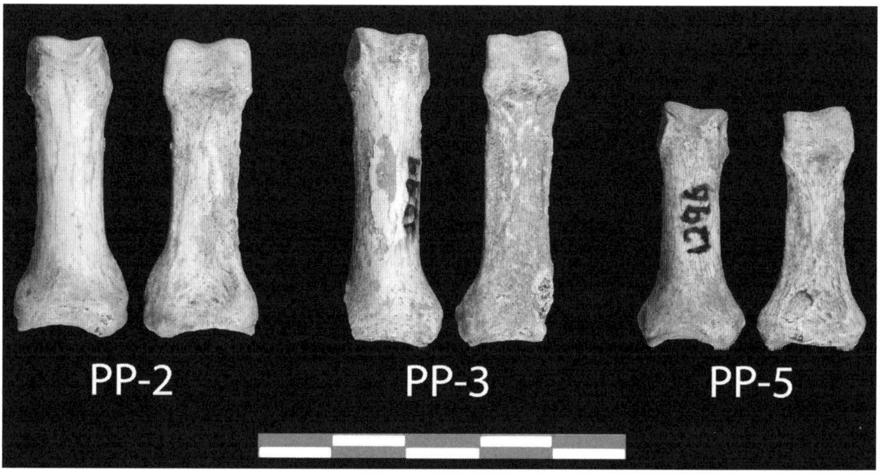

Figure 9-19. Dorsal and palmar views of the Tianyuan 1 manual proximal phalanges. The second (PP-2) and third (PP-3) are left; the fifth (PP-5) is right. Scale in centimeters.

2nd digit, there is a clear radial one but a very small ulnar one. On the 3rd phalanx, the 2 crests are moderately strong and symmetrical. On the 5th digit phalanx the ulnar crest is more marked than the radial one.

Neandertal proximal phalanges have been noted to have relatively wide proximal epiphyses (Musgrave, 1973). To assess the proportions of the Tianyuan 1 proximal phalangeal bases, the proximal maximum breadths are compared to the articular lengths for the second and fifth proximal phalanges (the third one is damaged on its proximoulnar corner) (Figure 9-20). In these distributions, the Neandertals are largely separate from the early modern humans (both Middle and Upper

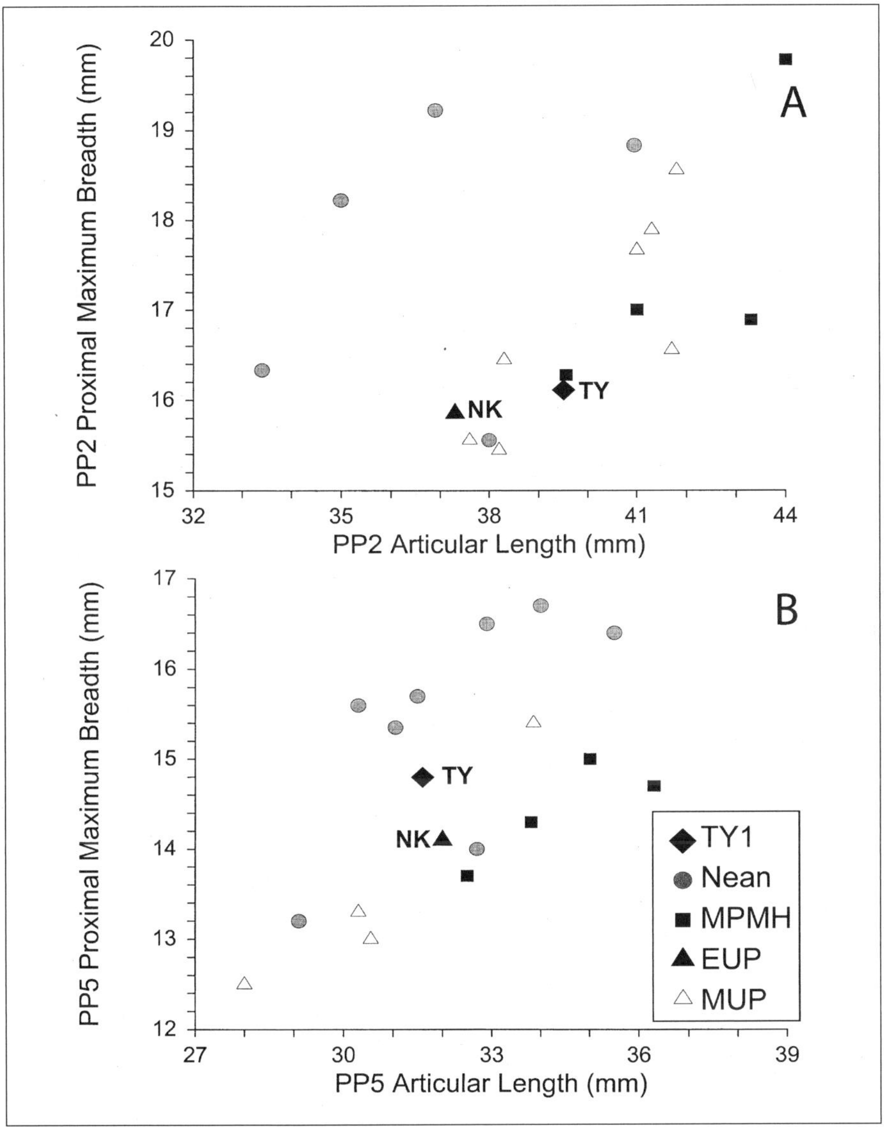

Figure 9-20. Bivariate plots of (*A*) second and (*B*) fifth proximal manual phalanx maximum base breadths versus their articular lengths for the Tianyuan 1 left proximal phalanges 2 and 5 (*TY*) and those of comparative samples. *TY:* Tianyuan 1; *Nean:* Neandertals; *MPMH:* Middle Paleolithic modern humans; *EUP:* western Early Upper Paleolithic; *MUP:* western Eurasian Mid Upper Paleolithic; *NK:* Nazlet Khater 2.

Paleolithic), even though there is one relatively gracile Neandertal in the second digit comparison (Shanidar 4) and two more gracile ones in the fifth digit comparison (La Ferrassie 2 and Shanidar 5). In the second digit assessment, Tianyuan 1, as well as the penecontemporaneous Nazlet Khater 2, are with the other early modern humans. In the right little finger comparison, Nazlet Khater 2 remains with most of the early modern humans but Tianyuan 1 is intermediate between the early modern humans and the majority of the Neandertals.

It is possible that the difference in base expansion between these two proximal phalanges reflects side asymmetry in the upper limbs. The second digit phalanx is left, and the fifth digit one is right. Given the elevated, right-side dominant humeral asymmetry, implying right-handedness, this difference between the phalanges might be a further reflection of this asymmetry. This applies especially since the comparative samples include right and left bones, albeit averaged for those individuals preserving both sides.

Middle and Distal Phalanges

There is one middle manual phalanx and one ulnar distal manual phalanx preserved for Tianyuan 1 (Table 9-15; Figure 9-21). They derive from the same digit, since their distal interphalangeal articulations have matching osteoarthritic degenerations (Chapter 12). Their distal interphalangeal articular surfaces are also a good match for each other, in terms of size and contours. As noted in Chapter 3, size comparisons with the distal articulations of the second and third proximal phalanges make it likely, but not conclusive, that they derive from the second digit. Their side remains indeterminate.

The middle phalanx is notable primarily for the modest development of the flexor digitorum superficialis tendon insertions on its palmar diaphysis. In addition, since most Neandertal middle phalanges have been shown to be relatively

Table 9-15. Osteometric dimensions of the Tianyuan 1 manual middle and distal phalanges, in millimeters

	Middle 2?	*Distal 2?*
Maximum length	24.9	17.2
Articular length	23.1	16.6
Midshaft height	5.5	4.0
Midshaft breadth	8.4	6.0
Proximal maximum height	9.7	pathological
Proximal maximum breadth	14.3	pathological
Proximal articular height	8.3	pathological
Proximal articular breadth	13.0	10.0
Distal articular breadth	pathological	9.0

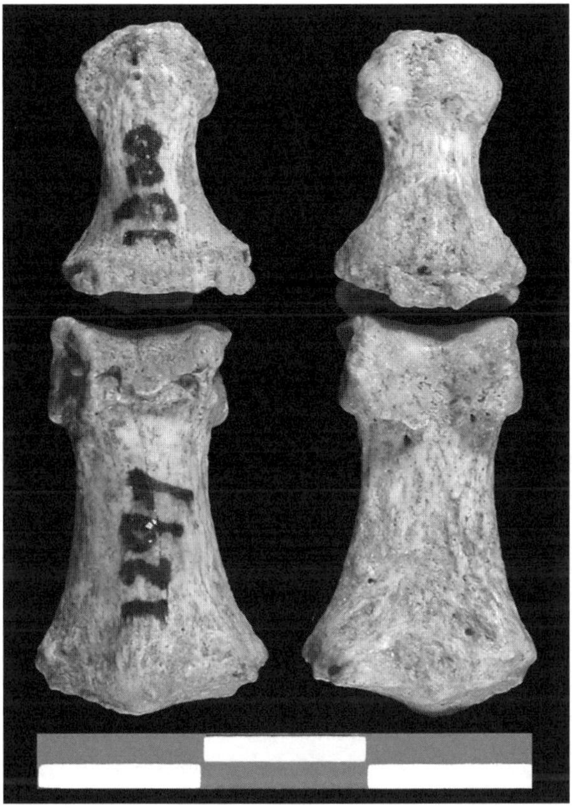

Figure 9-21. Dorsal (*left*) and palmar (*right*) views of the Tianyuan 1 second or third manual middle and distal phalanges. Scale in centimeters. The distal interphalangeal articulation has osteoarthritic degenerations.

broad compared to those of early modern humans (Musgrave, 1973; Trinkaus et al., 2007), the proximal epiphyseal maximum breadth of the Tianyuan 1 middle phalanx was compared to samples of second and third middle manual phalanges. Second and third middle phalanges were employed, given the ambiguity regarding the digital affinity of the Tianyuan 1 phalanx. If more than one of these phalanges was present for an individual, the right and left values for a digit (if available) were averaged, and then the 2nd and 3rd (average) values were averaged to provide a pooled value for that specimen. The resultant distribution (Figure 9-22) largely separates the Neandertals from the early modern humans, with two small Neandertals (La Ferrassie 2 and Palomas 92) slightly overlapping the early modern human range and the Initial Upper Paleolithic Saint-Césaire 1 value being relatively low for a Neandertal. The Nazlet Khater 2 mean phalangeal value is among the more robust of the early modern human phalanges, and three of the MPMH individuals (all but Qafzeh 8) fall in slightly above the MUP distribution. The Tianyuan 1 middle phalanx, however, is separate from all of the other early modern humans and is in the middle of the Neandertal distribution. Even more so

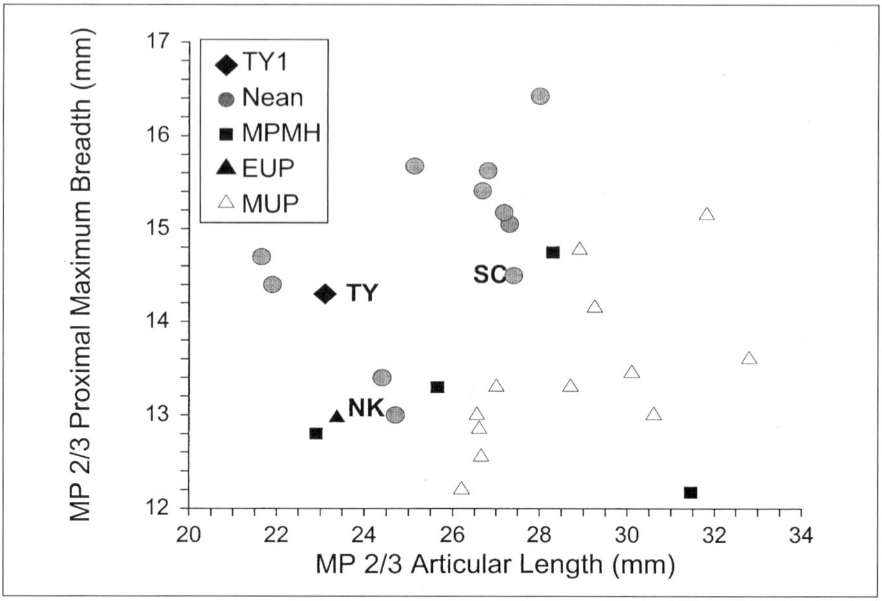

Figure 9-22. Bivariate plot of the second or third middle manual phalanx maximum base breadth versus its articular length for the Tianyuan 1 middle phalanx (*TY*) and those of comparative samples. *TY:* Tianyuan 1; *Nean:* Neandertals; *MPMH:* Middle Paleolithic modern humans; *EUP:* western Early Upper Paleolithic; *MUP:* western Eurasian Mid Upper Paleolithic; *NK:* Nazlet Khater 2; *SC:* Saint-Césaire 1 Initial Upper Paleolithic Neandertal.

than its fifth proximal phalanx, the Tianyuan 1 middle phalanx is archaic in this aspect.

The archaic nature of the middle phalanx breadth is echoed in the Tianyuan 1 distal phalanx. In particular, the apical tuft (or tuberosity) of the distal phalanx is large, is rounded, and lacks proximoradial and proximoulnar ungual spines. Although sometimes encountered in early and recent modern humans, this pattern of apical tuft morphology is characteristic of archaic *Homo,* including the Initial Upper Paleolithic Saint-Césaire 1 Neandertal (Musgrave, 1973; Trinkaus, 1983a; Rosas, 1985; Crevecoeur, 2002).

The expanded apical tuft of the Tianyuan 1 distal phalanx is evident in comparisons of its maximum breadth to phalangeal length (Figure 9-23A). In these comparisons, values for the middle three distal phalanges, when available, are averaged by individual, since it is difficult, even with a complete series of phalanges, to reliably assign the middle distal phalanges to digit. In the comparison of tuft breadth to phalangeal length, there is minimal overlap between the Neandertal distal breadth values and those of Middle or Upper Paleolithic modern humans, but the two distributions partially line up since the Neandertals on average have

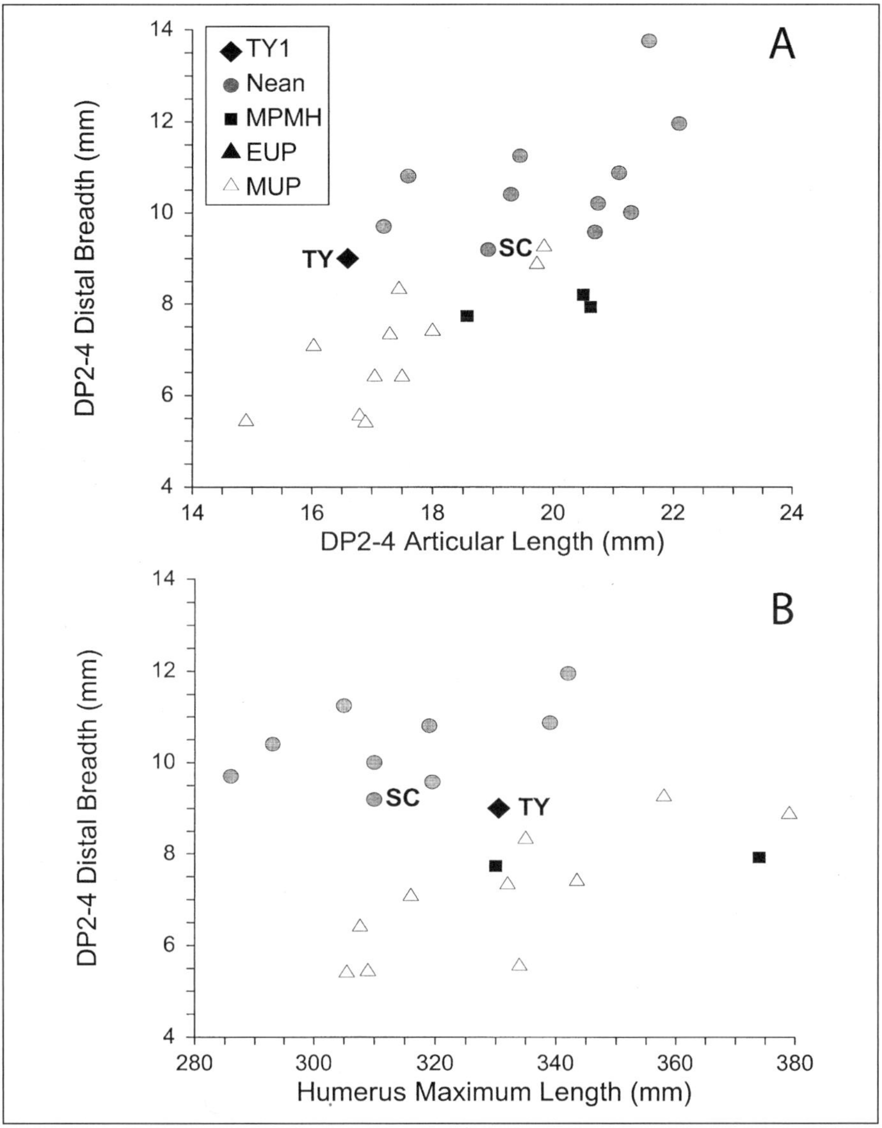

Figure 9-23. Bivariate plots of the second or third distal manual phalanx maximum apical tuft breadth (*A*) versus its articular length and (*B*) versus humeral maximum length for the Tianyuan 1 distal phalanx (*TY*) and those of comparative samples. *TY:* Tianyuan 1; *Nean:* Neandertals; *MPMH:* Middle Paleolithic modern humans; *EUP:* western Early Upper Paleolithic; *MUP:* western Eurasian Mid Upper Paleolithic; *SC:* Saint-Césaire 1 Initial Upper Paleolithic Neandertal.

longer distal phalanges. The Tianyuan 1 value is completely above the other early modern humans, especially separate from the MPMH values, and in line with the Neandertals. If the phalangeal distal breadths are instead compared to humeral length, thereby largely eliminating the effects of differential phalangeal length (Figure 9-23B), the Neandertals are completely separate from the early modern humans and Tianyuan 1 occupies a position between the two distributions. In this, note that the Upper Paleolithic Saint-Césaire 1 is among the other Neandertals, and the Middle Paleolithic Qafzeh phalanges are entirely with the MUP remains. Across primates, including recent humans, the breadth of the apical tuft reflects the size of the pulp and nail bed of the distal finger (Mitra et al., 2007).

It is therefore apparent that the distal phalanx of Tianyuan 1, along with the relative breadth of its middle phalanx and at least one of its proximal phalanges, retains the archaic *Homo* pattern and contrasts with the narrower fingers of Middle and earlier Upper Paleolithic modern humans. Unfortunately, distal phalanges are absent from Nazlet Khater 2 (Crevecoeur, 2008), and the few proximal phalanges associated with the Minatogawa remains are incomplete (Baba and Endo, 1982). There is an isolated distal hand phalanx from the eastern Asian earlier Upper Paleolithic of Pinza Abu, which has a narrow apical tuft (Sakura, 1981, 1985). However, it may derive from the fifth digit, in which case it would be undiagnostic as to archaic versus modern morphology (Trinkaus, 1983a). The same consideration applies to the isolated Middle Stone Age distal hand phalanx from the South African site of Sea Harvest (Grine and Klein, 1993).

Summary

The upper limb remains of Tianyuan 1 display a suite of characteristics that indicate their general similarity to other Late Pleistocene human arms and hands. In particular, the diaphyseal robusticity of the humeri, whether assessed through external diameters or cross-sectional parameters, falls in the middle of the overall Late Pleistocene distributions. This is found similarly for the humeral diaphyseal midshaft cross-sectional shape and for the relative development of the interosseus crest and adjacent diaphyseal shape of the radius. The musculoligamentous markings on the epiphyses and hand remains, however, tend to be modest in rugosity, and they are joined by the small muscle insertion areas on the proximal humeral diaphyses These indicators of arm robusticity are joined by the relative proportions of Tianyuan 1's left 2nd proximal manual phalanx base and the anteromedial orientation of the radial tuberosity.

At the same time, the humeral diaphyses and the scapular acromion processes show marked asymmetry, with the right side larger in both cases. For the one that can be compared across Late Pleistocene individuals, humeral diaphyseal asymmetry, the Tianyuan 1 level of asymmetry places it at the upper limits of the normal range of variation for the western Eurasian comparative samples. The one main contrast in this feature is with the Minatogawa sample, whose humeral diaphyses are exceptionally symmetrical. This arm asymmetry may be echoed as well in the modest robusticity of the left 2nd proximal manual phalanx and the more robust right 5th one. Yet, the radial diaphyses appear largely symmetrical. Other features are insufficiently bilaterally preserved to assess their levels of symmetry.

There are also a series of upper limb features that appear to be unrelated or at least indirectly related to robusticity, which have been shown to partially separate late archaic humans (especially Neandertals, given the marked dearth of late archaic human postcrania elsewhere) from early modern humans. Many of these features are also present among scattered remains of earlier (mostly Middle Pleistocene) archaic humans (see Trinkaus, 2006a). Of particular relevance here are the "modern" versus "archaic" features that separate late archaic humans from the early modern humans that preceded the Tianyuandong individual in time, namely the MPMH sample.

Among the apparently derived modern human features are the very wide scapular glenoid fossa (despite considerable overlap in this feature for left scapulae), the probable ventral axillary sulcus pattern of the left scapula, and the modest lateral curvature of the radial diaphysis. Among the more archaic features for the Tianyuan 1 upper limbs are the narrow medial pillar of the distal humerus, the large hamulus proximodistal length, the pronounced breadth of the middle manual phalanx, the rounded apical tuft of the distal manual phalanx, and the large breadth of that apical tuft. These archaic features of the hand bones are all the more relevant given the generally gracile muscular markings of the more proximal upper limb remains of Tianyuan 1. To these archaic features can be joined the more sagittal orientation of the metacarpal 2 facet on the capitate. It should be noted that Nazlet Khater 2, but not the MPMH humans, shares the long hamulus with Tianyuan 1.

Chapter 10

The Tianyuan 1 Lower Limb Remains

THE TIANYUAN 1 lower limbs retain most of a femur and of a tibia, partial elements of the other femur, the other tibia and a fibula, a talus and a calcaneus, four metatarsals, and two proximal pedal phalanges (see Figure 3-2). As with the upper limb remains, the long bones tend to be mostly diaphyseal, and the pedal remains are more complete. Nothing remains of the pelvis, nor of the patellae or the anterior tarsal bones. Nonetheless it is possible to estimate reliably the original lengths of a femur and a tibia and a femoral head diameter (Tables 10-1 and 10-4).

The distal halves of the femoral diaphyses are clearly abnormal, and the same or a related abnormality has affected the proximal posterior tibial diaphyses. Yet, the proximal thirds of the femora appear to be normal, as do the distal halves of the tibiae. These abnormalities are discussed in detail in Chapter 12, although some of the data relating to them are included in this chapter.

Femora

Proximal Femora

The portions of the proximal femora that relate to the hip region are limited to the subtrochanteric areas and the inferomedial necks on both femora, plus the greater trochanter and most of the neck on the right femur (Chapter 3; Figure 10-1). Since the proximal right femur is reassembled from separate pieces, little remains of the internal trabeculae. The observations are therefore related to overall size and orientation plus the pelvis-to-femur muscle insertions.

Neck-Shaft Angle

Even though the right femoral head is missing, it is possible from the original bone and the anteroposterior radiograph of the bone to estimate its femoral neck-

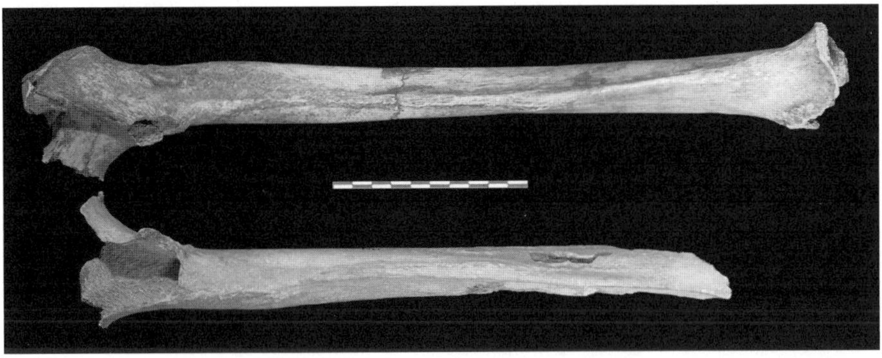

Figure 10-1. Posterior views of the Tianyuan 1 right (*above*) and left (*below*) femora. Proximal is on the left. Scale in centimeters.

shaft angle. The angle is definitely between 120° and 115°, and the estimated value of 118° (Table 10-1) is probably close to original value [bearing in mind that all femoral neck-shaft angle measurements have an intra- and interobserver error of ±1°–2° (Trinkaus, 1993b)]. Neck-shaft angles are developmentally plastic, decreasing with increasing loads on the hip region especially during preadolescence, with the angle becoming essentially fixed by the end of the 2nd decade of life (Houston and Zaleski, 1967; see Anderson and Trinkaus, 1998, and references therein).

Recent and Late Pleistocene human foraging samples have relatively low neck shaft angles, with most individual angles below 130° and sample means close to or below 125° (Trinkaus, 1993b; Anderson and Trinkaus, 1998; Table 10-2; Figure 10-2). In addition, the Mladeč 28 proximal femur had an angle ~124°, but it is too incomplete to provide a reliable value (Trinkaus, Smith, et al., 2006). The Late Pleistocene exceptions to this pattern are the Middle Paleolithic modern human (MPMH) sample (Trinkaus, 1993b) and Nazlet Khater 2 (Crevecoeur, 2008), and to a lesser extent the eastern Asian early modern human sample [the poorly dated KNM-ER 999 femur also has a high neck-shaft angle (Trinkaus, 1993c)]. Nazlet Khater 2 has abnormally short femora (Crevecoeur, 2008), and its neck-shaft angle probably reflects a developmental abnormality. The distribution for the eastern Asian sample reflects the high values for the Minatogawa sample (Baba and Endo, 1982) combined with lower values for the ZKD-UC 105 (126°) and 117 (127°) femora measured from Weidenreich's (1941) plates; Weidenreich (1941) provided a range of 123° to 137° (N = 4) for the ZKD-UC femora, with a mean of 129°. The high values for the Minatogawa and MPMH samples, as well as at least one of the ZKD-UC femora, are more difficult to explain; they match recent urban values and a minority of recent nonurban angles, and they therefore suggest little mobility and/or burden carrying for the juveniles of those populations (see Trinkaus, 1993b for the MPMH sample).

Table 10-1. Osteometric dimensions of the Tianyuan 1 femora, in millimeters except as noted

	Right	*Left*
Maximum length (M-1)	463.1[1]	—
Bicondylar length (M-2)	460.4[2]	—
Biomechanical length[3]	435.6[3]	—
Midshaft anteroposterior diameter (M-6)	36.8[4]	38.0[4]
Midshaft mediolateral diameter (M-7)	24.4[4]	24.9[4]
Midshaft circumference (M-8)	98.0[4]	100.0[4]
Subtrochanteric anteroposterior diameter (M-10)[5]	26.2	(26.5)
Subtrochanteric mediolateral diameter (M-9)[5]	33.6	31.5
Anterior curvature chord (M-27)	347.0	—
Anterior curvature maximum subtense	20.0	—
Femoral head anteroposterior diameter (M-19)[6]	(53.8)	—
Femoral head superoinferior diameter (M-18)[6]	(53.6)	—
Gluteal tuberosity breadth[7]	12.7	11.7
Hypotrochanteric fossa	absent	absent
3rd trochanter	present	—
Neck-shaft angle[8]	(118°)	—
Lateral condylar anteroposterior diameter (M-22)[9]	(75.0)	—

Note: Includes discrete trait presence/absence. Both of the middle to distal diaphyses are pathological, and it is unclear to what extent the measurements from them have been altered by the abnormalities. Measurements in parentheses are estimated.

1 Maximum length estimated from the direct distance between the proximal neck and the average of the two distal pseudo landmarks, the proximal anterolateral patellar surface margin and the posterolateral supracondylar margin (398.5 mm) using a linear regression based on a sample of recent human femora [MaxLen = ((TY1Len × 1.149) + 5.2); N = 35; r^2 = 0.980; estimated length = 463.1 ± 4.6 mm; SE_{est} = 1.00%].

2 Bicondylar length estimated from the estimated maximum femoral length using a least squares regression based on recent human femora [BicondLen = ((MaxLen × 1.012) – 8.5); N = 40; r^2 = 0.996; estimated length = 460.4 ± 6.4; SE_{est} = 1.39%; the SE_{est} combines the SE_{est} for maximum length (1.00%) with the SE_{est} for bicondylar length from maximum length (0.39%)].

3 Biomechanical length is from the proximal neck on the diaphyseal axis to the average of the distal condylar surfaces (Ruff and Hayes, 1983a). It is estimated for the Tianyuan 1 right femur using the distance from the preserved proximal neck to the average of the distal margins adjacent to the anterolateral patellar surface and posterolateral supracondylar surface (398.5 mm) using a linear regression based on a sample of recent human femora [BiomechLen = ((TY1Len × 1.086) + 2.7); N = 35; r^2 = 0.988; estimated length = 435.6 ± 3.3 mm; SE_{est} = 0.76%].

4 The midshaft cross-sectional geometry has been affected by the pathological abnormalities of the femoral diaphyses, minimally reducing the mediolateral dimensions and possibly exaggerating the anteroposterior ones. Although the measurements provided here are accurate measures of the bones, they do not represent the original "normal" proportions of the Tianyuan 1 femoral midshafts.

5 The subtrochanteric diameters are taken, as is appropriate, as a maximum mediolateral diameter across the gluteal buttress and then an anteroposterior diameter perpendicular to the mediolateral one.

6 The anteroposterior and superoinferior diameters of the femoral head were estimated from the lateral condylar anteroposterior diameter (posterior lateral condyle to anterior lateral patellar surface margin; 75 mm) using least squares regressions based on a pooled recent human (N = 41) and Late Pleistocene fossil human (N = 6) sample. Anteroposterior diameter = 0.641 × LatCondAP + 5.8, r^2 = 0.881; estimated diameter: 53.8 ± 1.7 mm; SE_{est} = 3.16%. Superoinferior diameter = 0.620 × LatCondDia + 7.1, r^2 = 0.850; estimated diameter: 53.6 ± 1.8; SE_{est} = 3.36%.

7 The maximum breadth of the gluteal tuberosity proper.

8 Given the absence of the femoral head, a fully verifiable neck-shaft angle cannot be taken. However, from the preserved portion of the proximal femur, as determined on both the bone and an anteroposterior radiograph of the proximal femur, it is apparent that it was ≤120° and ≥115°, and the value of 118° is the best estimate available. As with all neck-shaft angles, it has a reliability of ±1°-2° (Trinkaus, 1993b).

9 The anterior margin of the lateral patellar surface is slightly abraded, and therefore the lateral condylar anteroposterior diameter has been visually estimated as ~75 mm, adding 0.4 mm onto the preserved length of the lateral condylar piece (see Chapter 3). This estimate should be a minimum value, and given the relatively and absolutely large dimension of it, and of the femoral head diameters estimated from it, it should be conservative in any comparisons.

Table 10-2. Comparisons of Tianyuan 1 to Late Pleistocene samples for femoral osteometrics and third trochanter frequencies

	Neck-shaft angle (°)	*Gluteal tuberosity breadth (mm)*	*Third trochanter frequency (% (N))*[1]	*Pilastric index*	*Anterior curvature index*
Tianyuan 1	(118°)	12.2	present	151.7	5.8
East Asia EMH	129.7° ± 4.7° (6)	9.5	8.3 (6)	112.0 ± 21.1 (6)[2]	3.3 ± 0.3 (4)
West MUP	121.8° ± 6.5° (22)	9.7 ± 2.0 (13)	60.7 (14)	117.5 ± 10.8 (29)	4.3 ± 1.0 (11)
West EUP	~124°, 131.5°	8.0, 12.5	0.0 (1)	117.9, 118.8	4.0, 6.0
MPMH	133.2° ± 2.6° (6)	9.0 ± 2.0 (4)	0.0 (4)	124.2 ± 11.9 (8)	2.9, 5.8, 6.0
Neandertals	120.6° ± 4.9° (8)	12.6 ± 2.1 (13)	0.0 (9)	103.2 ± 8.7 (15)	4.7 ± 1.0 (6)
K-W P-value	0.001	0.014	—	0.002	0.233

Note: EMH: early modern humans; *MUP:* Mid Upper Paleolithic; *EUP:* Early Upper Paleolithic; *MPMH:* Middle Paleolithic modern humans. For the osteometrics, mean ± standard deviation (N) are provided. Kruskal-Wallis (K-W) P-values are across the comparative samples.

[1] The frequencies for the eastern Asian EMH and western MUP samples third trochanter presence reflect asymmetry on the Ohalo 2 and Minatogawa 3 femora. When a single femur is present for an individual, symmetry is assumed.

[2] The mean and standard deviation for the eastern Asian sample includes the Minatogawa femora and the two ZKD-UC ones for which casts exist. If Weidenreich's (1941) mean for the six ZKD-UC femora is averaged with the four Minatogawa ones, the mean becomes 114.8. Note that Weidenreich's (1941, Plate 24) drawings of the ZKD-UC femoral cross sections are distorted relative to his measurements.

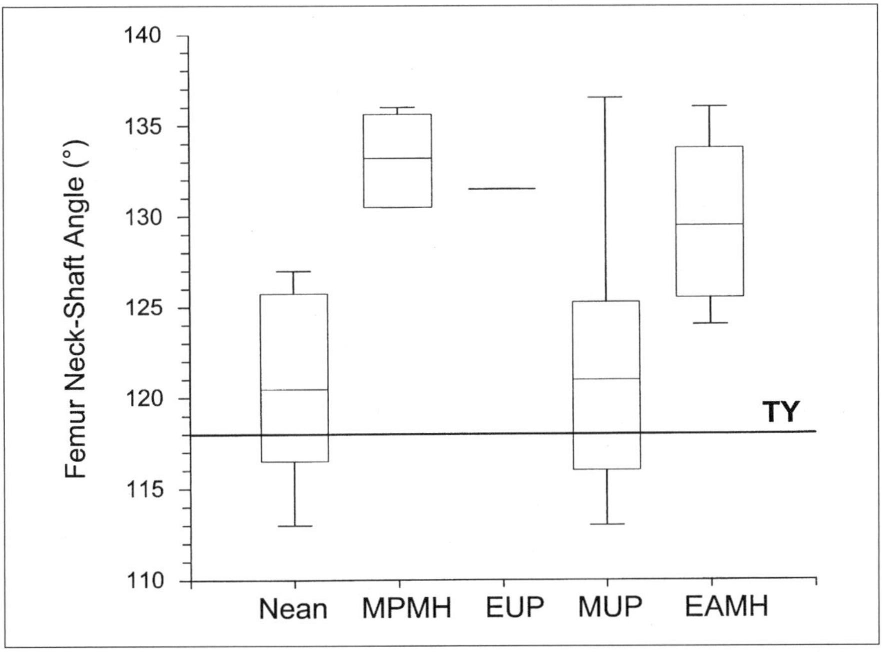

Figure 10-2. Box plot of femoral neck-shaft angles. *TY:* Tianyuan 1; *Nean:* Neandertals; *MPMH:* Middle Paleolithic modern humans; *EUP:* western Early Upper Paleolithic; *MUP:* western Eurasian Mid Upper Paleolithic; *EAMH:* east Asian modern humans. Tianyuan 1 value is ~118°.

The Tianyuan 1 angle is below the medians of all of the Late Pleistocene samples but well within the ranges of variation of the late archaic and Mid Upper Paleolithic (MUP) samples (Figure 10-2). It is also low but comfortably within the ranges of variation for recent human nonmechanized societies. The Tianyuan 1 angle is in accord with its general level of lower limb robusticity and therefore suggests that its level of mature locomotor loading was a continuation of a developmental pattern.

Trochanters, Gluteal Tuberosity, and Spiral Line

The partially preserved greater and lesser trochanters present few aspects of note (Figure 10-3). The left lesser trochanter preserves a strong medially deflected rim, which appears to have been originally present on the less complete right one. There is a deep right digital fossa with 3 small enthesopathies, but the right greater trochanter is otherwise relatively smooth.

The right femur has a large, posteriorly deflected, third trochanter that is largely smooth on its posterior surface but rugose and pitted on its lateral face. The left femur appears to lack a third trochanter, but it would have occurred on or adjacent to the proximoposterior break, and therefore its absence is not certain. Third trochanters are variably present in recent human samples (Hrdlička, 1937; Kate, 1966), and they probably reflect an ossification of the more proximal fibers of the gluteus maximus insertion into the proximal femur. Third trochanters are not currently known from western Eurasian Late Pleistocene archaic humans, MPMH or western Early Upper Paleolithic (EUP) femora (Table 10-2). However, one is present on one of the Minatogawa 3 femora, and they are relatively common among the western MUP humans. One is also present on one of the western Eurasian late Middle Pleistocene femora (Krapina 213).

The gluteal tuberosity is moderately rugose on both femora, across the tuberosity on the left one and particularly along a medial ridge below the third trochanter on the right one. The maximum breadth of the gluteal tuberosity is slightly larger on the right side, producing an average breadth of 12.2 mm. This value is close to the Neandertal mean and relatively large for early modern human femora, being matched or exceeded only by the EUP Mladeč 28 and MUP Cro-Magnon 4323A femora. However, when scaled to body mass times femur length (representing the baseline load and load arm for gluteus maximus), gluteal tuberosity breadths show a shift from late archaic to early modern humans with considerable overlap between the samples (Trinkaus, 2000; Figure 10-4). The three high outliers are the Cro-Magnon 4324 MUP human, and the La Ferrassie 2, and Shanidar 6 Neandertals. The relative Tianyuan 1 value, as well as the EUP Mladeč 27 one, falls securely with the other early modern humans. There are no hypotrochanteric fossae on the two gluteal tuberosities.

The spiral lines are both strongly marked. The left one in particular is rugose and raised from the proximal linea aspera almost to the anterior greater trochanter.

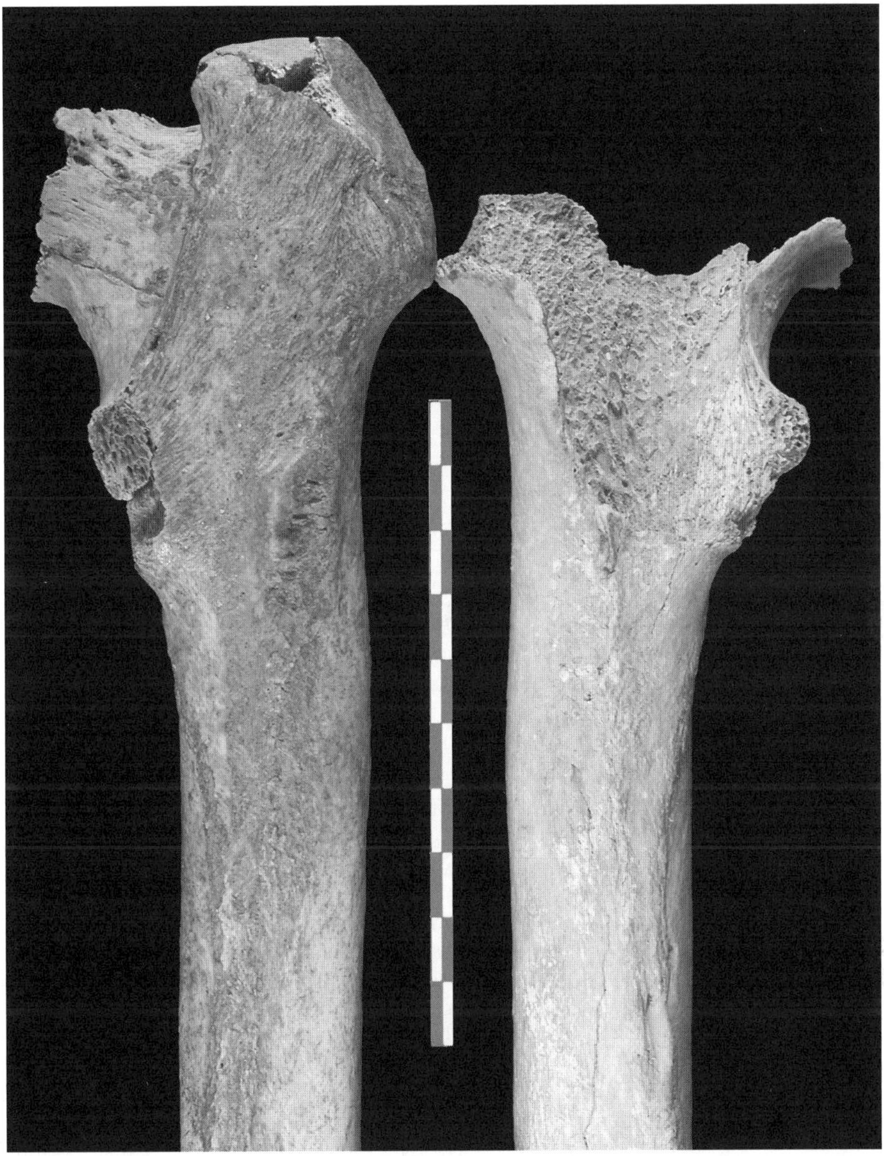

Figure 10-3. Posterolateral view of the Tianyuan 1 proximal femora, showing the gluteal tuberosities and the third trochanter on the right femur. Scale in centimeters.

Subtrochanteric Diaphyseal Proportions

There is variation among Late Pleistocene femora in the degree to which there is a gluteal buttress on the proximal diaphysis, a longitudinal swelling at the level of the gluteal tuberosity that is lateral or modestly posterolateral. It can vary from a gentle swelling to a ridge that is separated from the adjacent anterior and poste-

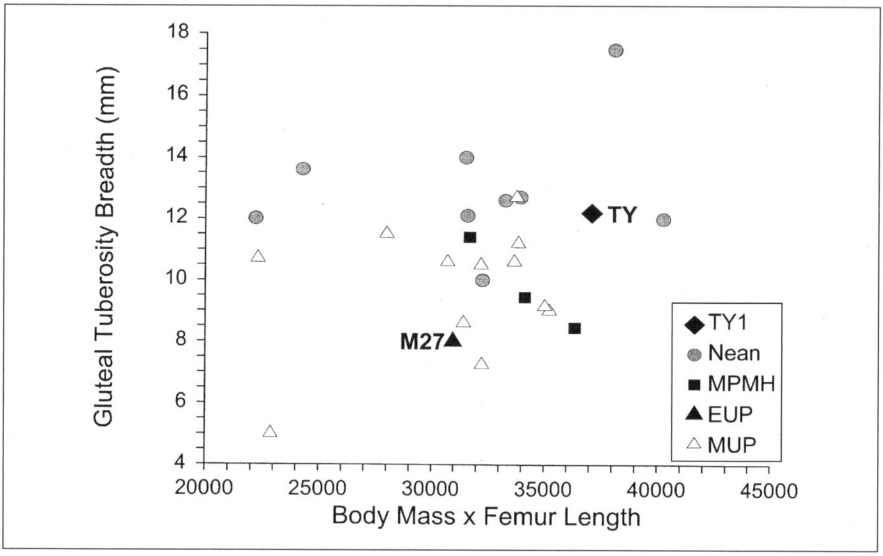

Figure 10-4. Bivariate plot of gluteal tuberosity breadth versus femur length times estimated body mass for Tianyuan 1 and comparative Late Pleistocene samples. *TY:* Tianyuan 1; *Nean:* Neandertals; *MPMH:* Middle Paleolithic modern humans; *EUP:* western Early Upper Paleolithic; *MUP:* western Eurasian Mid Upper Paleolithic; *M27:* Mladeč 27.

rior diaphyseal surfaces by longitudinal sulci. The Tianyuan 1 femora have modest gluteal buttresses (Figure 10-5). Anterolaterally the diaphyses are flattened as the anterior contours grade onto the gluteal buttresses. Posterolaterally there are slight longitudinal sulci, roughened and possibly exaggerated by the insertions of the gluteus maximus fibers.

The relative development of the gluteal buttress is partially quantified by the distribution of the approximately mediolateral diameter taken at the maximum extent of the buttress as a maximum diameter (usually along an anteromedial to posterolateral line through the buttress) relative to the approximately anteroposterior diameter taken perpendicular to this diameter. Usually quantified as the meric index [inappropriately called the "index of platymeria" (Weidenreich, 1941; Olivier, 1960; Bräuer, 1988)], this proportion is better visualized as a bivariate plot (Figure 10-6). As previously noted (Twiesselmann, 1961; Trinkaus, 2006d), Middle Paleolithic late archaic and early modern humans tend to have relatively round subtrochanteric femoral cross sections, whereas MUP modern humans have more mediolaterally elongated ones, reflecting their frequently large gluteal buttresses. The three EUP specimens span the entire range, with Mladeč 28 having a large gluteal buttress, Mladeč 27 having a relatively round cross section, and Nazlet Khater 2 (as with Skhul 5) having a greater anteroposterior diameter. In this, Tianyuan 1

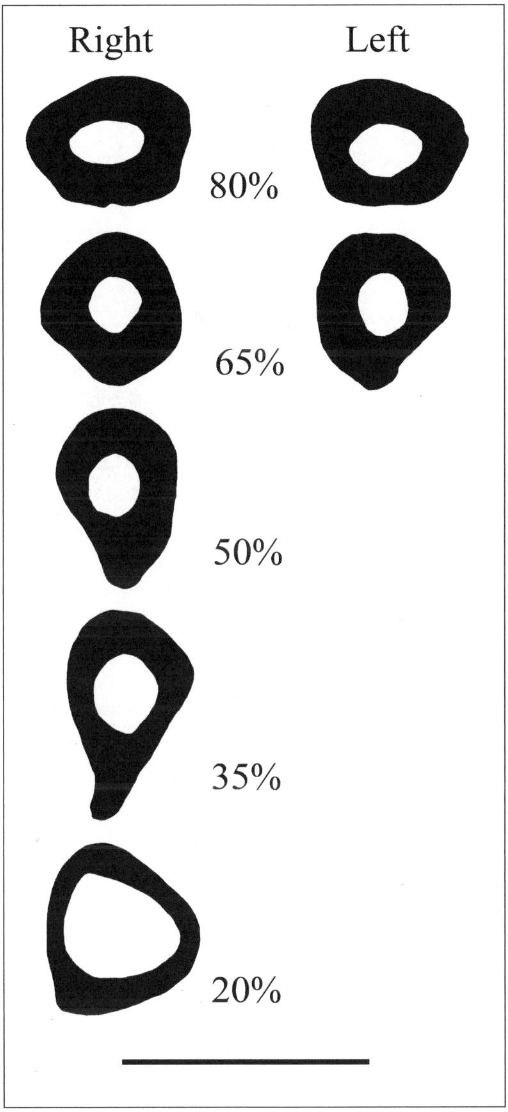

Figure 10-5. Diaphyseal cross sections of the right and left femora. Sections are in distal view, anterior is above, medial is to the inside, and 0% is distal. Scale bar equals 50 mm.

falls in the middle of the Neandertal distribution, close to the two other MPMH femora and to Mladeč 27, and at the top of the western Eurasian MUP and eastern EMH ranges of variation. Measurements from the two preserved ZKD-UC femoral casts place them very close to the Tianyuan 1 value [Weidenreich (1941) provided only a range (81.3—104.8) and a mean (86.1) for the meric index for 6 ZKD-UC femora, including the two providing cast data; these indices make them rather round in this cross section, since the index of the average diameters

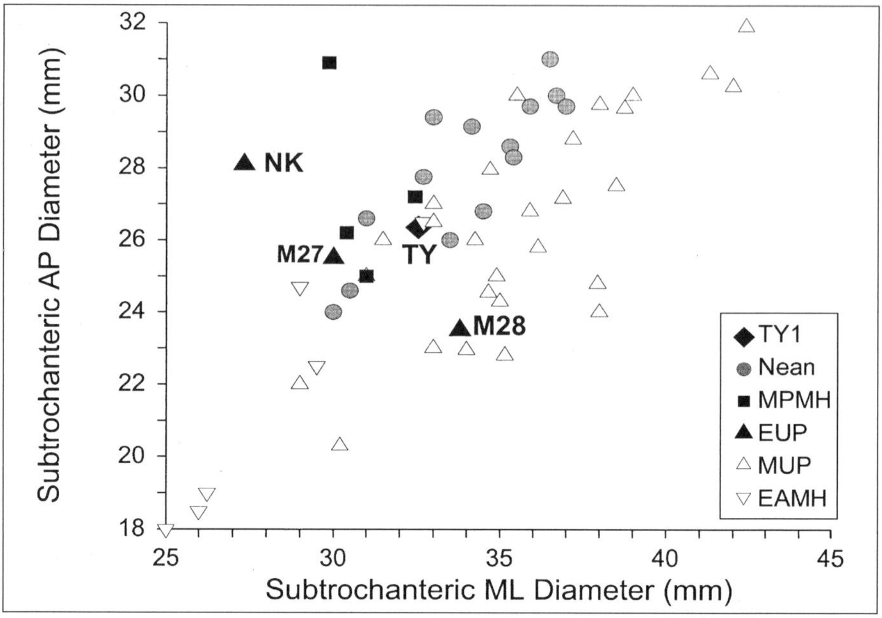

Figure 10-6. Bivariate plot of femoral subtrochanteric anteroposterior versus mediolateral diameters. *TY:* Tianyuan 1; *Nean:* Neandertals; *MPMH:* Middle Paleolithic modern humans; *EUP:* western Early Upper Paleolithic; *MUP:* western Eurasian Mid Upper Paleolithic; *EAMH:* eastern Asian modern humans; *NK:* Nazlet Khater 2; *M27:* Mladeč 27; *M28:* Mladeč 28.

for Tianyuan 1 (Table 10-1) is 81.0]. The Tianyuan 1 proximal femora are therefore relatively round compared to western Upper Paleolithic humans and the Minatogawa sample, but similar to contemporary and earlier Late Pleistocene femora (as well as to 2 ZKD-UC femora) in this feature.

Femoral Mid-Diaphysis

Midshaft Diaphyseal Shape

As noted above and described in Chapter 12, the mid-diaphyses and especially mid-distal diaphyses of Tianyuan 1 are abnormal in their cross-sectional shapes. The approximate midshaft, however, may be close enough to its original shape to provide some indication of its original nonpathological morphology and of its overall robusticity (Figures 10-5 and 10-7).

From the subtrochanteric region to approximately midshaft, there is the normal convergence of the spiral line and distal gluteal tuberosity on each midproximal diaphysis, with a prominent linea aspera descending from each of those muscle line convergences. The linea aspera is ~6.0 mm wide at midshaft on the right and

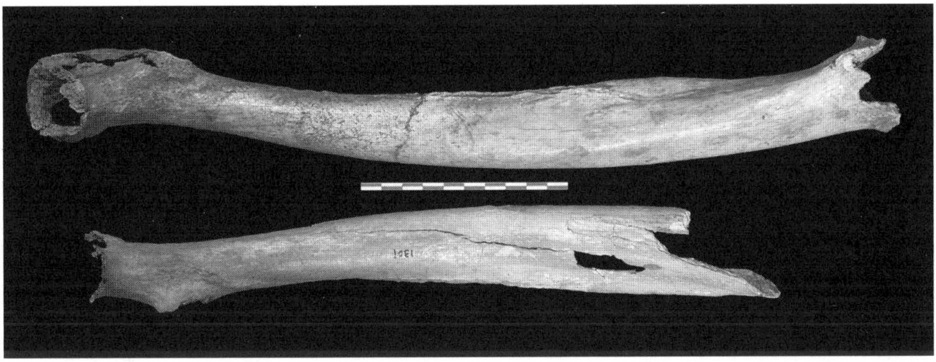

Figure 10-7. Medial view of the Tianyuan 1 femora. Proximal is on the left. Scale in centimeters.

~7.5 mm wide on the left femur. Each linea aspera is on a clear pilaster, demarcated laterally by a distinct longitudinal sulcus forming between the linea aspera and the convexly rounded lateral diaphysis. Medially the bone is bulbous along the linea aspera, especially on the right femur, but then on each one the medial surface is anteroposteriorly flat rather than the normal convexity of the medial diaphysis. The anterior surfaces are normal in being transversely convex, with the maximum anterior projection of the diaphyseal bone slightly medial on the diaphyses. Therefore, these femora exhibit the distinct pilastric morphology of early (and later) modern humans and contrast with the subcircular cross sections of late archaic human femora. However, the medial surfaces of the midshafts are unusually flat, probably related to the abnormalities of the mid-distal diaphyses.

Despite these irregularities of the femoral midshaft, it is possible to compare the Tianyuan 1 femoral cross-sectional shapes to other Late Pleistocene femora. The pilastric index of Tianyuan 1 (151.7) exceeds all of the other Late Pleistocene values. The closest ones are 139.0 (ZKD-UC AN67), 138.0 (ZKD-UC AN68), 137.3 (Skhul 5), and 136.9 (Ohalo 2). The Tianyuan 1 value is 5.6 standard deviations from the Neandertal sample, 3.2 standard deviations from the MUP sample, 2.3 standard deviations from the MPMH sample, but only 1.9 standard deviations from the variable eastern Asian sample (Table 10-2). A better reflection of its external proportions is a bivariate plot of these two midshaft diameters (Figure 10-8A), in which Tianyuan 1 remains separate from all of the other Late Pleistocene specimens, even though the two ZKD-UC femora for which there are casts approach it. The Mladeč 27 and Nazlet Khater 2 femora remain in the middle of the Late Pleistocene distribution, as does the late Neandertal from Saint-Césaire despite the lower position of most nonpilastric Neandertal femora.

Using midshaft cross-sectional geometry reduces most of the scatter among the Late Pleistocene specimens (Table 10-3; Figure 10-8B), even though it maintains most of the Neandertals at or below the early modern human distribution.

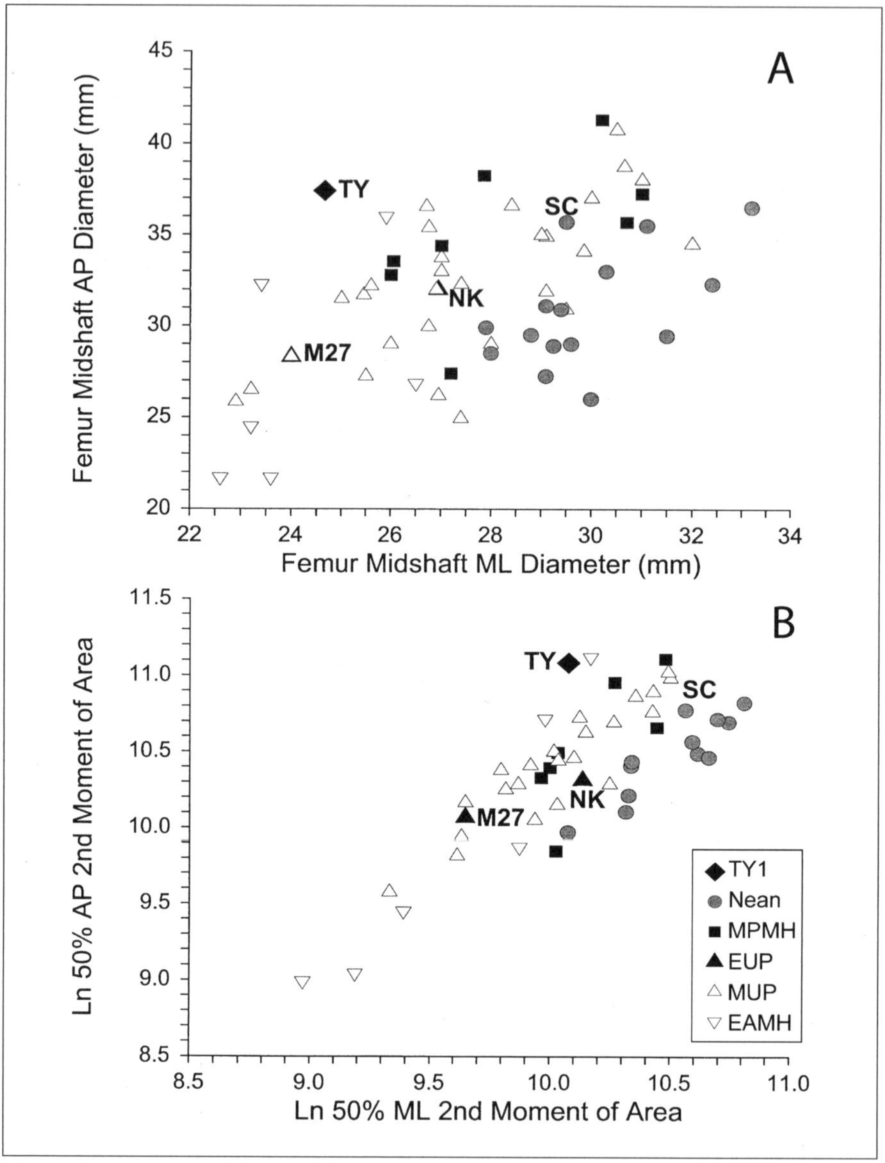

Figure 10-8. Bivariate plots of (*A*) femoral midshaft anteroposterior versus mediolateral diameters and (*B*) anteroposterior versus mediolateral second moments of area. *TY:* Tianyuan 1; *Nean:* Neandertals; *MPMH:* Middle Paleolithic modern humans; *EUP:* western Early Upper Paleolithic; *MUP:* western Eurasian Mid Upper Paleolithic; *EAMH:* eastern Asian modern humans; *NK:* Nazlet Khater 2; *M27:* Mladeč 27; *SC:* Saint-Césaire 1 Initial Upper Paleolithic Neandertal.

Table 10-3. Cross-sectional geometric parameters of the Tianyuan 1 femora

	Right	*Left*
20%		
Total area (mm^2)	906.4	—
Cortical area (mm^2)	495.1	—
Medullary area (mm^2)	411.3	—
Anteroposterior second moment of area (I_x) (mm^4)	(67,464)	—
Mediolateral second moment of area (I_x) (mm^4)	(41,012)	—
Maximum second moment of area (I_{max}) (mm^4)	67,464	—
Minimum second moment of area (I_{min}) (mm^4)	41,012	—
Polar moment of area (mm^4)	108,476	—
Orientation of I_{max} (theta)	(90°)	—
35% (pathological)		
Total area (mm^2)	758.4	—
Cortical area (mm^2)	591.5	—
Medullary area (mm^2)	166.9	—
Anteroposterior second moment of area (I_x) (mm^4)	(90,544)	—
Mediolateral second moment of area (I_x) (mm^4)	(24,982)	—
Maximum second moment of area (I_{max}) (mm^4)	91,093	—
Minimum second moment of area (I_{min}) (mm^4)	24,433	—
Polar moment of area (mm^4)	115,526	—
Orientation of I_{max} (theta)	(85°)	—
50% (pathological)		
Total area (mm^2)	700.7	—
Cortical area (mm^2)	592.4	—
Medullary area (mm^2)	108.3	—
Anteroposterior second moment of area (I_x) (mm^4)	(65,195)	—
Mediolateral second moment of area (I_x) (mm^4)	(23,759)	—
Maximum second moment of area (I_{max}) (mm^4)	66,018	—
Minimum second moment of area (I_{min}) (mm^4)	22,935	—
Polar moment of area (mm^4)	88,953	—
Orientation of Imax (theta)	(98°)	—
65%		
Total area (mm^2)	717.1	676.3
Cortical area (mm^2)	618.4	559.2
Medullary area (mm^2)	98.7	117.1
Anteroposterior second moment of area (I_x) (mm^4)	(48,760)	(46,314)
Mediolateral second moment of area (I_x) (mm^4)	(33,651)	(28,334)
Maximum second moment of area (I_{max}) (mm^4)	49,232	47,001
Minimum second moment of area (I_{min}) (mm^4)	33,178	27,648
Polar moment of area (mm^4)	82,410	74,649
Orientation of I_{max} (theta)	(100°)	(101°)
80%		
Total area (mm^2)	751.5	716.2
Cortical area (mm^2)	636.8	589.0
Medullary area (mm^2)	114.7	127.2
Anteroposterior second moment of area (I_x) (mm^4)	(39,607)	(36,648)
Mediolateral second moment of area (I_x) (mm^4)	(49,206)	(43,131)
Maximum second moment of area (I_{max}) (mm^4)	49,674	44,526
Minimum second moment of area (I_{min}) (mm^4)	39,139	35,253
Polar moment of area (mm^4)	88,813	79,779
Orientation of I_{max} (theta)	(12°)	(23°)

Note: At least the 35% and 50% cross sections were pathologically altered. Although sufficient external cortical bone exists on the left femur for midshaft osteometrics (Table 10-1), the bone is insufficiently preserved for cross-sectional geometry. Orientations are based on a sagittal plane through the linea aspera and the mediolateral middle of the diaphysis at midshaft; the values for I_x, I_y, and theta depend upon this approximation, and they are therefore placed in parentheses.

The ZKD-UC femora providing data from their casts are markedly anteroposteriorly hypertrophied, with the larger one closely approaching the otherwise very high position of Tianyuan 1. From these comparisons, it appears that the original proportions of the Tianyuan 1 femoral midshafts may not have been truly exceptional, given that they are approached by the ZKD-UC femora and a couple of other Late Pleistocene femora. However, the anteroposterior versus mediolateral proportions of the Tianyuan 1 femora were undoubtedly affected to some extent by its abnormalities, given the exceptional proportions of the mid-distal diaphyses (Chapter 12).

Femoral Diaphyseal Robusticity

It is nonetheless possible to use the Tianyuan 1 femoral midshafts to assess their robusticity. A plot of the polar 2nd moment of area versus femur length times estimated body mass (Figure 10-9) places Tianyuan 1 in the middle of the overall distribution. Minor adjustments in its position on the abscissa given estimation of its femoral length and body mass (Chapter 7) would have little effect on this assessment, given its position very close to the middle of the Late Pleistocene distribu-

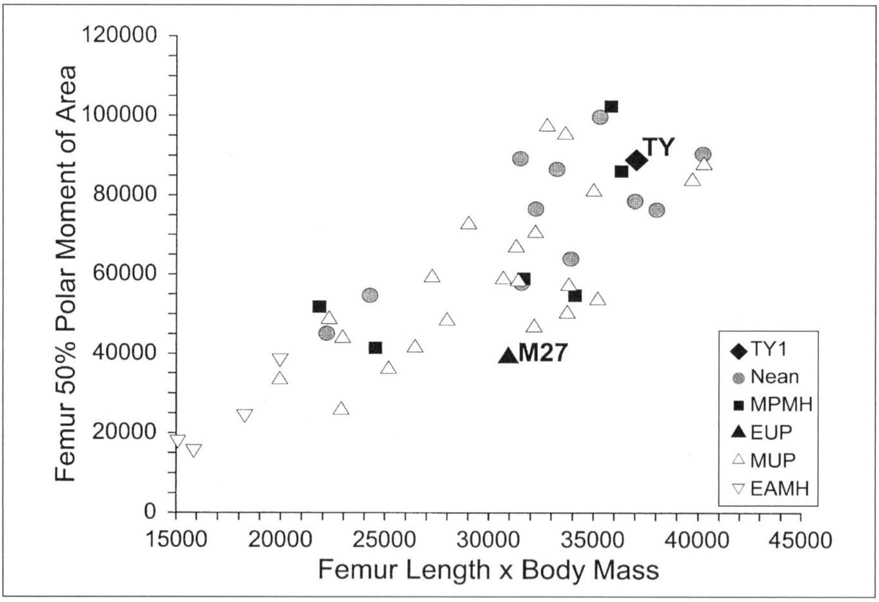

Figure 10-9. Bivariate plot of femoral midshaft polar second moments of area versus estimated body mass times femoral length. *TY:* Tianyuan 1; *Nean:* Neandertals; *MPMH:* Middle Paleolithic modern humans; *EUP:* western Early Upper Paleolithic; *MUP:* western Eurasian Mid Upper Paleolithic; *EAMH:* eastern Asian modern humans; *M27:* Mladeč 27. Given its abnormal femoral lengths, Nazlet Khater 2 is not included.

tion. Mladeč 27, despite being attributed linear body proportions (Trinkaus, Smith, et al., 2006), remains gracile. These data, despite the estimation of body mass and the abnormalities of the Tianyuan 1 femoral diaphysis, are sufficient to indicate a level of lower limb hypertrophy for Tianyuan 1 fully commensurate with the mobility characteristic of Late Pleistocene foraging populations.

Anterior Femoral Curvature

It is also possible to assess the degree of anterior diaphyseal curvature of the Tianyuan 1 right femur. Its curvature has the normal human form, with an even arc and the maximum projection of the diaphysis modestly distal of midshaft (Figure 10-7). Its index of curvature of 5.8 is relatively high, but it is matched (>5.7) by Fond-de-Forêt 1, Nazlet Khater 2, Ohalo 2, Předmostí 4, Qafzeh 9 and Skhul 6, only the first of which is a Neandertal (Table 10-2). As previously noted (Shackelford and Trinkaus, 2002), there is little difference across Late Pleistocene samples in femoral curvature, whether assessed by an index, a bivariate plot of subtense to chord (Figure 10-10), or the absolute subtense. The only exception is the low

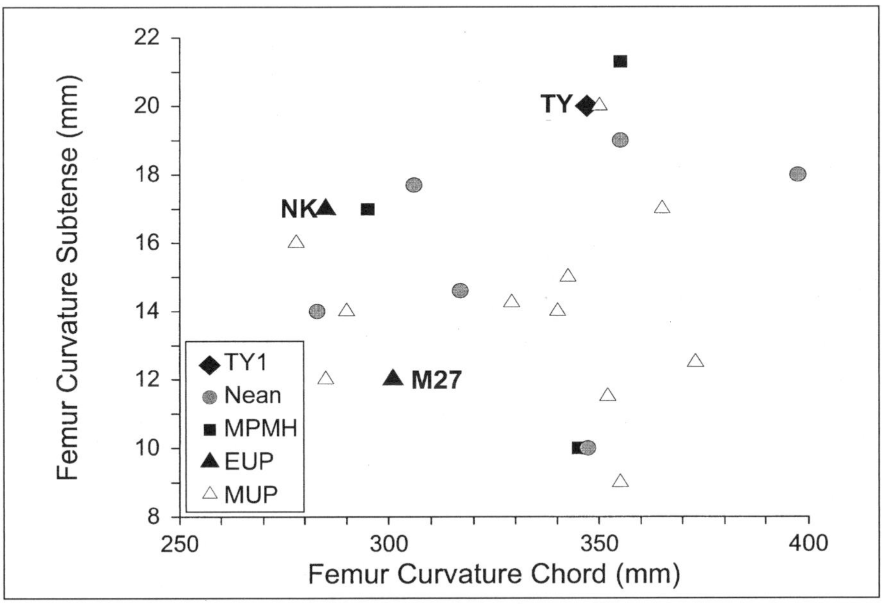

Figure 10-10. Bivariate plot of femoral anterior curvature subtense versus its (subtrochanteric to suprapatellar) chord. *TY:* Tianyuan 1; *Nean:* Neandertals; *MPMH:* Middle Paleolithic modern humans; *EUP:* western Early Upper Paleolithic; *MUP:* western Eurasian Mid Upper Paleolithic; *NK:* Nazlet Khater 2; *M27:* Mladeč 27. The position of Nazlet Khater 2 may be anomalous given its short femora.

values for the Minatogawa sample, for whom only indices are available (Baba and Endo, 1982).

Distal Femoral Morphology

The distal femoral epiphyses are represented only by the partial lateral condyle of the right femur with a portion of the patellar sulcus. There is little of note on it, other than that it provides a maximum anteroposterior diameter of ~75 mm (Table 10-1), which in turns provides a femoral head diameter estimate (Table 10-1) and from that an assessment of body shape and a body mass estimate (Chapter 7).

Tibiae

Diaphyses

Given both the pathological alterations of the right and left proximal posterior tibial diaphyses and the absence of the anterior portions, it is not possible to assess midproximal (nutrient foramen or cnemic level) shaft proportions or details of the original shaft contours. Considerations of the Tianyuan 1 tibial diaphyses are therefore limited to general assessments of shape and of midshaft robusticity.

Diaphyseal Shape

The proximal and midproximal diaphyses of the Tianyuan 1 tibiae are represented only by a small posterior section of the left one and the full posterior diaphysis of the right one (Figure 10-11). Both are altered pathologically (Chapter 12).

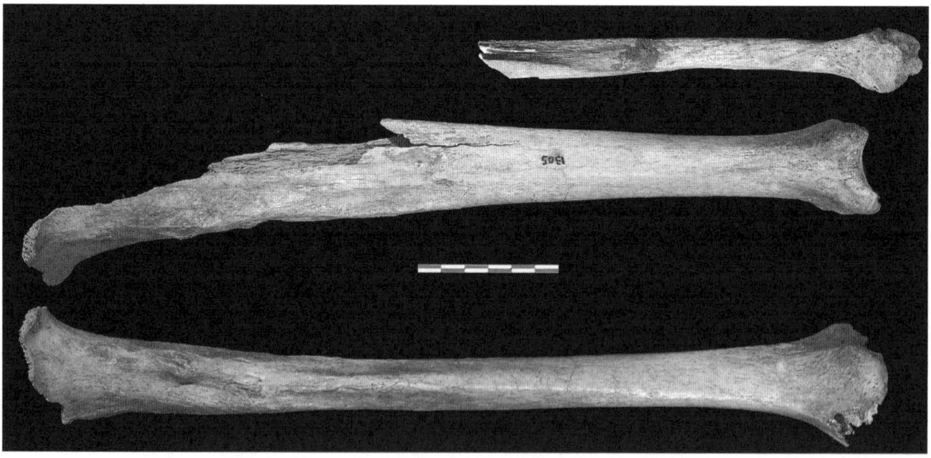

Figure 10-11. Posterior (*below*) and lateral (*middle*) views of the Tianyuan 1 right tibia and medial view (*above*) of the right fibula. Proximal is on the left. Scale in centimeters.

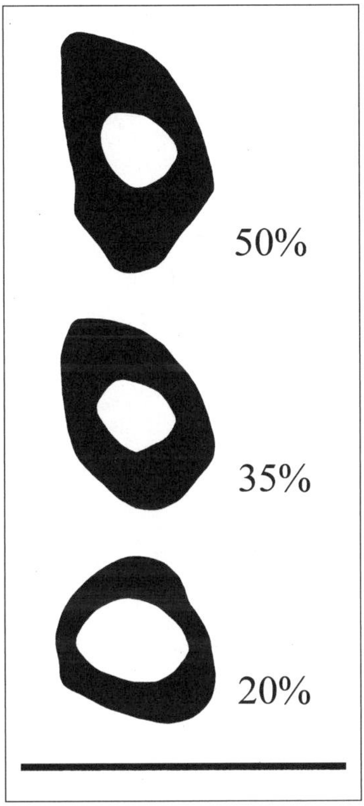

Figure 10-12. Diaphyseal cross sections of the mid and distal right tibia. Sections are in distal view, anterior is above, medial is to the right, and 0% is distal. Scale bar equals 50 mm.

On the more complete right tibia there is a prominent interosseus line that appears to be separate from the abnormal posterior diaphyseal surface. In the midproximal region there is the beginning of a large tibial pilaster that extends through midshaft (Figure 10-12). The pilaster in turn supports a well-demarcated but otherwise normal flexor line between the origins of tibialis posterior and flexor digitorum longus. The right diaphysis, despite its incomplete nature and altered condition, exhibits the concave surfaces anterolaterally, posteromedially, and posterolaterally that are common among modern human tibia (Figures 10-11 and 10-12); in this it contrasts with the more amygdaloidal cross-sectional shape characteristic of archaic *Homo* tibiae (Woo and Chia, 1954; Stringer et al., 1998; Trinkaus and Ruff, 1999; Churchill et al., 2000; Trinkaus, 2006d, 2009).

The overall diaphyseal shape combined with its prominent tibial pilaster results in midshaft external proportions that place it along the upper limits of Late Pleistocene human variation (Figure 10-13A), above the Middle Paleolithic Neandertals and modern humans, close to several western MUP humans, and exceeded in these proportions only by Předmostí 3. The more biomechanically relevant comparison of maximum (essentially anteroposterior) to minimum (approximately

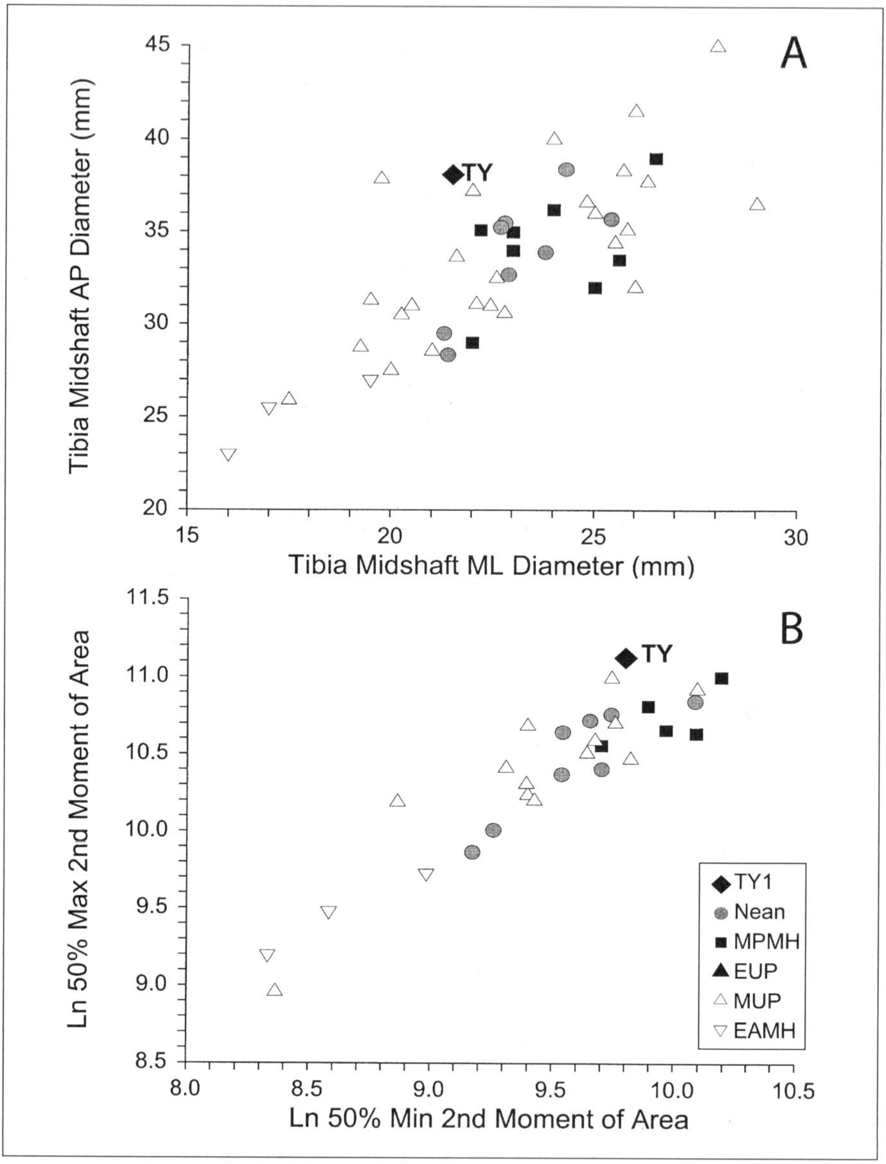

Figure 10-13. Bivariate plots of (*A*) tibial midshaft anteroposterior versus mediolateral diameters and (*B*) maximum (anteroposterior) versus minimum (mediolateral) second moments of area. *TY:* Tianyuan 1; *Nean:* Neandertals; *MPMH:* Middle Paleolithic modern humans; *EUP:* western Early Upper Paleolithic; *MUP:* western Eurasian Mid Upper Paleolithic; *EAMH:* eastern Asian modern humans.

mediolateral) second moments of area (Figure 10-13B) narrows the scatter of the Late Pleistocene tibiae but retains the position of the Tianyuan 1 tibia as the most pronounced in terms of anteroposterior to mediolateral midshaft proportions. It is possible that the anterior midshaft diaphyseal crack with its slight displacement of the anterior margin (Chapter 3) has modestly affected the results, but the anteroposterior dimensions were corrected for this distortion prior to these assessments (Table 10-4). The Tianyuan 1 tibia therefore remains at the top of the Late Pleistocene distribution in this feature.

Table 10-4. Osteometric dimensions of the Tianyuan 1 tibiae, in millimeters

	Right	*Left*
Maximum length (M-1a)	385.2[1]	—
Medial total length (M-1b)	375.1[2]	—
Lateral total length (M-1)	378.3[3]	—
Medial articular length (M-2)	360.9[4]	—
Lateral articular length (M-2)	364.0[5]	—
Biomechanical length[6]	362.5	—
Midshaft anteroposterior diameter (M-8)	(38.1)[7]	—
Midshaft mediolateral diameter (M-9)	21.5	—
Midshaft circumference (M-10)	97.0	—
Distal minimum circumference (M-10b)	79.0	—
Talar articular breadth	29.5	—
Lateral talar articular depth	31.3	—
Medial squatting facet height[8]	5.5	absent
Medial squatting facet breadth[9]	6.0	absent
Lateral squatting facet height[8]	10.7	9.1
Lateral squatting facet breadth[9]	14.1	15.0

1 The maximum length of the Tianyuan 1 right tibia was estimated from the preserved length from the distoposterolateral condylar margin to the middle of the trochlear surface (345.0 mm) using a least squares regression based on recent human tibiae [MaxLen = ((TY1Len × 1.064) + 17.8); N = 39; r^2 = 0.964; estimated length = 385.2 ± 5.2 mm; SE_{est} = 1.34%].

2 The medial total length of the Tianyuan 1 right tibia was estimated from the preserved length (see note 1) using a least squares regression based on recent human tibiae [MedTotLen = ((TY1Len × 1.038) + 17.0); N = 39; r^2 = 0.957; estimated length = 375.1 ± 5.5mm; SE_{est} = 1.46%].

3 The lateral total length of the Tianyuan 1 right tibia was estimated from the preserved length (see note 1) using a least squares regression based on recent human tibiae [LatTotLen = ((TY1Len × 1.028) + 23.5); N = 39; r^2 = 0.969; estimated length = 378.3 ± 4.6 mm; SE_{est} = 1.22%].

4 The medial articular length of the Tianyuan 1 right tibia was estimated from the preserved length (see note 1) using a least squares regression based on recent human tibiae [MedArtLen = ((TY1Len × 0.996) + 17.3); N = 39; r^2 = 0.967; estimated length = 360.9 ± 4.6 mm; SE_{est} = 1.28%].

5 The lateral articular length of the Tianyuan 1 right tibia was estimated from the preserved length (see note 1) using a least squares regression based on recent human tibiae [LatArtLen = ((TY1Len × 0.986) + 23.8); N = 39; r^2 = 0.974; estimated length = 364.0 ± 4.0 mm; SE_{est} = 1.11%].

6 Average of the medial and lateral articular lengths (Ruff and Hayes, 1983a).

7 The diaphysis measures 39.7 mm anteroposteriorly, but the lateral expansion crack is 1.6 mm wide at that level, hence the value of 38.1 mm.

8 The maximum height, approximately proximodistal, of the squatting facet from the edge of the talar facet.

9 The maximum mediolateral breadth of the squatting facet, along the edge of the talar facet.

Diaphyseal Robusticity

Plotting tibial midshaft polar second moments of area against the product of tibial lengths and estimated body masses produces no meaningful separation of the Late Pleistocene comparative samples (Table 10-5, Figure 10-14). Even the diminutive Minatogawa tibiae are in line with the western Eurasian samples and close to two small Neandertals. In this comparison, the Tianyuan 1 tibial diaphysis is one of the largest specimens providing data (adjacent to Skhul 5), but its position is unexceptional for a Late Pleistocene human. Adjusting its estimated body mass and tibial length upward, thereby making its body mass to stature proportions even more extreme than those employed here (Chapter 7), would place it closer to the middle of the Late Pleistocene distribution. Alternatively, adjusting body mass and bone length downward would make it one of the most robust human tibiae in the Late Pleistocene. Consequently, there is no evidence of any reduction in the

Table 10-5. Cross-sectional geometric parameters of the Tianyuan 1 right tibia

20%	
Total area (mm^2)	517.6
Cortical area (mm^2)	333.5
Medullary area (mm^2)	184.1
Anteroposterior second moment of area (I_x) (mm^4)	(22,914)[1]
Mediolateral second moment of area (I_x) (mm^4)	(15,714)
Maximum second moment of area (I_{max}) (mm^4)	24,052
Minimum second moment of area (I_{min}) (mm^4)	14,577
Polar moment of area (mm^4)	38,629
Orientation of I_{max} (theta)	110°
35%	
Total area (mm^2)	544.6
Cortical area (mm^2)	439.5
Medullary area (mm^2)	105.1
Anteroposterior second moment of area (I_x) (mm^4)	(33,973)
Mediolateral second moment of area (I_x) (mm^4)	(17,495)
Maximum second moment of area (I_{max}) (mm^4)	37,655
Minimum second moment of area (I_{min}) (mm^4)	13,813
Polar moment of area (mm^4)	51,468
Orientation of I_{max} (theta)	(113°)
50%	
Total area (mm^2)	666.3
Cortical area (mm^2)	555.2
Medullary area (mm^2)	111.1
Anteroposterior second moment of area (I_x) (mm^4)	(65,930)
Mediolateral second moment of area (I_x) (mm^4)	(20,110)
Maximum second moment of area (I_{max}) (mm^4)	68,031
Minimum second moment of area (I_{min}) (mm^4)	18,008
Polar moment of area (mm^4)	86,039
Orientation of I_{max} (theta)	(102°)

[1] I_x, I_y and theta are in parentheses, since the anteroposterior plane of the bone is approximated using the lateral plane of the midproximal diaphysis.

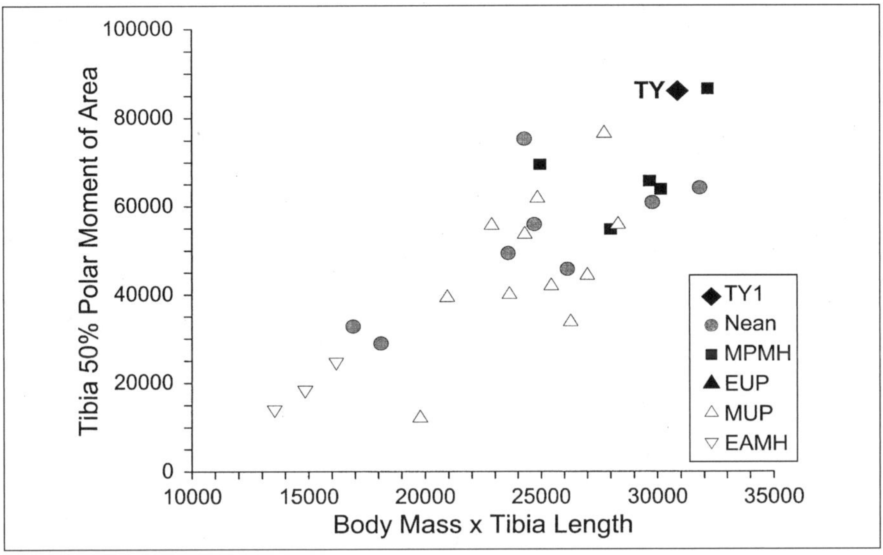

Figure 10-14. Bivariate plot of tibial midshaft polar second moments of area versus estimated body mass times tibial length. *TY:* Tianyuan 1; *Nean:* Neandertals; *MPMH:* Middle Paleolithic modern humans; *EUP:* western Early Upper Paleolithic; *MUP:* western Eurasian Mid Upper Paleolithic; *EAMH:* eastern Asian modern humans.

habitual loads placed on its legs (through mobility and/or burden carrying) as a result of its lower limb abnormalities.

Distal Epiphyses

The distal articulations of the Tianyuan 1 tibiae (Figure 10-15) are unexceptional, presenting anteroposteriorly concave trochlear facets with a gentle swelling for the midtrochlear sulcus. The left one retains part of the malleolus, and its articular facet flares medially at an angle of ~70° relative to the trochlear surface.

The right tibia has a small medial squatting facet and a substantially larger lateral one (Table 10-4). The left tibia lacks a medial squatting facet, but it has a lateral one similar in size to the right one. The pattern on the left tibia matches the facets on the neck of the left talus. Similar squatting facets are common among Late Pleistocene humans (Table 10-6). They are present, on the distal tibia and/or dorsal talar neck, on all of the tali prior to the Mid Upper Paleolithic except Amud 1 and Nazlet Khater 2, as well as on 3 Minatogawa skeletons and about half of the MUP individuals. Squatting facets, on the tibia and/or talus, have been documented in frequencies ranging from ~30% to ~80% for a variety of recent human samples across Eurasia; squatting facets tend to be rare mostly among modern urban populations

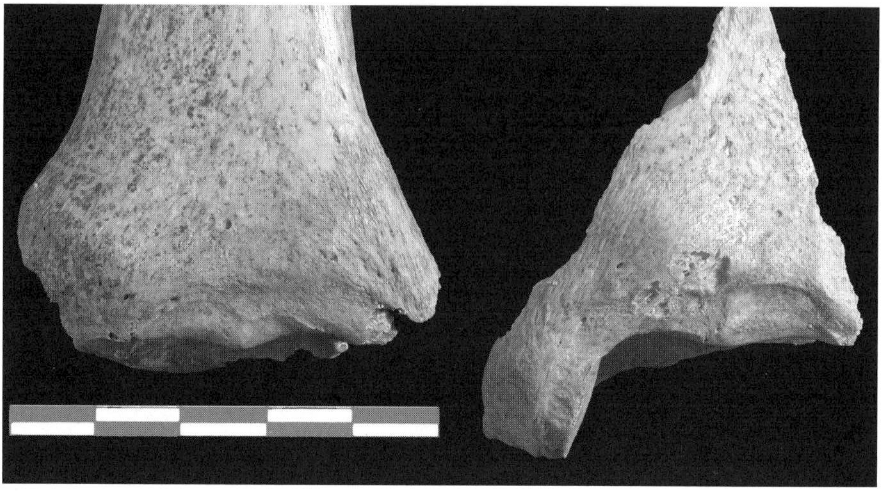

Figure 10-15. *Left to right:* Anterior views of the Tianyuan 1 right and left distal tibiae. Scale in centimeters.

Table 10-6. Articular facet variation for the talocrural, talocalcaneal, and metatarsal 1-2 articulations for Tianyuan 1 and Late Pleistocene samples

	Squatting facet presence (% (N))	*Sulcus tali facet presence (% (N))*	*Full fusion of the talar facets on the calcaneus (% (N))*	*Metatarsal 1-2 facet presence (% (N))*
Tianyuan 1	Present	Present	(Present)[1]	Absent
East Asia EMH	100 (3)	—	100 (3)	—
West MUP	53.3 (15)	55.6 (9)	81.8 (11)	0.0 (8)
West EUP	50.0 (2)	100 (1)	100 (1)	—
MPMH	100 (6)	25.0 (4)	100 (3)	66.7 (3)
Neandertals	90.0 (10)	33.3 (9)	75.0 (10)	50.0 (8)

Note: EMH: early modern humans; *MUP:* Mid Upper Paleolithic; *EUP:* Early Upper Paleolithic; *MPMH:* Middle Paleolithic modern humans. For the squatting facets, data are pooled for distal tibiae and dorsal tali by individual.

[1] The Tianyuan 1 fusion of the talar facets on the calcaneus is in parentheses, to indicate that, although they form one continuous curve, there is a slight division between the medial and anterior facets.

and Native Americans (Barnett, 1954; Baba, 1970; Trinkaus, 1975b). Tianyuan 1, with its matching tibial and talar facets, follows the pattern seen in most Late Pleistocene individuals.

Fibula

The Tianyuan 1 right fibula (Figure 10-11) preserves the diaphyses from slightly distal of midshaft to the distal malleolar facet. Using the mean estimate of the right tibial length of 385 mm provides an articular length estimate of 366.6 mm for the fibula (± 4.7 mm), based on a sample of recent human paired tibiae and fibulae ($r^2 = 0.958$, N = 50). The maximum preserved length of 189.5 is ~52% of this estimated length, indicating that the most proximal point (which is present only posterolaterally) is close to midshaft.

The midshaft has very prominent anterior and medial borders, especially anteromedially. The anterior surface is distinctly concave, and the medial surface is concave but less markedly so. The lateral surface is flat, and it then rounds onto the posterolateral surface and to a sharp posterior border. As such, the Tianyuan 1 fibula possesses the more common modern human pattern of having distinct and usually concave longitudinal sulci anteriorly and medially, with acute angles along the various borders. It contrasts with the usual pattern in archaic human fibulae, in which the surfaces are largely flat and the angles more obtuse (Heim, 1982; Trinkaus, 1983a). Yet, the longitudinal sulci are not as pronounced as those on some of the western Eurasian Upper Paleolithic fibulae (Trinkaus, 2006d).

The malleolus has a talar facet (articular height: 26.4 mm; articular depth: 21.7 mm) that is almost flat proximally, has a transversely rounded angle in the middle, and then is slightly concave vertically on its distal portion. There is little rugosity of the ligament attachments for the distal tibiofibular syndesmosis. The digital fossa is shallow, and the tip of the malleolus is at the same level as the distal end of the talar facet. The malleolus overall appears to deviate little laterally, such that the proximal portion of the talar facet is in a parasagittal plane parallel to the parasagittal plane of the diaphysis.

Posterior Tarsals

Talus

The essentially complete Tianyuan 1 left talus permits a series of observations on this articular cog between the tibia and fibula proximally and the subtalar skeleton distally (Figure 10-16; Table 10-7).

Talocrural Articulations

Dorsally, the trochlea presents a modest wedging, with a trochlear angle that is essentially on the means of the Middle Paleolithic samples and very similar to the Mladeč 30 value; Nazlet Khater 2 also appears (Crevecoeur, 2008) to have a low trochlear angle. The MUP sample has consistently higher angles, with a range of 8° to 18°, resulting in the significant difference across the Late Pleistocene samples (Table 10-8). The Tianyuan 1 trochlea has a shallow midline sulcus that fades out posteriorly and anteriorly and matches the rounded ridges on the distal tibiae. Anteriorly, there is a medial extension of the trochlea, along side of the anteriorly extending medial malleolar facet, but there is little anterior extension of the lateral trochlea.

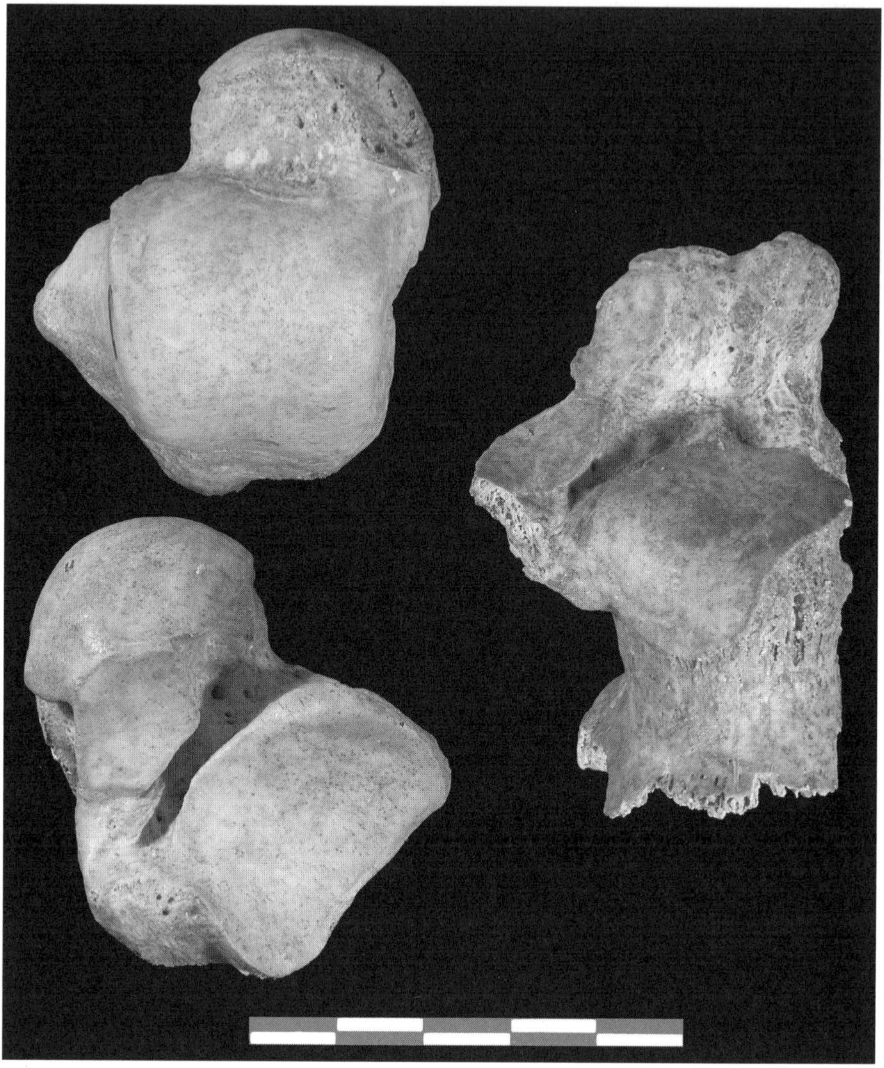

Figure 10-16. Dorsal (*upper left*) and plantar (*lower left*) views of the Tianyuan 1 left talus, and dorsal view (*right*) of the Tianyuan 1 right calcaneus. Scale in centimeters.

The malleolar surfaces have an intermediate degree of flare. This applies especially to the medial facet, but the lateral one flares more on its plantar third. Its articular to trochlear breadth proportions, in the context of Late Pleistocene comparative samples (Figure 10-17A), places it between the Neandertals with relatively wide articular breadths and the MPMH and Upper Paleolithic humans with narrower tali; Mladeč 30 and estimated values for Nazlet Khater 2 are completely within the early modern human distribution.

Table 10-7. Osteometric dimensions and articular discrete traits of the Tianyuan 1 left talus, in millimeters except as noted

Maximum (lateral) length (M-1a)	60.9
Length (M-1)	56.6
Articular height (M-3b)	24.5
Articular breadth (M-2b)	48.5
Trochlear length (M-4)	34.3
Trochlear breadth (M-5)	29.6
Head-neck length (M-8)	21.0
Head length (M-9)	35.3
Head breadth (M-10)	24.1
Posterior calcaneal facet length (M-12)	34.0
Posterior calcaneal facet breadth (M-13)	24.3
Trochlear angle[1]	7°
Neck angle (M-16)	20°
Torsion angle (M-17)	50°
Squatting facet—medial	absent
Squatting facet—lateral	present
Anterior and medial calcaneal facet fusion[2]	partial
Anterior medial malleolar extension	present
Anterior medial trochlear extension	present
Anterior lateral trochlear extension	absent
Sulcus tali facet[3]	present

[1] Angle in the horizontal plane between the medial and lateral margins of the trochlea.
[2] Degree of fusion between the anterior and medial facets for the calcaneus.
[3] A distinct articular surface on the sulcus tali, which is a rounding extension from the posterior calcaneal facet.

Table 10-8. Comparative talar angular measurements of the Tianyuan 1 left talus and Late Pleistocene samples

	Trochlear angle	*Neck angle*	*Torsion angle*
Tianyuan 1	7°	20°	50°
East Asia EMH	—	27°, 30°	30°, 31°
West MUP	11.6° ± 2.9° (10)	22.8° ± 2.4° (16)	35.0° ± 2.7° (9)
West EUP	8°	27°	45°
MPMH	7.0° ± 3.2° (7)	25.8° ± 2.3° (6)	38.4° ± 9.0° (5)
Neandertals	6.5° ± 3.4° (11)	24.6° ± 2.5° (11)	40.4° ± 4.8° (11)
K-W P-value	0.015	0.018	0.030

Note: EMH: early modern humans; *MUP:* Mid Upper Paleolithic; *EUP:* Early Upper Paleolithic; *MPMH:* Middle Paleolithic modern humans. Kruskal-Wallis (K-W) P-values across the comparative samples are provided.

The dorsal talar neck has a lateral squatting facet, but it lacks a medial one. In this, it matches the distoanterior left tibia, and as noted above, places it among the majority of the Late Pleistocene specimens in indicating habitual hyperdorsiflexion of the talocrural articulation, as in squatting. The anterior extension of the medial malleolar facet, a feature ubiquitous among Late Pleistocene tali (Trinkaus, 1975b, 2006d), is part of the same articular complex.

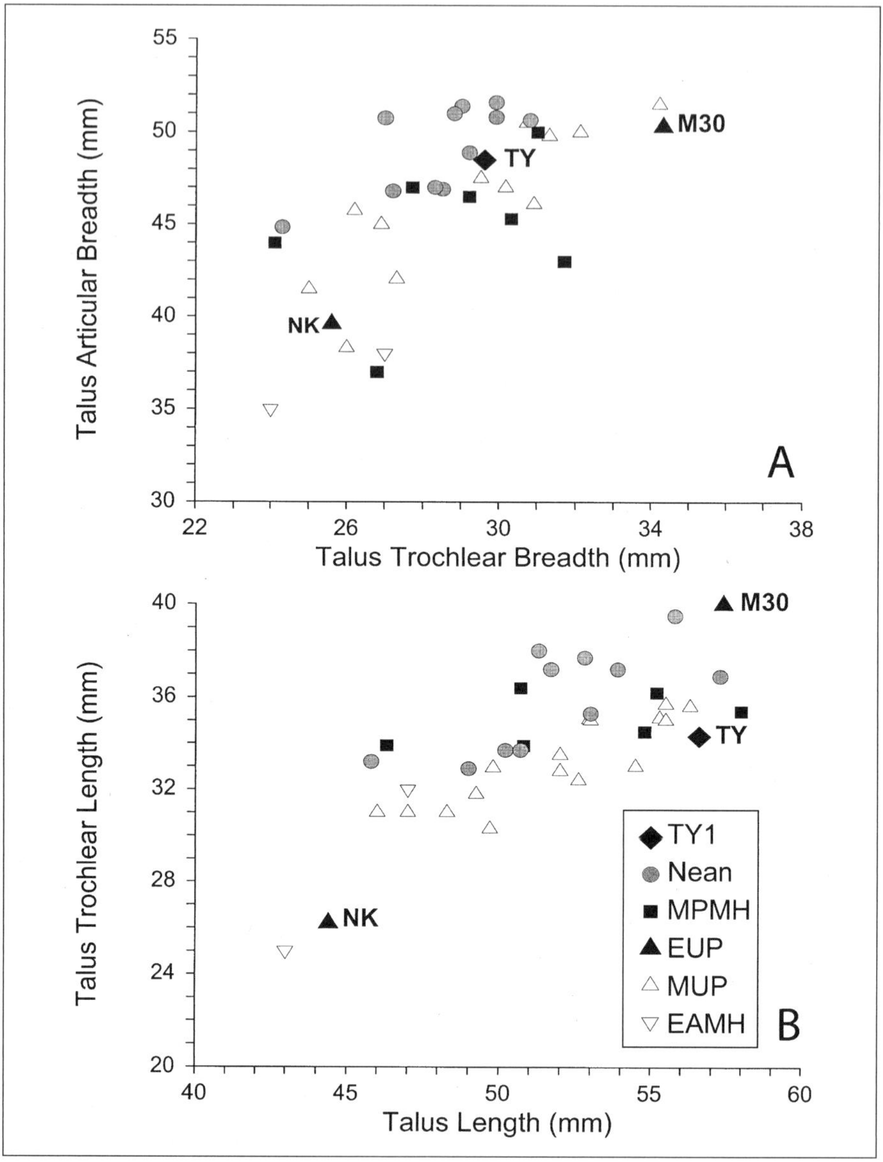

Figure 10-17. Bivariate plots of (*A*) talar articular breadth versus trochlear breadth and (*B*) talar trochlear length versus talar length. *TY:* Tianyuan 1; *Nean:* Neandertals; *MPMH:* Middle Paleolithic modern humans; *EUP:* western Early Upper Paleolithic; *MUP:* western Eurasian Mid Upper Paleolithic; *EAMH:* eastern Asian modern humans; *NK:* Nazlet Khater 2; *M30:* Mladeč 30. The articular breadth and talar length of Nazlet Khater 2 are approximate.

Talar Neck

The neck angle for the talus is relatively low, at the bottom of the Late Pleistocene ranges of variation and matched only by Amud 1, Nahal-Ein-Gev 1, and Dolní Věstonice 16 (Table 10-8). However, it is well within the ranges of variation of recent humans and distinct from the higher neck angles of large-bodied apes (Lisowski, 1967; Trinkaus, 1983b).

The Tianyuan 1 talus also has a relatively long neck (Figure 10-16). Since the neck and trochlear lengths are inversely correlated (Rhoads and Trinkaus, 1977), the neck length is best assessed by a comparison of trochlear to talar length (Figure 10-17B). As previously noted (Trinkaus, 1975a, 2006d), Neandertals have relatively short necks, or long trochleae; most Upper Paleolithic tali have relatively longer necks (or shorter trochleae); and the MPMH tali span the Late Pleistocene range of variation. Mladeč 30 falls with the Neandertals in this aspect (Trinkaus, Smith, et al., 2006), but estimated values for Nazlet Khater 2 (from Crevecoeur, 2008) place it among the MUP tali with longer necks. Tianyuan 1 has one of the shortest relative trochlear lengths, or longest relative neck lengths; it is matched only by the Skhul 6 talus.

Subtalar Articulations

The two notable features of the Tianyuan 1 plantar talus are its distinct sulcus tali facet and the angular distinction between its medial and anterior calcaneal facets. The former facet is an anterior rounding of the posterior calcaneal facet into the sulcus tali, and it articulates with the anterolateral dorsal calcaneus in hyperdorsiflexion of the foot (Trinkaus, 1975b). Sulcus tali facets are relatively abundant among Late Pleistocene humans, varying between one-quarter and one-half of the tali, plus the Mladeč 30 talus.

The Tianyuan 1 anterior and medial calcaneal facets are discrete surfaces, with a notch between them on the sulcus tali side and an elevated ridge between them on the medial side. The anterior one is slightly convex whereas the medial one is modestly concave. They form an angle of ~50° between them. The facets therefore conform to type B of Trinkaus (1975b), and type III of Arora et al. (1979). The left calcaneus is not preserved, but such clear separation of these facets is usually associated with a full separation of the anterior and medial talar facets on the calcaneus. Such a configuration of the anterior and medial facets is associated with more limited movement at the subtalar articulation (Drayer-Verhagen, 1993; Madhavi et al., 2008). It is moderately common among recent humans (Trinkaus, 1975b; Madhavi et al., 2008), but it is rare among Late Pleistocene humans (Trinkaus, 1975b; Table 10-6).

Talar Proportions

In addition to the assessment of relative talar breadths and lengths, it is possible, given the completeness of many Late Pleistocene tali, to assess the affinities of the Tianyuan 1 talus in a multivariate framework. This analysis was done using both a more complete set of ten linear measurements, and then to include a larger sample, a reduced set of 6 measurements, each with respect to the raw linear measurements and then with respect to the size-adjusted variables. The discriminant function analysis of these data using cross-validation (Table 10-9) provided relatively low percentages of correct assignment to the original samples. If only archaic versus modern human correct classification are considered, the percentages increase even though almost a third of them are misclassified. Despite these high error rates, the Tianyuan 1 talus is primarily aligned with either the MPMH or the MUP sample, and only in the analysis of the raw data of the full measurement set is it secondarily aligned with the Neandertals. That secondary Neandertal association is probably driven mostly by the large sizes of the Tianyuan 1 and many Neandertal tali, as reflected in their articular breadths and talar lengths. The penecontemporaneous Mladeč 30 talus, despite its relatively long trochlea, is consistently aligned with the MPMH sample.

The first two factors of a principal components analysis of 10 size-adjusted

Table 10-9. Discriminant functional affinities and posterior probabilities for the Tianyuan 1 talus dimensions

	Full set:[1] *raw measures*	*Reduced set:*[2] *raw measures*	*Full set: size corrected*	*Reduced set: size corrected*
Tianyuan 1 1st[3]	MPMH (0.742)	MUP (0.655)	MPMH (0.965)	MUP (0.794)
Tianyuan 1 2nd	Nean (0.190)	MPMH (0.327)	MUP (0.033)	MPMH (0.183)
% correct—all	42.9	54.5	39.3	36.4
% correct—LAH/EMH	67.9	69.7	67.9	69.7
Comparative N	28	33	28	33
Mladeč 30	MPMH (>0.999)	MPMH (0.994)	MPMH (0.999)	MPMH (0.916)

Note: MPMH: Middle Paleolithic modern humans; *MUP:* Mid Upper Paleolithic; *Nean:* Neandertals; *LAH:* late archaic human; *EMH:* early modern humans. Includes correct classification rates for the comparative samples across all 5 samples and between the late archaic humans (Neandertals) versus the pooled early modern human samples (LAH/EMH). Only the first two posterior probabilities are provided, since the others are markedly smaller. Results are provided for sets of 10 and 6 measurements, using raw measurements and size-corrected values (the latter following Darroch and Mosimann, 1985).

[1] The full measurement set includes talar length, articular height and breadth, head-neck length, and the lengths and breadths of the trochlea, head, and posterior calcaneal facet.

[2] The reduced measurement set includes talar length, head-neck length, and the lengths and breadths of the trochlea and head.

[3] For each set of talar measurements, the group with the highest and then second highest posterior probability for Tianyuan 1 is provided, followed in parentheses by the posterior probability of Tianyuan 1 for that sample.

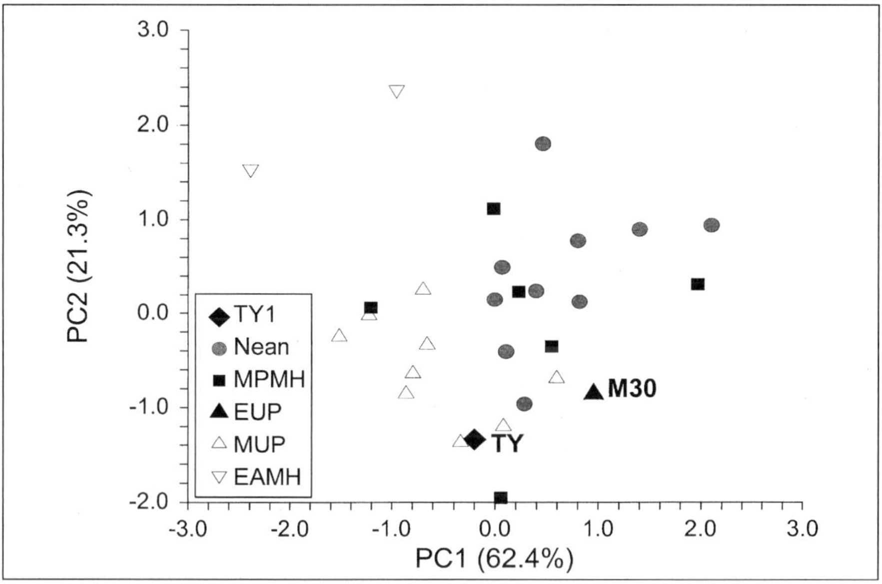

Figure 10-18. Bivariate plot of principal components (PC) 2 versus 1 for 10 size-adjusted linear talar measurements. *TY:* Tianyuan 1; *Nean:* Neandertals; *MPMH:* Middle Paleolithic modern humans; *EUP:* western Early Upper Paleolithic; *MUP:* western Eurasian Mid Upper Paleolithic; *EAMH:* eastern Asian modern humans; *M30:* Mladeč 30.

linear measurements (Figure 10-18) account for 83.7% of the variation. PCA 1 provides for substantial separation between the Neandertals and the Mid Upper Paleolithic modern humans, with the Middle Paleolithic modern humans encompassing the variation of the entire sample. The Mladeč 30 talus aligns with the MPMH and Neandertal tali. PCA 2 provides little separation between the samples. The Tianyuan 1 talus falls at the edge of the Neandertal variation on both axes, being relatively low compared to most tali on PCA 2. It is nonetheless close to several early modern human tali.

Calcaneus

The preserved right calcaneus (Table 10-10; Figure 10-16) retains principally its articular facets, the sustentaculum tali, and the medial body. The distal end of the medial process of the tuberosity is present, and it is relatively prominent plantarly and distally; there is no lipping on its distal margin for the plantar aponeurosis and abductor hallucis brevis. The origins for the long plantar ligament on the plantar body are moderately marked. There is a medially directed process of bone at the distal end of the plantar surface, probably for an hypertrophied origin for the short plantar ligament. In addition, there is an extension of bone from the dorsodistolateral corner of the calcaneus, extending distally over the dorsolateral corner of the cuboid facet; it should represent a partial ossification on the calcaneal side of the calcaneocuboid ligament. The articular facets are normal.

The sustentaculum tali is moderately projecting. An index of its sustentacular breadth to its body breadth (Table 10-11) places it among the higher of the Late Pleistocene calcanei. Its index is well below the Neandertal mean, but it is above the means of the three early modern human samples and especially distinct from the nonprojecting sustentacula of the Minatogawa calcanei. However, its index is close to those of two MPMH calcanei (Qafzeh 8 and 9) and exceeded by the index of Předmostí 10 and especially that of Dolní Věstonice 54 (Trinkaus et al., 2010). This sustentacular projection suggests a medially deviating talar head and neck, which may contrast with the low neck angle of the left talus.

The anterior and medial talar facets are separated by a slight transverse roughened line, but they otherwise form a continuous curve from their posteromedial to anterolateral extents. The anteromedial margin is abraded, and there is a concavity to the posterolateral margin of the facets, so it is not possible to determine if the fusion of the two facets is partial or complete. Their designation in Table 10-6

Table 10-10. Osteometric dimensions, in millimeters, and an articular discrete trait for the Tianyuan 1 right calcaneus

Medial breadth (M-2)	44.3
Sustentacular breadth (M-6)	15.8
Body breadth (M-2–M-6)	28.5
Posterior talar facet length	31.9
Posterior talar facet breadth	23.6
Talar articular facet fusion	(present)

Table 10-11. Comparative calcaneal sustentacular projection indices for the Tianyuan 1 right calcaneus and Late Pleistocene samples

	Sustentacular projection
Tianyuan 1	55.4
East Asia EMH	45.8, 47.2, 47.8
West MUP	51.0 ± 9.0 (11)
MPMH	51.6 ± 5.0 (4)
Neandertals	59.2 ± 12.3 (8)
K-W P-value	0.134

Note: EMH: early modern humans; *MUP:* Mid Upper Paleolithic; *MPMH:* Middle Paleolithic modern humans. Index = (sustentacular breadth/body breadth) x 100. Kruskal-Wallis (K-W) P-value across the comparative samples is provided. There are no sufficiently intact comparative EUP calcanei.

is therefore placed in parentheses. However, the even curve of the two facets contrasts with the sharply separated and angled calcaneal facets on the left talus. Moreover, the configuration of these facets on the calcaneus conforms to that seen in 83.9% of the Late Pleistocene tarsals (Table 10-6) and the majority of recent humans in especially nonurban samples (Trinkaus, 1975b; Shiraishi et al., 2007; Madhavi et al., 2008).

The contrasting patterns of talocalcaneal facets between the right and left Tianyuan 1 feet indicates a marked level of asymmetry. Similar asymmetry is present in the Shanidar 3 Neandertal (Trinkaus, 1983a), and it occurs between 4.5% and 14.0% in three recent human samples (Trinkaus, 1978). It is unclear what this asymmetry might mean in terms of pedal function, but it implies greater subtalar joint mobility on the right side.

Metatarsals and Pedal Phalanges

Metatarsals

The four preserved metatarsals (Table 10-12; Figure 10-19) present few features of note, given that there are few significant differences across Late Pleistocene samples in metatarsal morphology (Heim, 1982; Trinkaus, 1975a, 1983a, 2006d).

The first metatarsal has a moderately concave tarsal facet, which rotates from directly proximal on its dorsal half to slightly medial on its plantar half. It lacks an articular facet for the 2nd metatarsal, the arrangement for about two-thirds of

Table 10-12. Osteometric dimensions of the Tianyuan 1 metatarsals in millimeters except as noted

	1 Left	*3 Right*	*4 Right*	*5 Left*
Maximum length	—	78.0	—	69.8
Articular length	—	73.4	72.5	63.3
Midshaft height	15.0	9.9	9.8	7.5
Midshaft breadth	13.1	8.8	8.7	11.7
Proximal maximum height	30.8	21.9	—	14.9
Proximal maximum breadth	21.1	18.3	—	20.9
Proximal oblique maximum breadth				20.8
Proximal articular height	29.2	20.3	—	14.5
Proximal articular breadth	16.9	15.7	—	15.1
Distal height	—	17.8	17.2	8.9
Distal articular breadth	—	12.0	10.1	15.6
Proximal horizontal angle[1]	—	70°	—	61°
Torsion angle[2]	—	3°	—	6°

[1] Angle between the plane of the tarsal facet and the diaphyseal axis, measured in the horizontal plane of the bone.

[2] Angle between the dorsoplantar axis of the tarsal facet and the dorsoplantar axis of the head. A positive angle indicates a medial deviation of the dorsal metatarsal head.

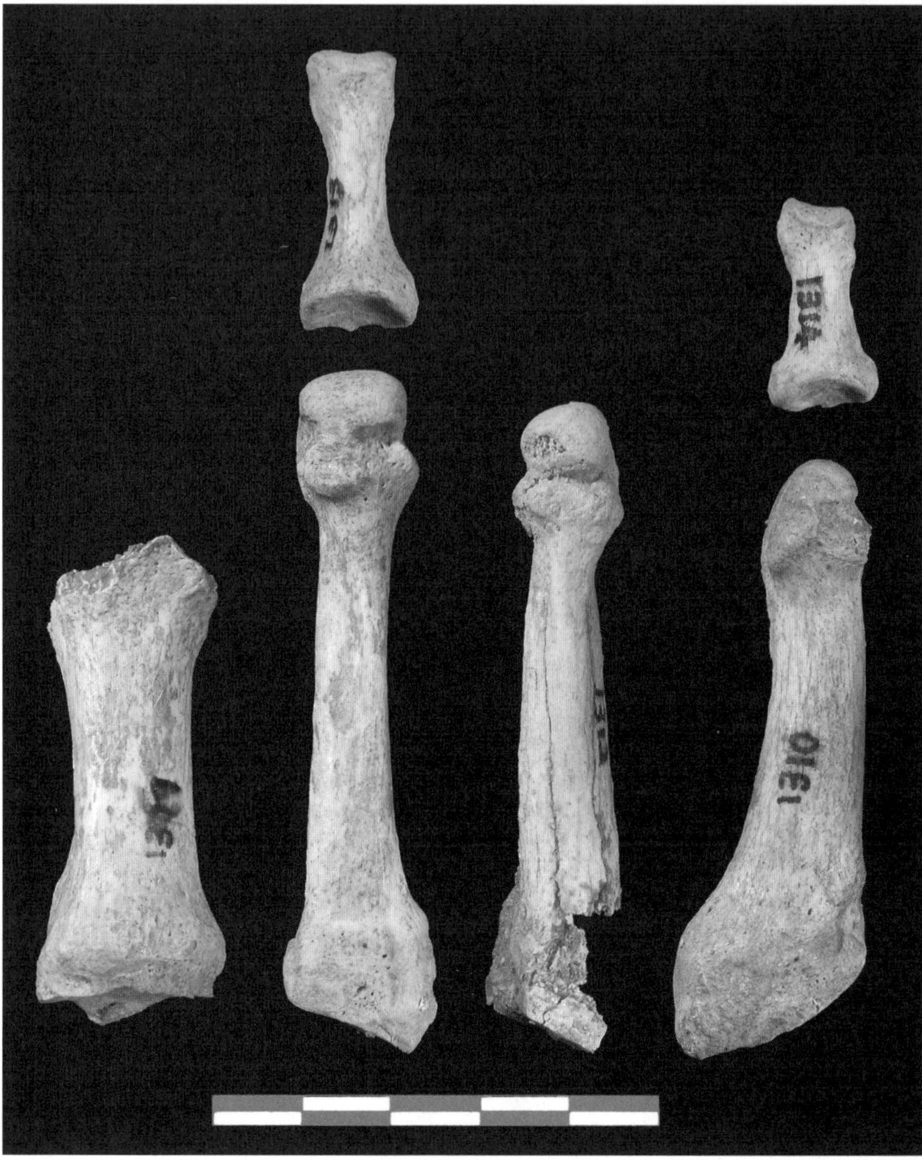

Figure 10-19. Dorsal views of the Tianyuan 1 metatarsals and proximal pedal phalanges. *Left to right:* metatarsal 1 left, metatarsal 2 right, metatarsal 3 right, and metatarsal 5 left. The more medial proximal phalanx is aligned with its probable digit 2, and the right proximal phalanx 5 is positioned above the left metatarsal 5. Scale in centimeters.

the Late Pleistocene humans and all of the MUP feet for which data are available (Table 10-6). The peroneal tubercle is marked by a slightly rugose concavity. The absence of the head prevents consideration of morphological reflections of hallux valgus.

The third metatarsal has a single and small medial facet for the second metatarsal and two larger ones laterally for the fourth metatarsal. The base of the fourth metatarsal is too eroded to assess its facet configurations. Both of these bones have clear but modest torsion for the transverse pedal arch, indicating a normal arch but suggesting a moderately low one. The fifth metatarsal similarly exhibits a low level of torsion. Musculoligamentous markings on the bones are universally weak.

Proximal Phalanges

Tianyuan 1 retains two complete proximal pedal phalanges (Figure 10-19; Table 10-13). One is from the right fifth ray, given its morphology, its length, and the right deviation of its trochlea. The other one is probably right based on its minimal head deviation, and it most likely derives from the second ray, given its length relative to the metatarsals (Chapter 3). The proximal articular height is similar to its articular breadth, which supports a second ray designation, since the metatarsal facets tend to be become relatively broader laterally.

Both of the phalanges have metatarsal facets which are oriented dorsoproximally, rather than directly proximally, indicating habitual dorsiflexion of the meta-

Table 10-13. Osteometric dimensions of the Tianyuan 1 proximal pedal phalanges in millimeters except as noted

	2 (or 3)	5
Maximum length	32.5	25.0
Articular length	29.2	21.4
Midshaft height	6.9	5.2
Midshaft breadth	6.6	7.1
Proximal maximum height	12.9	10.9
Proximal maximum breadth	13.7	12.8
Proximal articular height	11.8	9.7
Proximal articular breadth	11.8	10.0
Distal height	7.0	6.7
Distal articular breadth	10.4	9.0
Proximal vertical angle[1]	72°	76°
Proximal horizontal angle[2]	88° right	79° right
Torsion angle (both dorsal to right)[3]	11°	4°

[1] Angle, in the parasagittal plane of the bone, of the tangent to the metatarsal surface relative to the diaphyseal axis. An angle $< 90°$ indicates a distal deviation of the dorsal margin.

[2] Angle, in the horizontal plane of the bone, of the tangent to the distal trochlear surface relative to the diaphyseal axis.

[3] Angle between the plantar surface of the phalangeal base and the plantar surface of the distal trochlea.

tarsophalangeal articulations during heel-off. The fifth ray phalanx has a shaft that is broader than wide, which is the normal arrangement for that bone. The second one, however, is higher than it is wide, the more common arrangement among recent humans (Trinkaus and Hilton, 1996), one absent from the Neandertals (N = 8), but the variably present configuration for early modern humans (MPMH: 50.0%, N = 4; MUP: 63.6%, N = 11).

Of more interest is the relative robusticity of the proximal phalanx 2, since there is a consistent reduction in the diaphyseal robusticity of the middle proximal pedal phalanges between the Middle Paleolithic modern humans and Neandertals on the one hand and the western Eurasian Mid Upper Paleolithic modern humans on the other hand (Trinkaus, 2005b; Trinkaus and Shang, 2008). Given the lack of reduction in overall leg robusticity across these groups, and hence in mobility levels and loading patterns (Trinkaus, 2006d; Figures 10-10 and 10-15), the reduction in middle pedal phalangeal robusticity is inferred to indicate a marked reduction in the mechanical effect of ground reaction force on the lesser toes, and hence an increase in the habitual use of protective footwear (see Trinkaus, 2005b).

Given uncertainties in assigning isolated middle proximal phalanges to digit, comparative data are pooled for phalanges 2 to 4, averaging the values when more than one phalanx is available. In addition, to assess robusticity, the polar second moments of area for phalangeal midshafts were computed using an ellipse formula (O'Neill and Ruff, 2004) and modeling the phalanges as solid beams. A resultant bivariate plot of the polar second moments of area (Figure 10-20) provides the previously documented separation of the Middle Paleolithic and Mid Upper Paleolithic pedal phalanges, with minimal overlap in the distributions. The Tianyuan 1 second proximal phalanx is, if anything, gracile compared to even the most gracile of the MUP sample. Adjusting the body mass value downward, given the uncertainties in its estimation (Chapter 7), would still maintain the Tianyuan 1 pedal phalanx well within the gracile MUP range and separate from the Middle Paleolithic late archaic and early modern humans. Assuming that this phalangeal diaphyseal reduction reflects reduced habitual loads on the pedal phalanges, and given that the Tianyuan 1 femoral and tibial diaphyseal robusticity is similar to that of other Late Pleistocene humans, the gracility of its second proximal pedal phalanx supports the habitual use of footwear by this individual (see Trinkaus and Shang, 2008).

Summary

The lower limb remains of Tianyuan 1, despite the pathological alterations of the mid-distal femoral diaphyses and the midproximal tibial diaphyses, provide considerable data on the morphology of these limbs. This information is in addition to the inferences concerning body size and proportions (Chapter 7).

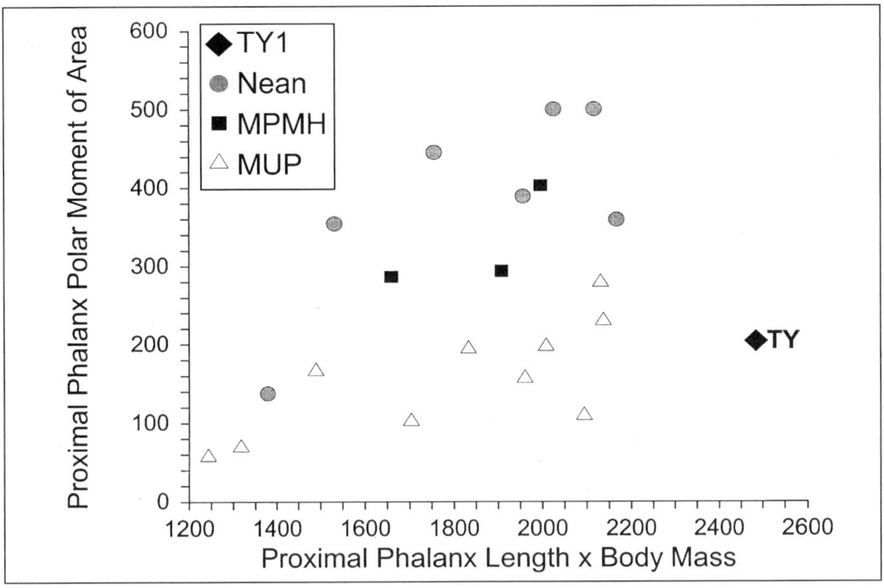

Figure 10-20. Bivariate plot of middle proximal pedal phalangeal polar second moments of area versus phalanx length times estimated body mass. Second moments of area were computed modeling the diaphyses as solid beams. For specimens preserving more than one of the proximal pedal phalanges 2 to 4, the second moments of area and the lengths of the available phalanges were averaged for that individual. Inclusion of individuals lacking a body mass estimate increases the sample sizes and maintains the same general pattern (Trinkaus and Shang, 2008). *TY:* Tianyuan 1; *Nean:* Neandertals; *MPMH:* Middle Paleolithic modern humans; *MUP:* western Eurasian Mid Upper Paleolithic.

In a comparative morphological sense, the lower limbs of Tianyuan 1 are clearly aligned with those of early modern humans. This is reflected in the distinct femoral pilaster, the presence of longitudinal sulci on the lateral tibial diaphysis, the anterior and medial fibular diaphysis, and the elongation of the neck of the talus. Even though the degree of expression of these traits may have been functionally influenced, the pattern of their presence is shared with other early modern humans, especially those from the earlier Upper Paleolithic of western Eurasia.

The femora and tibiae exhibit levels of robusticity that are shared with late archaic humans (Neandertals) and Upper Paleolithic modern humans. In a developmental context, the low femoral neck-shaft angle of Tianyuan 1 places it among the lower angles of the Neandertals and MUP humans, and completely separate from the Middle Paleolithic modern humans, the Minatogawa eastern Asian early modern humans, and the (pathological) Nazlet Khater 2 skeleton. This angle reflects elevated locomotor loading levels during development. The same elevated locomotor activity was continued through maturity, as is reflected in the robusticity values for the femoral and tibial midshafts, despite the alterations of the muscular insertions on those diaphyses. Interestingly, however, its scaled gluteal tuberosity development is among the early modern humans and distinct from the Neandertal values.

These morphological affinity and functional hypertrophy indicators are combined with clear evidence for postures that produced hyperdorsiflexion of the talocrural articulation (i.e., squatting), an extremely gracile middle pedal proximal phalanx indicating reduction in ground reaction forces through some form of mechanical insulation (i.e., footwear), and a marked asymmetry of the talocalcaneal articulations suggesting differential mobility of the right and left subtalar articulations.

Chapter 11

Sex and Age-at-Death of Tianyuan 1

ASSESSMENTS OF the overall place of the Tianyuan 1 skeletal remains in Late Pleistocene paleobiological variation can benefit from knowledge of the biological sex and the age-at-death of the individual. Even though primary indicators of these parameters are not preserved on the remains, it may be possbile to gain an approximate assessment of them from indirect indicators.

Biological Sex of Tianyuan 1

The biological sex of the Tianyuan 1 partial skeleton can be primarily assessed from its overall body size and mandibular morphology in comparison to other Late Pleistocene humans. The complete absence of the pelvis means that the primary sexual characteristics of the skeleton (Brůžek, 2002) are unknown, and the absence of the cranium means that secondary sexual characteristics of that region (Walker, 1994, 2008) cannot be assessed.

Body Size Indicators

As detailed in Chapter 7 with respect to long bone lengths, Tianyuan 1 was the tallest of the currently known eastern Asian Late Pleistocene humans but average in size compared to early modern humans from western Eurasia (see Figures 7-1 and 7-2). Its estimated long bone lengths exceeded all of the direct and estimated measurements of long bone lengths for the limited samples of eastern Asian early modern humans (Weidenreich, 1941; Baba and Endo, 1982), and they exceed those for more recent Late Pleistocene humans from mainland eastern Asia (Shackelford, 2005). However, there are Late Pleistocene modern humans from Australia whose statures were in the same range as that of Tianyuan 1 (Bowler and Thorne, 1976; Webb, 1989). Moreover, if the Tianyuan 1 long bone lengths are compared to those of early modern humans from western Eurasia, they fall in the middle of the overall range of variation, at the bottom of the male distributions, and among some of the taller females.

The lower limb weight-bearing articulations of Tianyuan 1 are at the top of the Late Pleistocene range of variation. Its estimated femoral head diameter is exceeded only by those of two Neandertals and approached by two other Neandertals and two MUP humans, five of which are pelvically sexed as male (see Figure 7-3). In talar length, the Tianyuan 1 value is exceeded or matched by one pelvically sexed male Neandertal, and by one talus from each of the Middle Paleolithic modern human, Early Upper Paleolithic, and Mid Upper Paleolithic samples (see Figure 7-3). However, the femoral head diameter is statistically indistinguishable, given estimation standard errors, from that of the pelvically sexed female late Middle Pleistocene Jinniushan 1 archaic human skeleton (Rosenberg et al., 2006).

Given these considerations, body size indicators, especially lower limb articular dimensions, might support a male diagnosis for the Tianyuan 1 sex. But its long bone lengths in a Late Pleistocene context are undiagnostic, and the hip articular dimensions of the, albeit archaic, female Jinniushan 1 remains make a male diagnosis on the basis of those joint sizes less secure.

Mandibular Indicators

There have been various attempts to determine sex from the mandible, despite interpopulational variation in mandibular form even among recent humans (see Steele and Bramblett, 1988; White and Folkens, 1991; Walker, 1994, 2008). The general pattern, however, appears to be a more prominent mentum osseum (or mental eminence) among males (Walker, 1994), a more open (or obtuse) gonial angle among females (Steele and Bramblett, 1988), and more gonial flare among males (White and Folkens, 1991). It remains unclear to what extent these criteria might apply to the Tianyuan 1 remains, given the absence of a contemporaneous eastern Asian reference sample. Nonetheless, these criteria can be assessed with respect to recent human and early modern human mandibular variation.

The mentum osseum rank of Tianyuan 1 of 4 (see Table 6-3; Figure 6-4) corresponds to Walker's (1994) rank 3 ("ambiguous sex") or possibly 4 ("probably male"). The gonial angle of the right corpus and ramus, ignoring the rounded gonial area and measuring the angle between the inferior corpus anterior of the ramus and the posterior ramus above the gonial rounding, is ~112° (see Figure 6-2). This angle places it in Steele and Bramblett's (1988) male category, since it is well below 125°. The gonial angle, however, turns markedly medially (see Figure 6-1), a criterion that should make it female (White and Folkens, 1991). These considerations therefore provide one "ambiguous to probably male," one "male," and one "female" designation.

The mental eminence form may or may not indicate gender among these Late Pleistocene humans. Western Eurasian early modern humans overwhelmingly exhibit mentum osseum rank 4 (see Table 6-3), whether pelvically male or female (e.g., Brůžek et al., 2006). However, more recent eastern Asian early modern humans are evenly distributed between ranks 4 and 5, and at least one of those individuals with rank 5 (Minatogawa 1) is pelvically sexed as male. In addition, of the three mandibles with rank 5 or 4/5 and preserved rami (Minatogawa 1 and 5 and ZKD-UC 101) two have flaring gonial areas. But the purportedly female ZKD-UC 104 mandible also exhibits some gonial flaring. Furthermore, the Minatogawa 1 and ZKD-UC 101 mandibles have relatively open gonial angles, whereas ZKD-UC 104 has a low gonial angle. Given this, the modest development of the mental eminence on the Tianyuan 1 mandible, in an eastern Asian early modern human context, might be taken to indicate "probably female" status rather than "ambiguous sex," but the gonial criteria do not follow a clear pattern.

The morphology of the Tianyuan 1 mandible is therefore best considered as indeterminate with respect to biological sex. Given the extensive overlap in these features among pelvically sexed Late Pleistocene and recent human mandibles, and the inconsistent associations of the trait forms within mandibles, it would take a clustering of all of its features at one end of the variation to provide a secure sex diagnosis for this individual. This is not the case.

Age-at-Death of Tianyuan 1

There is little doubt that Tianyuan 1 was fully mature at the time of death. The right M_3 was fully formed with complete closure of the root apices, in occlusion for an extended period of time, and then supererupted following antemortem tooth loss (Chapters 5 and 12). All of the synchondroses of the second cervical vertebra were fully closed and obliterated (Chapter 8). The preserved epiphyses (distal humerus, femoral trochanters, distal tibiae and fibula, manual phalanges, metatarsals, and pedal phalanges) were all completely closed and obliterated (Chapters 9 and 10). The symphysis between the manubrium and the first sternal segment was completely open, but it is common for it to remain so throughout adulthood (Trotter, 1934; McKern and Stewart, 1957).

The only age-related degenerative processes evident on the remains are the moderate osteoarthritis on the second cervical vertebra, the first to second stenebral fusion, and the occlusal attrition of the preserved dentition. The osteoarthritis of the manual interphalangeal articulation is likely to be activity related, and the abnormal changes to the distal femoral and proximal tibial diaphyses are apparently idiopathic and not likely to be age related (Chapter 12).

The degenerative changes of the axis suggest a reasonably advanced age-at-death, but since such changes are also related to the levels and frequency of biomechanical loads on the cervical vertebral column, such changes may indicate little more than an age-at-death unlikely to be less than the fourth decade of life.

As noted in Chapter 8, the fusion of the sternebral segments is normally complete by the end of the third decade postnatal. The complete fusion of the first and second sternebrae therefore indicates a probable minimum age-at-death in the fourth decade.

The advanced dental occlusal attrition of Tianyuan 1 (see Figures 5-1 to 5-3) can be compared to those evident among Late Pleistocene and Holocene individuals skeletally aged by other methods (e.g., Brothwell, 1963; Trinkaus and Thompson, 1987; Hillson et al., 2006) and to the levels of attrition evident among recent high latitude foragers (e.g., Davies and Pedersen, 1955; Moorrees, 1957). Based on these frameworks, the occlusal wear of the Tianyuan 1 teeth, especially of the premolars and the M_1 and M_2 with almost complete exposure of occlusal dentin (all except for a small area on the M_2), suggests an age into the fifth decade postnatal and possibly into the sixth decade. The moderate wear gradient for the Tianyuan 1 mandibular molars implies that these levels of attrition accumulated relatively slowly compared to many Late Pleistocene and recent humans; therefore they suggest an age-at-death toward to the higher end of this range. The antemortem tooth loss of at least the lower left side and the upper right side (Chapter 12) might, however, have accelerated the occlusal attrition on the remaining teeth, which might counteract the effect on age-at-death estimation suggested by the low wear gradient.

Independent of the absolute age-at-death of Tianyuan 1, the amount of wear on the postcanine dentition, that which would be unaffected by differential levels of paramasticatory use of the anterior dentition, places this individual among the older of the known late archaic and early modern human remains. Its degree of attrition is matched by several Neandertals (Arcy-Hyène 8, Artenac 1, Banyoles 1, La Ferrassie 2, and Shanidar 1, 3, 4, and 5) and MUP humans (Brno 2, Caviglione 1, Dolní Věstonice 3, Pavlov 1, Sunghir 1). Its degree of attrition was exceeded only by the La Ferrassie 1 Neandertal and the Dolní Věstonice 16, Ohalo 1 and 2, and ZKD-UC 101 early modern humans (not counting individuals with extensive antemortem tooth loss).

Summary

These considerations suggest that the Tianyuan 1 partial skeleton derived from an individual who was, for a Late Pleistocene human, relatively but not excessively advanced in age at the time of death. It is tempting to assign a male sex to the remains, given the large dimensions of its lower limb articulations and implied high body mass. However, the long bone lengths are inconclusive with respect to broader (not just eastern Asian) Late Pleistocene human stature, and the elements of the mandible that might be sexually diagnostic are similarly inconclusive.

Chapter 12

Paleopathology of the Tianyuan 1 Partial Skeleton

THROUGHOUT THE chapters on the individual skeletal elements of Tianyuan 1, there are many references to pathological abnormalities. Some of these are briefly described since they relate to the paleobiology of concern in those chapters, but detailed description and diagnosis of them is presented here. Some of the relevant details and illustrations are in Chapters 5 through 10.

Developmental Stress Indicators

Dental Enamel Hypoplasia

Extensive wear of the I_2–M_2 and the loss of most of the M^{1-2} buccal crown means that only the right M_3 has sufficient crown for an assessment of dental enamel hypoplasia (DEH) or developmental defects of the enamel (see Figures 5-1–5-3). The M_3 has a small pit at the occlusal edge (Figure 12-1), on the buccal side of the mesiobuccal cusp, 4.7 mm from the cervix. Given M3 crown formation in the late first and early second decade postnatal (Smith, 1991), this defect probably formed between the ages of 9 and 11 years of age.

A variety of localized or systemic insults, including especially dietary deficiencies and infections, can produce such a minor DEH (Kreshover, 1960; Sweeney et al., 1971; Goodman et al, 1991; Hillson, 1996). However, M3 DEH is less frequent than on the more mesial molars among recent humans [in the rare cases when they are recorded (e.g., El-Najjar et al., 1978)] and among Late Pleistocene early modern humans (Trinkaus, Hillson, et al., 2006). They are nonetheless the most common location for molar defects among the Neandertals (Ogilvie et al., 1989). But the overall frequencies of all of the analyzed Late Pleistocene samples are within recent human ranges of variation (Ogilvie et al., 1989; Brennan, 1991; Guatelli-Steinberg et al., 2004; Trinkaus, Hillson, et al., 2006).

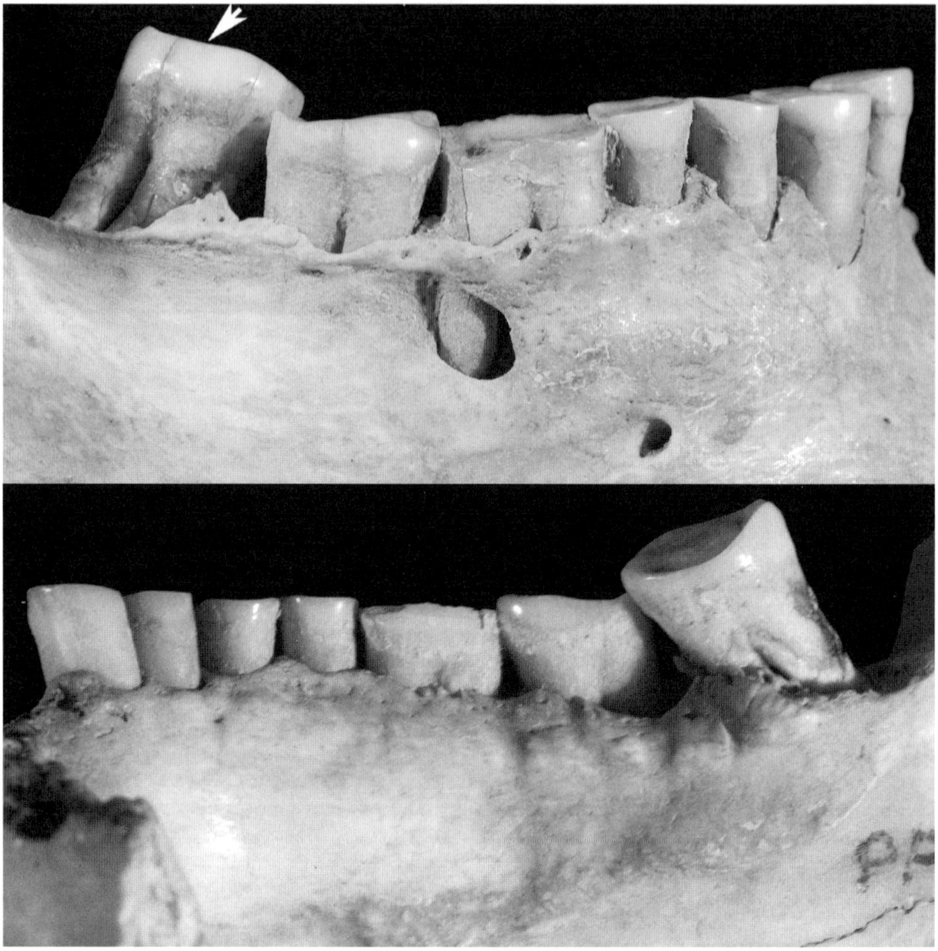

Figure 12-1. Buccal (*top*) and mesial (*bottom*) views of the Tianyuan 1 right mandibular dentition and mandibular corpus. The arrow points to the minor hypoplasia on the M_3.

Growth Arrest (Harris) Lines

There are relatively few sufficiently intact metaphyseal regions on the Tianyuan 1 appendicular remains for which the presence of growth arrest (transverse; Harris) lines (Aufderheide and Rodríguez-Martin, 1998: p. 422) can be assessed. Moreover, the moderately advanced adult age of the individual (Chapter 11) suggests that many originally present growth arrest lines would have been remodeled and removed. It is nonetheless possible to assess the presence of these features on the distal left humerus, the distal right tibia, and the proximal left metatarsal 1 (Figure 12-2).

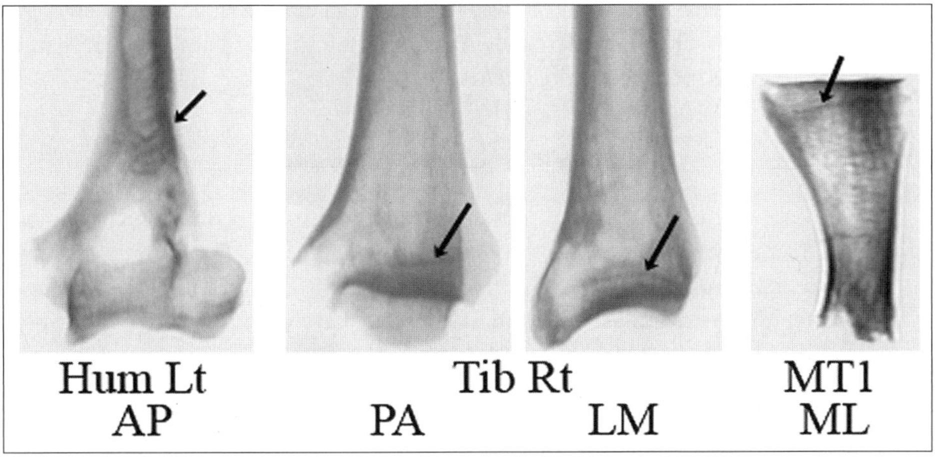

Figure 12-2. Radiographs of the Tianyuan 1 distal left humerus in anteroposterior (AP) view, of the distal right tibia in posteroanterior (PA) and lateromedial (LM) views, and the left metatarsal 1 in mediolateral (ML) view. Arrows point to the linear radiodensities, which may represent transverse (Harris) lines.

The distal humerus has about six mediolateral lines of increased radiodensity across the distal diaphysis, proximal of the olecranon fossa area. These lines are distant from the original metaphyseal line, and it is unclear whether they represent growth arrest lines or merely trabeculae associated with the supraolecranon area. On the distal tibia, however, there is a distinct line at the level of the proximal medial malleolus above the lateral half of the trochlear surface in the posteroanterior view but extending the full depth of the metaphysis in the lateromedial view (Figure 12-2). It is farther from the trochlear surface posteriorly and laterally. There is a similar dorsoplantar line across the proximal metatarsal 1 metaphysis, extending across the full dorsoplantar metaphysis and slightly farther from the medial cuneiform facet plantarly. These lines probably represents a growth arrest during adolescence given the mid to late second decade fusion of the distal tibial and proximal first metatarsal epiphyses (Scheuer and Black, 2000). They were probably of modest duration but sufficient in intensity for lines to form that persisted for a few decades of life. It is possible that the lines reflect the same insult as the trivial hypoplasia on the M_3, given the probable ages of formation of the two sets of defects, but it is more likely that the M_3 defect represents a slightly earlier stress event.

Growth arrest lines have been identified on a number of immature and mature Late Pleistocene human remains (Brennan, 1991; Trinkaus et al., 2002; Cowgill et al., 2007). However, the persistent difficulties of consistently identifying them on incomplete remains, their uncertain association with stress episodes, and the remodeling of them during life (Gindhart, 1964; Macchiarelli et al., 1994; Lewis and Roberts, 1997) makes any inference difficult, beyond stating that Tianyuan 1 went through a stress episode in adolescence.

Developmental Asymmetry

Several of the Tianyuan 1 postcranial remains exhibit varying degrees of bilateral asymmetry. The axis (C2) is asymmetrical in a number of aspects of its laminae, spinous process, craniocaudal facet positioning, and foramina transversaria (Chapter 8). The scapulae, although incomplete, have asymmetrical spinous processes (see Figure 9-1). The humeral diaphyses are among the most asymmetrical of the Late Pleistocene specimens, exceeded only by the (almost certainly) pathological Barma Grande 2 specimen in cortical area and polar moment of area comparisons (see Figure 9-8). The talocalcaneal articular facets are markedly asymmetrical, given the inferred facet configurations for the right talus and left calcaneus, and there may have been asymmetry in the degree of talar neck medial deviation (Chapter 10). Few other bilateral postcranial proportions are available, but the radial diaphyses appear largely symmetrical; the femoral diaphyses are symmetrical; and the distal tibiae appear to be similar. The partial mandible with its antemortem tooth loss (Chapter 6) cannot be assessed for asymmetry, although both the right mandibular molars and the left maxillary molar indicate normal maxillary overbite in the postcanine dentition.

The humeral diaphyseal asymmetry may be largely due to right-handedness of the individual, since similar (if less pronounced) humeral diaphyseal asymmetry is evident in a number of other Late Pleistocene humans (Trinkaus et al., 1994; Churchill and Formicola, 1997), and talocalcaneal facet asymmetry is present in Shanidar 3 and some recent humans (Trinkaus, 1978, 1983). Cervical vertebral asymmetry is present among recent humans (see Figure 8-3; Meseke et al., 2008). But the level of asymmetry throughout the preserved axial and appendicular skeleton suggests a more systemic effect. Correlations (but not etiologies) have been noted between various cervical, craniofacial and appendicular asymmetries (e.g., Korbmacher et al., 2007); the occlusal asymmetries sometimes correlated with cervical and shoulder asymmetries (Solow and Sandham, 2002; Korbmacher et al., 2007) are not present on Tianyuan 1, as indicated by its molar wear.

The degree of asymmetry of Tianyuan 1 is therefore probably not sufficient to warrant its designation as pathological. It is sufficient to be notable and to pose questions as to whether the different aspects of asymmetry might be behaviorally and/or developmentally correlated.

Dental and Alveolar Abnormalities

Dental Caries

There is no evidence of dental caries, occlusal, cervical, or radicular. If any caries occurred during the individual's lifetime, they were minor, and any evidence of them was removed through dental attrition.

Hypercementosis

There is moderate hypercementosis around the apical halves of the right M_3 roots, which is more pronounced on the lingual side of the distal root (see Figure 5-3). There is no evidence of hypercementosis on the exposed portions of the other preserved teeth.

The etiology of hypercementosis is unknown, but it may be due to a variety of localized or systemic causes, including dietary stress, supereruption of the teeth, localized trauma, traumatic occlusal stress, and malpositioning of the teeth, or more generally to conditions that affect the periodontal tissues (Gorlin and Goldman, 1970; Corruccini et al., 1987; Hillson, 1996). Given the anomalous occlusal position of the Tianyuan 1 M_3, some combination of these localized conditions probably promoted the hypercementosis.

Alveolar Process Degenerations

On the right side the alveolar bone mesial of the M_2 has only minor degenerations (see Figure 5-2). The I_1–I_2 to M_1–M_2 interdental septa show little or no porosity, and the lingual and buccal alveoli mesial of the M_1 appear normal. On the lingual side of the M_1 there are spicules of bone that rise up along the exposed root. Adjacent to the M_2, the buccal alveolar process is normal. However, lingually the M_2 alveolar bone recedes distally, and there is a space 2.5 mm mesiodistal between the distal M_2 root and the M_2–M_3 interdental septum, related to the elevated position of the M_3 (see Figures 6-1, 6-2, and 12-1). Despite its altered position, the M_3 alveolar margins show little abnormality (see Figures 5-2 and 12-3). The buccal margin is sharp, albeit with bone rising mesially to meet the elevated mesial root. The lingual side is similar, although it bulges slightly as part of the minor mandibular torus. The exposed socket shows little porosity.

Radiographically, the alveolar bone around the I_2 to P_4 roots and around the M_2 roots is normal and conforms closely to the margins of the roots. However, there is space around both roots of the M_1 and a space below the apex of the mesial root of the M_1. The bony resorption around the M_1 roots is associated with a buccal fenestration adjacent to the distal M_1 root (Figure 12-1). It is an irregular ovoid opening, with the long axis distosuperior to mesioinferior. Its height (perpendicular to the alveolar plane) is 5.0 mm, and its breadth is 4.9 mm, with the superior margin 3.7 mm below the alveolar margin. The mesial and inferior margins are sharp, but the distal one is rounded. There is no sign of porosity around the opening.

There is a modest mandibular torus lingually. It begins at the mesial M_1, is distinct under the M_2, and then presents a clear nubbin under the M_3. The last projects 2.6 mm, is mesiodistally 8.3 mm, and at the mesial M_3 is vertically 5.3 mm. It has long been discussed whether mandibular tori are an epigenetic trait or the result of biomechanical stress on the alveolar bone (lingually below and buccally above). It may be epigenetic in terms of individual predisposition, but mechanical stress may be necessary for its development (Yang, 1995; Al Quran and Al-Dwairi, 2006; Ihunwo and Phukubye, 2006; Ohzeki and Igarashi, 2007). Given the association of the Tianyuan 1 torus with extensive occlusal attrition, occlusal forces probably contributed to the development of the feature.

Supereruption of the Dentition

The extensive occlusal attrition and reductions in crown heights of the teeth (see Table 5-1) has resulted in considerable continuous eruption [supereruption or dentoalveolar extrusion (Costa, 1982; Whittaker et al., 1982; Newman, 1998; Rosen and Forman, 1999)] of the Tianyuan 1 teeth (see Figures 5-2 and 12-1). Since occlusal attrition and associated supereruption are normal life cycle processes, and not pathological ones, they are discussed in Chapter 5. It is principally the exceptional supereruption of the right M_3 that is of concern here.

The M_3 was elevated and mesially tilted relative to the M_2. The mesial M_3 occlusal margin is 4.2 mm above the distal M_2 occlusal margin, and the vertical midline axis of the buccal M_3 is ~60° relative to the alveolar plane (or tilted ~30° mesially). As such, the mesial margin of the M_3 is above the distal M_2. There is only a modest rounding of the distolingual M_2 occlusal margin. There is no apparent wear facet on the mesial M_3 cervix, where it contacts the distal M_2, but postmortem damage to the M_3 mesial root would obscure any such facet unless it were large. In addition, there are matching interproximal wear facets on the distal M_2 and the mesial M_3 (Chapter 5). The supereruption occurred sufficiently prior to death for the alveolar bone around the M_3 root to conform to the M_3 root in its supererupted position; it is unclear how long this would take. In addition, there appears to have been some supereruption of the M_2 relative to the M_1, given the greater cervix to alveolar crest (C/AC) heights of the M_2 (see Table 5-1).

Since supereruption is a normal process when occlusal forces from opposing teeth are insufficient to maintain the tooth at its original level, the supereruption of the M_2, and especially of the M_3, should have been secondary to right maxillary antemortem tooth loss. It is also possible that the minimal rounding of the distal occlusal margin of the M_2 was a result of little or no occlusal bite force on the M_3 subsequent to its supereruption.

Antemortem Tooth Loss

These degenerative changes in tooth position and the alveolar bone are associated with antemortem tooth loss (Figure 12-3). In particular, the right I_1 to the left P_3 were lost well before death. The teeth are absent, their sockets are completely closed, and the alveolar margin is pinched with minimal porosity along the supe-

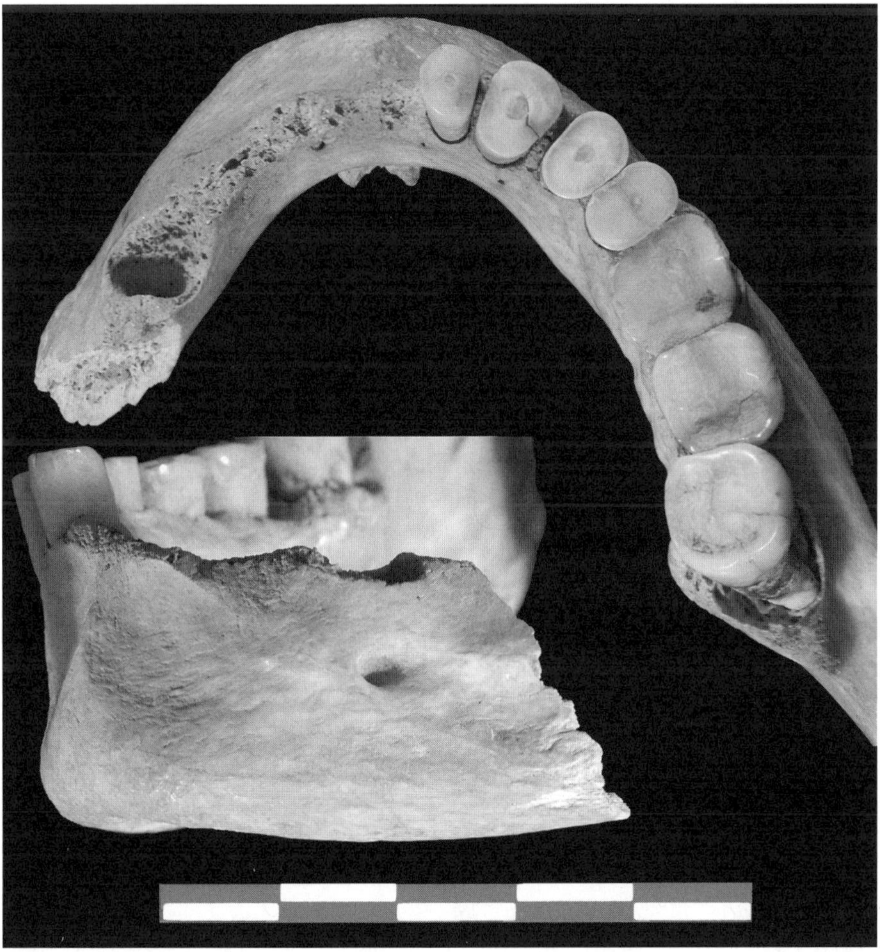

Figure 12-3. Occlusal (*top*) and anterolateral left (*bottom*) views of the Tianyuan 1 mandibular corpus and dentition. M_3 is in its socket in its supererupted position. Scale in centimeters.

rior edge. Radiographically, there is no evidence of the sockets, with the trabecular bone continuing up to the alveolar crest uninterrupted; the bone had fully remodeled subsequent to the tooth loss. There is a mesial interproximal facet on the right I_2, indicating that at least the right I_1 was originally present.

The left P_4 was in place at the time of death, since its socket remains (Figure 12-3). However, the socket is tilted mesiosuperior to distoinferior, with porosity along its mesial side. The mesial edge of the socket is above the mental foramen. Given that the two mental foramina are symmetrical with respect to the symphyseal midline (Chapter 6), and that the right mental foramen is below the P_4–M_1 interdental septum, it is likely that the P_4 drifted mesially as it tilted subsequent to the loss of the more mesial left mandibular teeth. It would not have been functional at the time of death.

The left M_1 is absent, and the region of the M_1 alveolus should have been in middle of the fossilization break on the left side. However, there is no sign of the mesial M_1 socket. Distal of the P_4 socket there is 4.4 mm of interdental septum, 4.5 mm of buccal alveolar bone, and 2.5 mm of lingual alveolar bone to the fossilization break. Either the M_1 was located much farther distally, which is unlikely, or the M_1 was lost antemortem with resorption of its root sockets.

In addition, the modest supereruption of the right M_2 and especially the marked supereruption of the right M_3 indicate that the opposing maxillary teeth must have been absent prior to death. This would have involved the right M^3, and it probably involved the right M^2 given the mesial tilt of the M_3.

Consequently, at least nine teeth were present at death (left M^{1-2}, left P^4, and right I^2 to M^3), even if some of them were not functionally occlusal. At least six teeth were lost antemortem (right I_1 to left P_3, left M_1, and right M^3) and probably a seventh (right M^2). Some of the antemortem tooth loss occurred well before death, in particular the right I_1 to left P_3 plus the left M_1. However, the loss of the right maxillary molar(s) may have occurred only briefly before death, sufficiently prior to death for the supereruption to have occurred but insufficiently for marked attritional rounding of the distal M_2 crown. It may well be that the right I^2 to M^1 were in occlusion, given the absence of supereruption of the I_2 to M_1.

Assuming a maximum of seven teeth lost antemortem and a maximum of fifteen teeth present at death (the nine indicated plus the right maxillary ones), this provides a rate of 32% (N = 22) antemortem tooth loss. If one limits the percent of tooth loss to teeth for which there is direct evidence, but including the right M^3 given the marked supereruption of the right M_3, one has 6 lost teeth and nine present, or a slightly higher rate of 40% (N = 15).

Given the absence of evidence for periodontal inflammation, caries, more than trivial dental enamel hypoplasia, and hypercementosis beyond that of the supererupted M_3, it is not clear what promoted the antemortem tooth loss of Tianyuan 1. It could be a case of facial trauma, but if so, it would be difficult to account for the lower left and upper right tooth loss without any evidence of trauma to the mandibular corpus. It is also unlikely to be the result of occlusal wear, since the preserved teeth, despite their advanced occlusal and interproximal wear, remained fully functional.

Osteoarthritis and Enthesopathies

Cervical Vertebral Osteoarthritis

Only one of the Tianyuan 1 vertebrae was recovered, a largely complete axis (C2) (Chapter 8). It presents three indications of degenerative processes (see Figure 8-1).

The first degeneration is an osteophyte on the ventral and right sides of the dens. The articular facet for the anterior arch of the atlas is normal, and the extra bone extends cranial of it. Ventrally, the additional bone extends up the cranial tip and beyond it, creating a dorsal hollow between it and the original tip of the dens. The osteophyte has a transverse dimension of 3.0 mm and projects cranially 2.1 mm. On the right side there is an extra lip of bone projecting dorsally next to the ventral one, 3.0 mm dorsoventral and 4.0 mm craniocaudal. The lip itself points craniodorsomedial. This additional bone is an osteophyte within the alar ligament, especially on the right side. Since the alar ligaments extend from the dens cranially and laterally to the medial sides of the occipital condyles, they stabilize the vertical rotational axis of the C2 with respect to the occipital bone. The osteophyte on the Tianyuan 1 dens indicates strain on these ligaments, especially on the right side.

The second C2 degeneration involves the left caudal articular facet. It has a beveled dorsal edge, up to 3.7 mm wide, at a ~5° angle to the rest of the otherwise normal facet. The beveled edge is polished with transverse mini ridges, resembling minor eburnation of the articular margin. The other facets are normal.

The third change involves a right side osteophyte on the dorsal body for the posterior longitudinal ligament. The bony processes are between the cranial facets, 4.4 mm mediolateral, ~5.0 mm craniocaudal, and projecting 2.2 mm into the spinal canal. The left side of the dorsal body is normal, as is the ventral body surface. The caudal intervertebral disk surface has small body projections, clustered on the sides of the surface. The disk surface, including the rings, is otherwise normal. The changes are a normal part of aging and are not necessarily pathological.

Mild degenerative changes on the C2/C3 articulations are common on the vertebral columns of older individuals, and the frequency of degeneration of C2/C3 facets is about average for the cervical vertebral column overall (Merbs, 1983; Bridges, 1994; Dawson and Trinkaus, 1997). Comparative data for alar ligament osteophytes are only available for an Inuit sample (Merbs, 1983), which show ~20% incidence of osteoarthritis on the dens portion of the atlantoaxial articulation. However, unless it was associated with unusual cephalic movements (vigorous and/or frequent rotation of the head) or trauma, these axial degenerations imply more widespread vertebral osteoarthritis (at least cervical) for Tianyuan 1.

Appendicular Osteoarthritis and Enthesopathies

The limb remains present several minor abnormalities. These occur in the context of largely normal articulations, in terms of both subchondral bone and surrounding ligamentous attachments. These normal articulations include the left scapular glenoid, the right distal humerus, the distal tibiae, the distal right fibula, and most of the articulations of the largely complete individual hand and foot bones. At the same time, most of the muscular or ligamentous attachments are free of superfluous ossifications.

There is trivial articular margin lipping on the caudal third of the ventral glenoid margin of the left scapula and extending around its caudal margin (see Figure 9-2), on the dorsal capitate (see Figure 9-16) and triquetral facets of the left hamate bone (see Figure 9-17), and on the dorsolateral cuboid facet of the right calcaneus (see Figure 10-16). None of the upper limb articular margin changes are pronounced, although the ligamentous ossification on the calcaneus is more marked. None of them involve the subchondral bone surface.

One articulation exhibits an osteoarthritic alteration (see Figure 9-21), the distal interphalangeal joint of the (probably second) manual ray. The distal trochlea of the middle phalanx is normal, but there are osteophytes around the subchondral bone. These changes are matched by those on the proximal end of the distal phalanx, which has osteophytes around the articulation but normal subchondral bone.

There is an enthesopathy for the pronator teres insertion on the lateral diaphysis of the left radius. It is a raised area, 24.8 mm proximodistal and 5.0 mm in breadth, It blends gradually with the adjacent subperiosteal bone posteriorly and distally, but it is separated from the adjacent bone by a groove anteriorly and proximally. There is a rugosity of the surface along the anterior margin. As such, it corresponds to category 3 of Galtés et al. (2008) for age and activity related alteration of the pronator teres insertion on the radius.

The digital fossa of the proximal right femur has three small spines in its superior portion, ossifications of portions of the obturator muscle tendons.

Femoral and Tibial Diaphyseal Alterations

The most pronounced pathological alterations of Tianyuan 1 affected the mid and distal femoral diaphyses and, to a lesser extent, the posterior proximal tibial diaphyses. Most of the changes involve the attachments for muscles along the linea aspera and the soleal line, but there are also alterations in the mid and distal femoral diaphyseal cortical bone distribution.

Femoral Diaphyseal Alterations

On the more complete right femur (see Figures 10-3, 10-5 and 10-7), the diaphyseal cross-sectional shape and muscle markings appear normal from the trochanters to the midproximal shaft, approximately to the level at which the proximal femoral muscle insertions converge into a discrete linea aspera (~65% of biomechanical length, measured from the distal end). The bone has an even anterior curvature with the maximum anterior extent distal of midshaft, a normal arrangement.

As described in Chapter 10, near midshaft there is a pronounced linea aspera, 6.0 mm wide, on a relatively normal pilaster. There is a lateral longitudinal sulcus along the pilaster, as is commonly present. Medially, there is no sulcus; the bone curves convexly anteriorly from the linea aspera around the posteromedial shaft. The surface then becomes flattened on the midmedial side, to then round onto the anterior surface. The medial flattening of the diaphysis gives the impression of a narrow midshaft. However, the mediolateral second moment of area of the midshaft is moderately large for a Late Pleistocene human femur, but its anteroposterior second moment of area is unusually so (see Figure 10-8). The one specimen that approaches it is the ZKD-UC AN67 femur.

Going distally, the anterior surface remains normally transversely convex, if narrow, and the anteromedial side slopes off posteriorly. Medially, the bone remains flat, and the flattened surface gradually rotates posteromedially. The pilaster becomes markedly pronounced distal of midshaft, beginning 240–250 mm distal of the proximal neck and extending ~95 mm along the posterolateral margin of the diaphysis. Given a biomechanical length (proximal neck to average of the distal condyles) estimate of 435.6 mm, it therefore begins ~55% of the distance distally from the proximal neck and ends ~80% of the distance distally from the neck. The crest (Figure 12-4) is rounded on its posterior edge, ~9 mm wide proximally and ~7 mm wide in its middle and near its distal end. Proximodistally, it has an even posterior convexity. The crest has a largely smooth surface, but it appears to lack the longitudinal fiber orientation of diaphyseal cortical bone, resembling partially remodeled woven bone.

The abnormal crest splits the linea aspera into two at its proximal end, and then the medial side bifurcates again (Figure 12-4). The lateral portion of the linea aspera runs along the edge of the crest, where it meets the more normal bone of the distal posterolateral diaphysis. One branch of the medial linea aspera runs along the medial side of the crest, close to its posterior margin. The other, and more evident, portion of the medial linea aspera curves convexly anteriorly as it

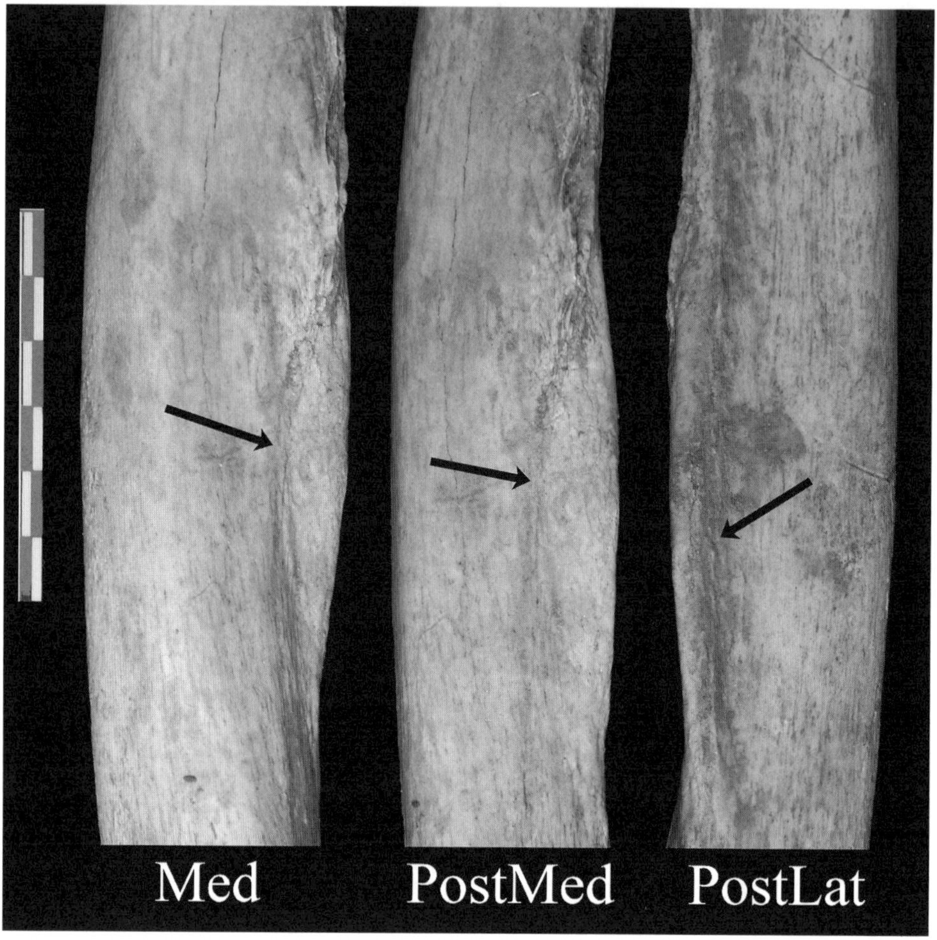

Figure 12-4. *Left to right:* Medial (Med), posteromedial (PostMed) and posterolateral (PostLat) details of the mid-distal right femoral diaphyseal abnormal crest and muscle attachment lines. Arrows point to the anterior edges of the muscle lines that are divided around the posterior bone crest. Scale in centimeters.

goes distally. The result is an almond-shaped area ~42 mm long and up to 10 mm wide, between the more posterior medial linea aspera and the flattened but otherwise normal medial cortical bone surface.

The left femoral diaphysis is less complete, but with minor variations it closely mirrors the pattern seen in the right femur. Near midshaft, the linea aspera is 7.5–8.0 mm wide, the anterior and lateral surfaces are rounded, but the medial surface exhibits the same flattening. The distal pilaster rises into a similar crest of new bone, rounded posteriorly and fading proximally into the more normal bone at the same proximodistal level. Laterally, the linea aspera is proximally on the crest and then deviated to be ~8 mm from the posterior margin of the crest. Medially, the medial portion of the distal linea aspera retains a line just to the medial side of the crest, but its major portion goes along the medial base of the crest, along the anterior margin of what is the almond-shaped surface on the better preserved right femur. The crest is ~9 mm wide proximally (close to midshaft), ~6 mm wide in its middle, and then ~5.5 mm wide distally. In its middle, it projects ~11 mm.

The mid-distal diaphyseal crest produces a pronounced anteroposterior component to the diaphysis (see Figures 10-5 and 10-7). The anteroposterior second moment of area (I_x) at 35% of its biomechanical length, approximately at maximum crest development, exceeds those of all other Late Pleistocene femora (Figure 12-5). The Mladeč 27 femur is anteroposteriorly elongated relative to other Late Pleistocene femora, but it is far less so than the Tianyuan 1 femur.

Radiographically, in lateral view, the right femoral endosteal contour along the medullary cavity appears normal, and the irregularity appears mostly as additional, more radiotranslucent bone laid down over the otherwise reasonably intact cortical bone (Figure 12-6). The less complete and more eroded left femoral dis-

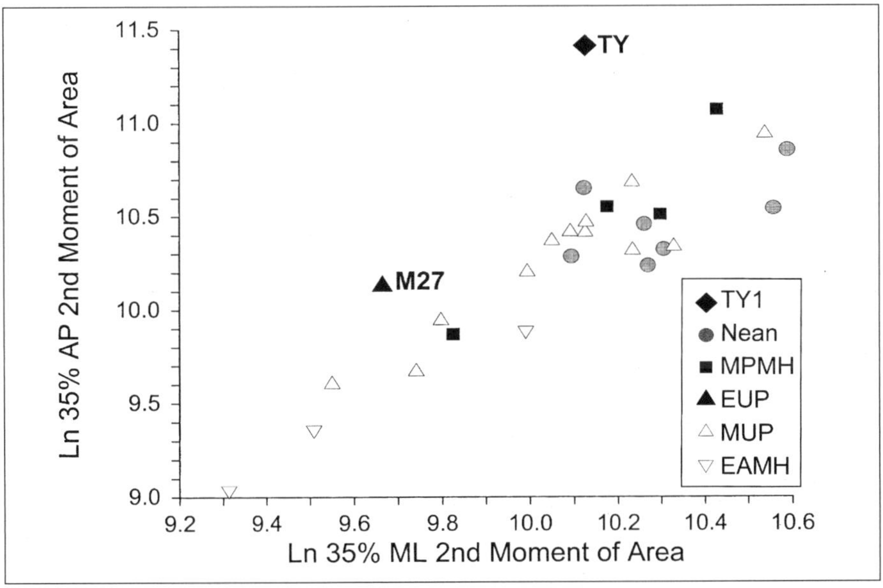

Figure 12-5. Bivariate plot of the anteroposterior (I_x) versus mediolateral (I_y) $\ln_e$ second moments of area for the femoral mid-distal shaft (35%). *TY:* Tianyuan 1; *Nean:* Neandertals; *MPMH:* Middle Paleolithic modern humans; *EUP:* western Early Upper Paleolithic; *MUP:* western Eurasian Mid Upper Paleolithic; *EAMH:* eastern Asian modern humans; *M27:* Mladeč 27.

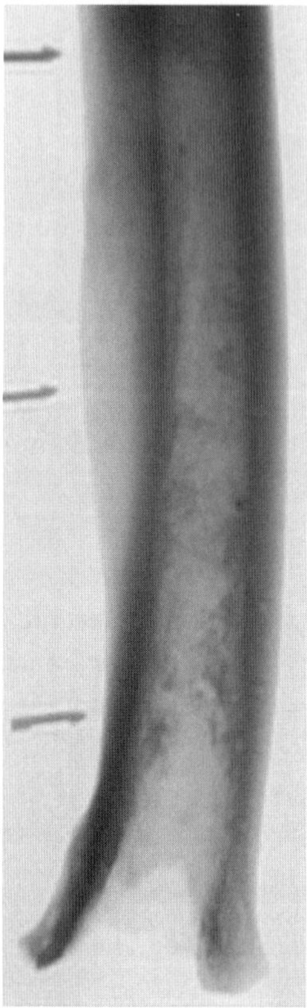

Figure 12-6. Radiograph of the Tianyuan 1 right distal femoral diaphysis in lateromedial view. Lead arrows along the posterior diaphysis indicate, from proximal to distal, the locations of the 50%, 35%, and 20% cross sections.

tal diaphysis appears similar radiographically. There are no additional trabeculae within the medullary cavities.

As a result, the symmetrical abnormalities of the posterior distal femoral diaphyses appear principally as a growth of new bone, extending posteriorly along what would otherwise be the gradual distal fading out of the pilaster. This resulted in the dividing of the muscle insertion lines that make up the distal linea aspera, including those of vastus medialis, vastus lateralis, the short head of biceps femoris, and adductor longus. It is likely that the insertion lines for the two vastus muscles ended up medial and lateral respectively of the new crest of bone, but it is unclear how the attachments of the other muscles might relate to the rearranged muscle insertion lines. The resultant changes in mechanical strain patterns may well have induced biomechanical remodeling of the femoral midshafts, including the flattening of the medial diaphyseal surfaces.

Tibial Diaphyseal Alterations

The distal epiphyses and the distal half of the right tibial diaphysis appear normal in their external configurations, articular surfaces, and the distribution of cortical bone (see Figures 10-11, 10-12 and 10-15). The same applies to the more anterior portions of the right proximal diaphysis, little which is preserved. However the soleal lines on both tibiae, but especially on the right one, are irregular (Figure 12-7).

The right soleal line is weakly marked proximoposterolaterally, and it continues as a faint line medially. Below the line along the posterolateral angle of the diaphysis, there is a raised area, 50–55 mm long and ~12 mm wide. Extending distally from the swelling are three prominent and sharp crests, with the smallest one lateral. The crests are ~18 mm long. Medial of the crests, there is a midposterior vertical sulcus, extending distally from the soleal line ~53 mm. The floor of the sulcus appears as though it is the original subperiosteal bone surface, and the sulcus has been created by swellings of new

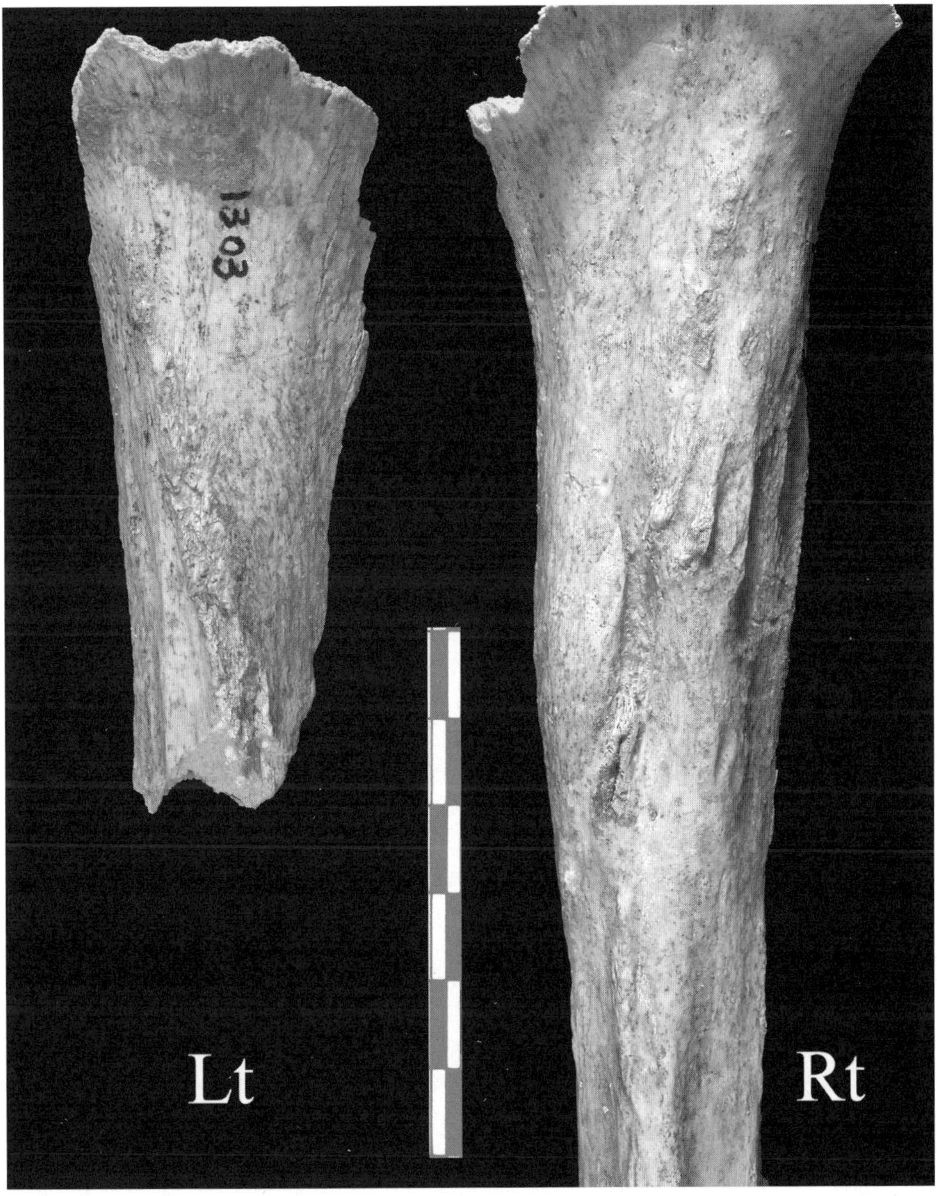

Figure 12-7. *Left to right:* Posterior views of the Tianyuan 1 left and right proximal tibial diaphyses showing the irregular muscle attachments around the soleal lines. Scale in centimeters.

bone, laterally and medially of it. There is then a posteromedial swelling of new bone laid down with clear borders, but most of the surface has become smoothed over. The whole of the posteromedial swelling is almond shaped, proximodistally ~67 mm and mediolaterally 20.4 mm as a chord and 26.0 mm as an arc. There is a midposterior lip of bone into the vertical sulcus, and a small crest of bone close to where the distomedial end of the soleal line should be. In the distal end of the sulcus there is a double spine, 17.1 mm long and 5.8 mm wide at its distal end. Then distal of the sulcus and below the swellings, there is a morphologically normal but pronounced flexor crest, between the origins for flexor digitorum longus and tibialis posterior, leading down to a marked tibial pilaster.

The abnormalities of the right proximal tibial diaphysis consist of two longitudinal swellings of bone, one along the medial soleal line and one distal of the lateral soleal line. The medial swelling is mostly smooth, but the lateral one has three sharp, pronounced crests.

The left proximal tibia consists of a small section of bone. The soleal line is normal in terms of position and the absence of adjacent swellings. However, the line is rugose, broad, and raised, and the middle portion of it has lipping directed distolaterally. The soleal line is 7.3 mm wide at its proximal end and 6.0 mm wide at its distal end.

Etiological Interpretation

We have been unable to determine satisfactorily the etiology of these abnormalities of the Tianyuan 1 femoral and tibial diaphyses. (1) They are not associated with osteoarthritis, to the extent documented by the right femoral condyle and the distal tibial articulations or the associated pedal remains. There are only minimal proximal femoral and pedal connective tissue ossifications. (2) They are not associated with indications, articular or diaphyseal, of any loss of locomotor function. Indeed, the assessments of midshaft robusticity of the right femur and tibia (Chapter 10) place them in the middles of the Late Pleistocene distributions for lower limb diaphyseal robusticity, indicating the maintenance of normal Late Pleistocene levels of locomotion, burden carrying, and other foraging activities for this individual. (3) Radiographically, the cortical diaphyseal bone appears normal, and the medullary cavities are unaltered. (4) There is no sign of diffuse subperiosteal porous new bone. (5) They are largely bilaterally symmetrical (see Figures 10-1 and 10-7) , despite asymmetries elsewhere in the skeleton. (6) The degree and pattern of anterior femoral curvature is normal, especially for a Late Pleistocene human. The lateral curvature of the radius is normal and modest, and neither the humeri nor the tibia exhibit curvature.

There are numerous pathological conditions, many involving infectious diseases or dietary deficiencies, that can affect these locations of the femoral and tibial diaphyses (Ortner and Putschar, 1981; Aufderheide and Rodríguez-Martin, 1998; see summary Table 1 in Fennell and Trinkaus, 1997). However, almost all of the ones usually identified skeletally involve lytic lesions and/or diffuse porous new bone deposition. Neither of these conditions applies to the Tianyuan 1 abnormalities. The only other conditions that we have located, which might be related to it, are paradiaphyseal calcific tendinitis and tumoral calcinosis. But paradiaphyseal calcific tendinitis (Fritz et al., 1994; Flemming et al., 2003) and tumoral calcinosis (Seeger et al., 1990) involve the laying down of irregular porous bone along the diaphyseal muscular attachments and are associated with cortical bone resorption. These conditions are not evident on the Tianyuan 1 diaphyses.

Consultation with paleopathological colleagues has provided cautious suggestions, including immature diaphyseal fractures with remodeling, an infantile postural behavior or a neonatal localized functional anomaly producing unusual remodeling, childhood rickets with extensive remodeling of the deformities, and new bone formation from overuse in a mature individual. In several of these scenarios, the femoral diaphyseal changes are seen as the primary abnormality and the tibial muscular attachment alterations would be secondary effects from compensatory overuse of the talocrural plantarflexors. An alternative explanation for the tibial abnormalities would be impingement on the muscles and their tibial origins from the distal femoral crest during hyperflexion of the knee in squatting, a habitual behavior indicated by the distal tibiae and talus (Chapter 10).

With respect to localized traumatic injury, there are no traces of earlier callus formation subperiosteally or endosteally on the femora or the tibiae, nor any increase in the radiodensity or cortical thickness of the bone in the regions of the irregularities. It is possible that the individual sustained an injury to the legs sufficiently early in childhood, such that most of the evidence of a fracture or muscle attachment tears would have been resorbed and/or remodeled through growth and the couple of decades of adult life of Tianyuan 1. In such a scenario, the injuries would be principally on the mid-distal femoral diaphyses, and the tibial changes would be secondary. It would nonetheless be remarkable to have such injuries occur bilaterally in the same places proximodistally on the femora, producing virtually identical reactions on both femora, and similarly associated changes on both tibiae.

Alternatively, an unusual posture or a functional anomaly in the legs in infancy could have produced changes in the curvature of the mid-distal femora, promoting compensatory posterior diaphyseal bone deposition as a mechanical buttress to changed anteroposterior strain patterns through the diaphysis. As with fractures, the tibial changes would be seen here as secondary. Such an etiology is more likely to be bilaterally symmetrical, if localized. However, the anterior curvature of the femoral diaphyses is not unusual for a Late Pleistocene human, and none of the other Late Pleistocene human femora (late archaic or early modern) with similar levels of anterior curvature has even a suggestion of the mid-distal posterior crest of bone such as is seen on both Tianyuan 1 femora.

An additional possibility is a childhood vitamin D deficiency resulting in rickets, with subsequent bony deposition to mechanically strengthen the bowed diaphyses and then secondarily a reduction in the original curvatures through growth remodeling. Certainly, at 39° N latitude, Tianyuandong is sufficiently north to reduce the potential for previtamin D_3 synthesis from UVB radiation (Jablonski and Chaplin, 2000), although the isotopic evidence for fish consumption (see Appendix 2) could well have compensated for any reduction in dermally synthesized vitamin D. Again, the changes in the proximal tibiae would reflect compensatory muscle activity for the resultant abnormal loading patterns during locomotion and the mid-distal femoral crests would be biomechanical reinforcements against the altered strain patterns in the diaphyses. If this were the case, one would have to argue that the rachitic bowing only affected the mid-distal femoral diaphyses since the arm bones and the lower leg bones exhibit no abnormal curvatures and there are no irregularities of the preserved metaphyseal regions. This is contrary to the usual pattern of childhood rachitic bowing, in which most of the long bones are affected, especially the weight-bearing ones (Ortner and Mays, 1998; Mays et al., 2006). One would also have to argue for sufficient remodeling of the diaphyses subsequent to the rickets to produce the normal level and pattern of curvature evident in the femora.

The last possibility is based on the observation that the mid-distal posterior crests on the femoral diaphyses appear superficially and radiographically to consist of nonlamellar bone, laid down over otherwise largely intact diaphyses with their original cortical bone distributions minimally altered other than along and immediately adjacent to the posterior flanges of bone. In that case, the changes would have occurred relatively soon before death, For that scenario to work, one would have to postulate a posture or behavior that would place abnormally high bending moments on the distal femoral diaphyses and overwork the talocrural plantarflexors. Squatting, such as is documented by the Tianyuan 1 talocrural articulations (Chapter 10), could be such an activity. However, any abnormal activity of Tianyuan 1 is not likely be an activity that overloaded the patellofemoral articulation, since the preserved lateral condyle and patellar surface on the right femur exhibit no osteoarthritic degeneration, such as routinely occurs with habitual overloading of a flexed knee (Zhang et al., 2004; Tangtrakulwanich et al., 2007).

Therefore, even though each one of these potential etiologies appears to be at odds with some aspects of the Tianyuan 1 skeletal remains, it is not possible to completely discount them. In order to have a firm diagnosis of these Tianyuan 1 lesions, one would have to demonstrate that one condition fits the evidence significantly better than the others. We are not aware that this is currently possible. Nor do we want to simply describe this set of abnormalities as idiopathic.

Although the differential diagnosis of the Tianyuan 1 lower limb abnormalities is a paleopathological puzzle, the remains clearly indicate that this individual sustained these lesions and was able not only to survive with them present for an extended period of time but also maintain a normal level of activity as a Late Pleistocene human forager. Should we be able to determine precisely the etiology of these femoral and tibial lesions, it might be possible to further assess the risk factors, survivability, and any secondary consequences that would not otherwise be evident on the skeletal remains.

Summary

Tianyuan 1 presents a suite of abnormalities. Some of these reflect the accumulation of minor lesions through life. They include cervical vertebral and manual interphalangeal osteoarthritis, minor tendinous and ligamentous ossifications, antemortem tooth loss with associated supereruption, alveolar changes and hypercementosis, and trivial dental enamel hypoplasia. The primary changes are the bilaterally symmetrical cresting and muscular attachment alterations of the distal femoral and proximal tibial diaphyses, whose etiology we have not been able to determine.

The first set of lesions are relatively common among both Pleistocene *Homo* and recent foraging populations, and they represent moderate risk of overuse lesions and perhaps minor trauma. The femoral and tibial lesions are apparently more systemic, and as such they join a growing set of nontrivial developmental or degenerative lesions documented for Pleistocene *Homo*. Moreover, they join a growing set of abnormalities of Pleistocene human remains whose ultimate etiologies remain uncertain (e.g., Trinkaus, 1983a; Buzhilova, 2000; Formicola and Buzhilova, 2004; Dočkalová and Vančata, 2005; Trinkaus, Ruff, and Conroy, 1999; Trinkaus, Hillson, et al., 2006; Mafart et al., 2007). More important, without a secure diagnosis of the Tianyuan 1 lower limb abnormalities, it is not possible to assess the degree to which they may have affected function, and hence the behavioral consequences of their presence.

Chapter 13

Paleoanthropological Implications of the Tianyuan 1 Human Remains

GIVEN ITS geological age, geographical position, and preservation of multiple elements, the Tianyuan 1 partial skeleton has the potential to provide insight into several aspects of Late Pleistocene human paleobiology and population processes. As noted in the introduction, one specimen, no matter how complete and no matter how diagnostic, cannot by itself resolve ongoing issues concerning human evolution and especially modern human emergence. Individuals do not evolve, populations do. But on a more global basis one individual can help reinforce a pattern seen elsewhere, if by no other means than increasing the sample size for an already-documented phenomenon. Moreover, a single individual can help refute a commonly held interpretation if it provides secure evidence that is counter to the predictions of that point of view. These considerations apply to both populational and paleobiological implications of the Tianyuan 1 human remains.

It is in this context that these remarks about the broader implications of the Tianyuan 1 remains are provided. Any of these inferences can be modified as more remains are discovered, securely dated, and appropriately analyzed. Nonetheless, given the extremely limited state of the eastern Eurasian, securely dated Late Pleistocene (especially from marine isotope stage [MIS] 3) human fossil record, the paleontological evidence from Tianyuan 1 provides a major addition and a point of future departure for our understanding of this time period in human emergence to the east of the 50th parallel.

Populational Considerations

Morphological Affinities of Tianyuan 1

From the beginning of its presentation here and in previous publications (Tong, Shang, et al., 2004; Trinkaus, 2005a; Shang et al., 2007; Trinkaus and Shang, 2008), Tianyuan 1 has been referred to as an early modern human. This designation is based on the presence on its remains of a series of derived modern human morphological characteristics (see Trinkaus, 2006a), both with reference to the available Late Pleistocene late archaic humans (albeit dominated by the Neandertals) and with respect to earlier (Middle and Early Pleistocene) archaic *Homo* remains.

Modern Human Aspects of Tianyuan 1

The most obvious of these derived characteristics are found within the mandible (Chapter 6). In particular, the mandible presents a distinct mental eminence, a protruding tuber symphyseos with adjacent lateral concavities (incisura mandibulae) relative to the contour of the anterior mandibular symphysis. Its chin area lacks prominent lateral tubercles, but these tubercles are generally absent from both Middle Paleolithic and earlier Upper Paleolithic modern humans. The anterior symphyseal configuration is joined by an absence of a planum alveolare and of a transverse symphyseal torus, and by a marked reduction in the lateral corpus thickness. In addition, the constellation of preserved ramal discrete traits, including a low coronoid process with a symmetrical mandibular notch, an open mandibular foramen, and the absence of an enlarged superior medial pterygoid tubercle, align it with early modern humans despite some overlap between the ranges of variation of late archaic and early modern humans in these features.

The upper limbs (Chapter 9) are notable principally for their modest musculoligamentous markings and moment arms, especially on the arm bones but also to some extent on the hand remains. This includes in particular the modest pectoralis major and deltoid tuberosities on the humerus, the anteromedial orientation of the radial tuberosity (even though that configuration is found among the Neandertals), the minimal radial lateral curvature, and the modest hamulus projection. These aspects of the upper limb are joined by the probable presence of a ventral sulcus pattern on the scapula, the dominant pattern among recent humans even though it is absent from Late Pleistocene Neandertals and western Eurasian early modern humans; it is also apparently the ancestral *Homo* pattern. The left scapula has a very wide glenoid fossa, a generally derived modern human feature in the right glenohumeral articulation, even though it is less distinctive in the left shoulder.

In the lower limb (Chapter 10), the femora, despite their pathological alterations, are distinctly pilastric, and this pattern is joined by the presence of clear longitudinal sulci on the tibia and fibula. Both of these features are distinctive of modern humans relative to archaic *Homo*. Farther distally, the talus has a relatively long neck.

These features combine to reinforce the taxonomic "diagnosis" of Tianyuan 1 as an early modern human, given its possession of several distinctive recent human morphological features. Of these aspects, only the anterior symphyseal morphology and the femoral pilastric diaphysis are truly uniquely derived among recent humans, but the remainder of these morphological configurations are dominant among recent humans and occur rarely or in isolation among both earlier and later archaic *Homo.*

Archaic Human Aspects of Tianyuan 1

Several aspects of the Tianyuan 1 skeleton make it unusual as a modern human and either align it with archaic *Homo* or place it in an intermediate position between archaic *Homo* and early modern humans. This is at least with respect to archaic *Homo* as known from the western Old World given the dearth especially of postcranial elements (other than the Middle Pleistocene ones known from Zhoukoudian Locality 1 and Jinniushan) from eastern Eurasia. One can only assume, until it is established otherwise, that these purportedly archaic skeletal features were generally present in archaic *Homo;* this assumption is supported by the archaic natures of the postcrania (and craniofacial remains) from the Middle Pleistocene Chinese Jinniushan and Zhoukoudian Locality 1 remains, as well as of earlier archaic *Homo* remains elsewhere in the Old World (Trinkaus, 2006a).

There are several upper limb features that are rather archaic. The humerus has a narrow medial pillar; the metacarpal 2 facet on the capitate is relatively sagittally oriented; there is a pronounced proximodistal length of the hamulus; the middle hand phalanx has a wide base; and the distal phalangeal tuberosity lacks ungual spines and is relatively wide. The hamulus hypertrophy is especially marked, given that it occurs in the otherwise more gracile left upper limb of Tianyuan 1.

There is also an unusual aspect of the Tianyuan 1 skeleton that distances it especially from the Middle Paleolithic modern humans. Even though Tianyuan 1 has a moderately high crural index, entirely separate from those of the Neandertals, it has exceptionally large lower limb (weight-bearing) articulations for its stature. Its estimated femoral head diameter to femur length proportions are outside the ranges of variation of any other sample of early modern humans and cluster with those of large, male, European Neandertals (Chapter 7). It also resembles the inferred proportions of the large, female, late Middle Pleistocene Jinniushan skeleton. Since the femoral head reflects body mass and femoral length reflects stature, this implies an individual with a markedly stocky body, one normally associated with colder climates and hence higher latitudes. It appears to be an archaic *Homo* characteristic, one that was lost among the Middle Paleolithic modern humans and their western Old World apparent descendants.

Intermediate Aspects of Tianyuan 1

These modern and archaic features are joined by a suite of features that place Tianyuan 1 morphologically between the known ranges of variation or in the proportional zones in which there is modest overlap of the late archaic and early modern human ranges of variation. In the dentition, although the overall dental metric proportions tend to align Tianyuan 1 more with early modern humans (especially the Middle Paleolithic modern humans), they are intermediate between the Neandertals and modern humans in relative anterior to posterior dental proportions. In the mandible, the position of the mental foramen is in the overlap zone of the reference samples, and originally (when the individual was a young adult) the mandible probably exhibited a small retromolar space, a feature occasionally found among early modern humans and commonly present among the Neandertals. In the hand skeleton, there is variation in the relative breadths of the phalangeal bases, with the left second proximal one aligning with modern humans but the right fifth proximal one being intermediate between the samples. In the lower limb, the relative malleolar surface projections of the talus are intermediate between these reference samples.

Summary of Morphological Affinities

It is best to consider the Tianyuan 1 skeletal and dental remains as indicating primarily a derived modern human morphological pattern. Yet, it possesses several features that would be common among late archaic humans and exceptional among early modern humans, as well as several aspects that place it in an intermediate morphological position between these reference samples. Although some of the skeletal features listed have some degree of developmental plasticity, especially with respect to appendicular robusticity, many of these features appear early in development and are reasonably interpreted as more closely reflecting the individual's genotype than its behavioral patterns. Moreover, even though a few of the upper limbs features (e.g., hamulus size, phalangeal base breadth, and apical tuberosity breadth) likely reflect musculoligamentous and/or articular hypertrophy, they are associated in the Tianyuan 1 upper limb with rather modest muscle markings. As a consequence, the Tianyuan 1 skeletal remains are best seen as modern without being entirely modern. They have a sufficient number of derived modern human features to be associated with the modern human morphological pattern, but they retain a greater than expected (were it a purely modern human) number of archaic and intermediate features. In many of these more archaic aspects of Tianyuan 1, it is particularly with respect to the earlier Middle Paleolithic modern humans that it contrasts.

Eastern Asian Populational Implications

As noted in the introduction, there is a long history of paleoanthropologists trying to decipher the population history of eastern Asia from the Middle and Late Pleistocene skeletal remains of the region, beginning with Weidenreich's (1939a,

1943, 1947) assessment of the Zhoukoudian–Upper Cave (and Locality 1) remains and continuing through a series of authors during the last several decades. This has been particularly so with respect to the appearance of modern human morphology in the region, but it has also included issues of Middle Pleistocene continuity and especially continuity between early modern and recent eastern Asian populations.

Previous Assessments

Many of the authors considering the appearance of modern human morphology in the region (e.g., Wolpoff et al., 1984; Habgood, 1992; Pope, 1992; Wu, 1992, 2004, 2006; Trinkaus, 2005a; Cartmill and Smith, 2009) have argued that the fossil data support and/or are compatible with a model in which there was considerable population continuity between eastern Asian late archaic humans and early modern humans in the region. At the same time, multivariate analyses of the ZKD-UC crania have been used to argue for a lack of continuity with the establishment of modern humans in eastern Asia (e.g., Bräuer, 1992; Kamminga, 1992; Bräuer and Mímisson, 2004), despite comments (e.g., Bräuer, 1992; Kamminga, 1992; Brown, 1993) that the samples of late archaic humans and, prior to the discovery of Tianyuan 1, early modern humans in eastern Asia are too scanty and poorly dated to make secure conclusions about the population processes in that region with the emergence of modern humans. Yet, others (e.g., Wolpoff et al., 1984; Pope, 1992; Wu, 2004) have noted the persistence of eastern Asian facial features across this human evolutionary transition, ones that are hard to explain without some degree of population continuity. This is reinforced by the ubiquitous presence of shovel-shaped incisors in the region Shovel-shaped incisors are an ancestral *Homo* feature, but they were replaced by the derived chisel-shaped incisor pattern among the earlier Middle Paleolithic modern humans (McCown and Keith, 1939; Vandermeersch, 1981; Tillier, 1999), thereby making chisel-shaped incisors the ancestral pattern for early modern humans (Trinkaus, 2005a, 2007a). The presence of shovel-shaped incisors in early and recent east Asian modern humans can therefore not be seen merely as a retained ancestral trait, but must be seen either as derived relative to the MPMH modern humans or, more likely, acquired through gene flow with regional archaic humans.

Earlier considerations of population continuity in eastern Asia, dating back to Weidenreich (1947), saw it in terms of regional continuity in the context of relatively normal levels of gene flow across Eurasia. With the gradual recognition that the earliest anatomically modern humans (albeit with variability and some archaic facial features) appeared in eastern Africa and temporarily in extreme southwest Asia during the late Middle and initial Late Pleistocene (MIS 6 and 5) (Vandermeersch, 1981; Day and Stringer, 1982; Trinkaus, 1984; White et al., 2003; McDougall et al., 2005; see Trinkaus. 2005a and references therein), the predominant model shifted from one of "regional continuity" to one of "assimilation" (or "dual ancestry"). In other words, the issue is no longer one of documenting whether there was strict population continuity in eastern Asia between late archaic humans and early modern humans. The issue has become one of assessing whether there was significant regional archaic human input into the subsequent populations of early modern humans as the latter dispersed across Eurasia in the middle of MIS 3 (e.g., Wu, 2004; Trinkaus, 2005a; see also Wolpoff et al., 2001).

Population Implications of Tianyuan 1

The Tianyuan 1 skeletal remains serve several roles. First, they document that by ~40,000 cal BP in mainland eastern Asia, as with peninsular southeastern Asia (Barker et al., 2007), there were populations of people who exhibited a long suite of the derived modern human characteristics previously present in eastern Africa and southwest Asia. Second, they strongly support the interpretation that there was substantial gene flow or population dispersal of recognizable modern humans eastward across Eurasia; one would have to argue otherwise for the parallel derivation of a suite of derived morphological characteristics. Third, the persistence of archaic features in the Tianyuan 1 skeleton, archaic features already lost in the presumably ancestral Middle Paleolithic modern humans, makes it highly unlikely that this eastward dispersal of modern human biology occurred without populational input from regional late archaic humans into those early modern human populations.

Further Considerations

It is important to reiterate the framework for assessing the populational implications of the Tianyuan 1 skeletal and dental morphology. As previously stated (Trinkaus, 2005a, 2007a), the ancestral modern human sample consists principally of eastern African and southwest Asian Middle Paleolithic–Middle Stone Age modern human remains, primarily from the sites of Herto, Omo-Kibish, Qafzeh, and Skhul. One could expand this sample to include MIS 5 and 4 African remains from northwest Africa (Aterian remains) and south Africa (MSA remains), which would shift the range of variation of that sample modestly (Chapter 4). Since Tianyuan 1 lived, within dating accuracy, at the same time as all of the other earliest MIS 3 Eurasian modern humans (~40,000 BP), it is only with respect to the substantially earlier (MIS 6–4) modern humans, as well as earlier archaic humans generally, that the polarities of its morphological patterns can be assessed. It could not have inherited traits from younger populations, and the approximately contemporaneous Early Upper Paleolithic remains (especially from Nazlet Khater and Oase, given shared preservation of skeletal elements) are unlikely to be representative of that ancestral group.

It is admitted that the dearth of late archaic humans in eastern Eurasia, and particularly the absence of a number of relevant aspects of the Tianyuan 1 skeleton, weakens any populational inferences made from Tianyuan 1. However, if one admits that the Middle Paleolithic modern human sample (however broadly or narrowly defined) is the relevant comparative sample, one still has to explain the presence of archaic features in the otherwise modern Tianyuan human remains.

One could dispute, on a feature-by-feature analysis of the Tianyuan 1 skeletal and dental remains, that these archaic features are plastic and hence with limited populational valence, probabilistically within the expected range of variation of the ancestral Middle Paleolithic modern humans, and/or within the range of variation of those ancestral modern humans but that limited samples and sample sizes mean that we do not have them recorded in the fossil record. However, in order for these lines of argumentation to hold, one would have to plead a special case for each of the above-mentioned features. Since all inferences of populational processes in the past are probabilistic statements, whether based on fossil remains or extrapolated back from the biology of recent humans, it becomes a question of which scenario is more probable. Is this mosaic of modern and archaic features in the Tianyuan 1 remains best explained as a consequence of simple and total replacement of regional eastern Asian archaic humans by in-dispersing modern humans? Or is this mosaic more easily, with a higher probability, explained as the result of some degree of admixture between these potentially ancestral populations? Given the likely independence of features (e.g., humeral pillar thickness and distal phalangeal tuberosity shape are unlikely to be pleiotropic), the assimilation scenario appears to be the more probable one.

Summary of Eastern Asian Populational Implications

Given the currently available fossil sample sizes, and especially a sample of one for the earliest eastern Asian modern humans, we cannot know the amount of admixture/assimilation that might have occurred. However, assuming that the preservation and discovery of Tianyuan 1 was random with respect to ancestry, the chances of finding the observed morphological mosaic, had admixture been truly trivial, would be vanishingly small.

The Broader Context

As noted above (especially Chapter 4), Tianyuan 1 is roughly contemporaneous with a small but growing sample of the earliest MIS 3 modern humans across the Old World. These include the Niah Cave partial skeleton from Sarawak (Brothwell, 1960; Barker et al., 2007); the Hofmeyr cranium from South Africa (Grine et al., 2007); the Nazlet Khater 2 skeleton from Egypt (Crevecoeur, 2008); the Peştera cu Oase, Cioclovina, and Muierii remains from Romania (Trinkaus et al.,

2003; Soficaru et al., 2006, 2007; Rougier et al., 2007); the Mladeč remains from the Czech Republic (Teschler-Nicola, 2006); and a series of immature mandibles, skeletal fragments, and isolated teeth from sites in France (e.g., Vallois, 1958; Henry-Gambier et al., 2004; Henry-Gambier and Sacchi, 2008; Ramirez Rossi, 2009). Although all of these remains present suites of distinctive, derived modern human characteristics and have the basic morphological gestalt of recent humanity (no one would refer to them as other than modern humans), many of them have been noted to exhibit archaic and/or regional features not present in the presumably ancestral Middle Paleolithic modern human sample (see esp., Vallois, 1958; Frayer et al., 2006; Grine et al., 2007; Trinkaus, 2005a, 2007a; Rougier et al., 2007; Crevecoeur, 2008; Ramirez Rossi et al., 2009; Crevecoeur et al., 2009).

The comparative morphological analysis of the Tianyuan 1 skeleton serves to reinforce a pattern that has been emerging over the past decade. Namely, among the earliest modern humans of MIS 3, the time of the real dispersal of modern human biology across the Old World, there is widespread evidence for some degree of assimilation of regional, archaic human populations. As with Tianyuan 1 and the eastern Asian population processes, the degree of assimilation, or the relative contributions to this "dual ancestry," are unknown and probably unknowable. Moreover, these data and inferences tell us little (or nothing) about subsequent human population dynamics or the ancestry of recent human regional groups. But they tell us a lot about the nature of those early—30,000- to 50,000-year-old-humans—and how they related to their late archaic predecessors across the Old World.

Structural Considerations

The discussion of populational considerations has largely assumed that most of the various anatomical features of interest on the Tianyuan 1 remains are largely independent and can be used to document an archaic/modern human morphological mosaic in this individual, comparable to the ones that have been documented in approximately contemporaneous early modern humans from the western Old World. Taking into consideration that some of these features are functionally relevant, if only with respect to skeletal hypertrophy, and that some of them may have a degree of developmental plasticity, it appears reasonable to regard most of them as epigenetic markers that can be used, in a general fashion, to shed light on Late Pleistocene population processes.

But discussions of the Tianyuan 1 morphological mosaic, as well as those of other early modern humans, begs the question of structural integration through the skeleton. Despite well-appreciated individual variation, it is not difficult to describe the contrasting morphological gestalts of late archaic and early modern humans (see Trinkaus, 2006a). Indeed, this has been done with varying degrees of detail and success since Boule (1911–1913). If indeed there are contrasting morphological gestalts, and the associations of various archaic features or of modern features are biological and not merely coincidental, then one could query how, biologically, one arrives at the kind of morphological mosaic observed in Tianyuan 1 (or for that matter, Oase1 and 2, Nazlet Khater 2, Hofmeyr 1, Muierii 1, the Mladeč sample, the Les Rois sample, etc.).

There is no simple answer to this question since by and large the underlying relationships between these skeletal (and dental) features are unknown. However, posing the question takes us beyond questions of sex, reproduction, and ancestry.

Paleobiological Considerations

The Tianyuan 1 skeleton has other implications that go beyond who was having sex with whom in the Late Pleistocene. After all, sex happens; that it occurred should not be surprising. The functional anatomy and paleopathology of this individual, as well as its ancestry, have broader implications for its behavior and that of its population.

The Behavioral Side of Ancestry

These assessments of the ancestry of the Tianyuan 1 individual, and of other, roughly contemporaneous, early modern humans, provide insight into the degree of behavioral contrast between late archaic humans and early modern humans during the transitional phase of MIS 3. From the conclusion that the assimilation model best accounts for the morphological pattern of Tianyuan 1, and of other early modern humans, one can infer that the late archaic and early modern humans saw each other as potential mates within a social context. Although behavioral differences may have existed, and certainly there were differences in appearance given contrasts in facial form, they must have shared the same basic kinds of social, technological, symbolic, and communication systems. This implies that whatever differences existed must have been subtle.

This inference of behavioral similarity does not necessarily imply that they were all fully behaviorally modern, in the sense of being within the ranges of variation of the behavior of ethnohistorically documented foraging societies. They were not Holocene humans. They may have been very close to Holocene hunter-gatherers in their behavioral spectra, or they may have had considerably less elaborated sociocultural systems. The only implication is that the differences between the behavioral patterns of late archaic and these early modern humans should have been modest.

Strength and Endurance

Mobility

The lower limb remains of Tianyuan 1 are almost universal in indicating that this was a highly mobile individual throughout its life. This is reflected in the femoral and tibial midshaft robusticity assessments, which should reflect both childhood mobility and burden carrying, as well as the persistence of those patterns throughout adult life (Pearson and Lieberman, 2004; Ruff, 2006; Ruff et al., 2006; Cowgill, 2008; Shaw and Stock, 2009a). The high level of immature locomotor activity is further reflected in the very low femoral neck-shaft angle of Tianyuan 1, a trait that reflects the loading levels of the hip region during development (Anderson and Trinkaus, 1998). In both of these aspects, Tianyuan 1 is unexceptional for a Late Pleistocene human since mobility seems to have been a necessary aspect of their existence, both among late archaic and early modern humans (Trinkaus et al., 2001; Trinkaus, 2007b).

Indeed, the persistence of lower limb diaphyseal robusticity (Chapter 10), despite the marked pathological changes in the femora and tibiae and their associated musculature (Chapter 12), reinforces the importance of maintaining mobility despite physical impairments. Similar persistence of robusticity despite lower limb deformities is evident among late archaic (Shanidar 1) and Mid Upper Paleolithic (Dolní Věstonice 15, Sunghir 3) individuals (Trinkaus, 1983a; Trinkaus et al., 2001; Formicola and Buzhilova, 2004).

Manipulation Strength

At the same time, the upper limbs of Tianyuan 1 present a mixed picture of arm and hand hypertrophy. The humeri are not particularly strong, in terms of their diaphyseal robusticity or muscle attachments. The radii have little diaphyseal curvature, but a strongly marked insertion for pronator teres. There is marked asymmetry of at least the shoulder and upper arm regions, as is reflected in the scapular acromial processes and especially the diaphyseal cortical bone hypertrophy of the humeri. This asymmetry does not appear in the radial diaphyses, but it may be present in the greater relative base breadth of the right fifth proximal phalanx compared to that of the left second proximal phalanx. Moreover, the base hypertrophy of the middle hand phalanx and the tuberosity enlargement of the distal phalanx (both unfortunately not assignable to side) suggest some degree of at least distal upper limb robusticity. Unfortunately, the lack of archeological remains at Tianyuan Cave, or strictly contemporaneous eastern Asian Paleolithic sites, makes it impossible to assess whether this pattern of upper limb robusticity could be correlated with significant shifts in manipulative behavior.

In addition, the mandibular dentition has occlusal attrition from the I_2 to the

M_3 that exhibits only a modest wear gradient across the molars and little differentiation in degrees of wear between the anterior teeth and first two molars. Among archaic humans, at least in the western Old World, an individual with the overall advanced occlusal wear of Tianyuan 1 would be expected to have markedly more pronounced anterior attrition (Trinkaus, 1983a, 1992).

Toes and Shoes

To these considerations of lower and upper limb hypertrophy can be added the marked gracility of the middle proximal pedal phalanx. As has been supported by patterns across relatively shod and unshod recent human populations (Trinkaus, 2005b), gracile lateral pedal phalanges in the context of otherwise robust lower limbs implies that those lateral toes were buffered from habitual ground reaction forces. The only normal mechanism to do so, given the normal plantarflexion of those toes during heel-off and hence ground reactions forces across them, is some form of protective sole placed under them to diffuse the ground reaction forces.

This pattern of robust legs but gracile lateral toes is ubiquitous among Mid Upper Paleolithic western Eurasian modern humans (Trinkaus, 2005b), including those with archeologically documented footwear (Trinkaus and Shang, 2008), but the earliest of them is ~31,000 cal BP. Earlier, Middle Paleolithic (archaic and modern) humans, as well as Middle Pleistocene ones, have robust proximal pedal phalanges in concert with their strongly built legs. The gracile toe of Tianyuan 1 is therefore best explained as a consequence of the habitual use of mechanically protective footwear, similar to that seen in the Mid Upper Paleolithic. The presence of this pedal pattern on Tianyuan 1 extends the antiquity of this pattern back an additional 9,000 years. Unfortunately, none of the other ~35,000 to 40,000 BP early modern humans preserve sufficient pedal remains for a similar assessment.

Stress and Survival

These functional considerations are combined with marked abnormalities of the femoral and tibiae, and a series of minor accumulated insults (Chapter 12).

Wear and Tear

The majority of the pathological lesions on Tianyuan 1 are the result of minor insults accumulated during the lifetime of this individual. This includes the osteoarthritis of the upper cervical vertebrae and the distal interphalangeal articulation, as well as the minor ossifications of connective tissue on the scapula, radius, hamate, femur, and calcaneus. The cervical and interphalangeal osteoarthritis would have been painful during movements of those joints, but neither one is likely to have impeded movement. There is no evidence of subchondral bone degeneration in any of the weight-bearing articulations. Unless the osteoarthritis can be attributed to minor trauma, there is no evidence of healed trauma on the individual.

In general, given the probably older middle age of the individual at death (Chapter 11), the remains are relatively free of accumulated minor degenerations. By this age-at-death, almost all of the sufficient complete Neandertals exhibit some level of serious degeneration (Trinkaus, 1983a; 2007b). The level of such accumulated minor to moderate insults is markedly lower among the samples of Mid Upper Paleolithic individuals (Trinkaus, Hillson, et al., 2006).

Antemortem Tooth Loss

Tianyuan 1 is notable for its degree of antemortem tooth loss, tooth loss that is evident in the resorbed sockets of the right I_1 to left P_3 and M_1 and in the super-eruption of the right M_3 (Chapters 5 and 12).

Antemortem tooth loss is well documented among later Pleistocene archaic humans, with Aubesier 11, La Chapelle-aux-Saints 1, Shanidar 5, and especially Guattari 1 exhibiting pronounced tooth loss (Sergi, 1974; Trinkaus, 1983a, 1985; Lebel and Trinkaus, 2002). There is even one case of the survival in the Early Pleistocene of complete antemortem tooth loss (Lordkipanidze et al., 2005). Antemortem tooth loss is largely absent from early modern humans, even in cases of marked occlusal wear similar to that of Tianyuan 1 [e.g., Brno 2, Caviglione 1, Dolní Věstonice 3 and 16, Pavlov 1, Předmostí 3, Sunghir 1, and ZKD-UC 101 (Matiegka, 1934; Weidenreich, 1939a; Jelínek, 1959; Kozlovskaya and Mednikova, 2000; Trinkaus, Hillson, et al., 2006)]; it is minimally present in the Minatogawa sample (Hanihara and Ueda, 1982). Levels of antemortem tooth loss similar to that seen in Tianyuan 1 are not unusual in recent high latitude human foraging populations (Moorrees, 1957; Costa, 1980).

Lower Limb Abnormalities

The most impressive of the Tianyuan 1 abnormalities are the alterations of the distal femoral diaphyses and the proximal tibial diaphyseal muscular attachments. As discussed in Chapter 12, the etiology of these bilaterally symmetrical changes is not apparent. The several possibilities, which individually appear unlikely given the nature of the lesions and their associations in the skeleton, are infantile deformities, childhood rickets, a childhood trauma, or adult abnormal use, each with extensive healing and remodeling through life. If rickets were the case, then it would represent the only case known from the Pleistocene, even among higher latitude individuals subjected to low levels of solar radiation and hence of vitamin D synthesis. If a childhood trauma were the case, it would be less spectacular, since serious injury to Middle Paleolithic individuals that occurred during childhood is well documented (e.g., Feldhofer 1, Qafzeh 11, Shanidar 1). The other possibilities are more difficult to assess. However, none of these etiologies appears likely. The etiological ambiguity of these abnormalities is unfortunate, since it negates the possibility of assessing the degree to which they may have affected function, and hence the behavioral consequences of their presence.

Evidence of serious developmental and/or degenerative lesions is relatively common among Pleistocene *Homo* specimens. These lesions include substantial dentoalveolar loss of function in Dmanisi 3444/3900, Broken Hill 1, Aubesier 11, La Chapelle-aux-Saints 1, and Guattari 1 (Carter, 1928; Sergi, 1974; Trinkaus, 1985; Lebel and Trinkaus, 2002; Lordkipanidze et al., 2005). They include more than trivial trauma, frequently with partial loss of function, in Hulu 1, Krapina 34.7, La Chapelle-aux-Saints 1, Feldhofer 1 and 2, Shanidar 1 and 3, Saint-Césaire 1, Qafzeh 11, and Dolní Věstonice 3 and 15 (Trinkaus, 1983a, 1985; Dawson and Trinkaus, 1997; Tillier, 1999; Zollikofer et al., 2002; Mann and Monge, 2006; Smith et al., 2006; Trinkaus, Hillson, et al., 2006; Shang and Trinkaus, 2008). There are a number of cases of developmental abnormalities, including Salé 1, Singa 1, Kebara 2, Qafzeh 12, Dolní Věstonice 3, 15 and 16, and Sunghir 3 (Duday and Arensburg, 1991; Hublin, 1991; Spoor et al., 1998; Tillier, 1999; Formicola and Buzhilova, 2004; Trinkaus, Hillson, et al., 2006). And there are several cases of degenerations, infections, or dietary deficiencies that were survived sufficiently long for skeletal reactions to have taken place, including KNM-ER 1808, KNM-ES 11693, Kocabaş 1, La Ferrassie 1, Kiik-Koba 1, Shanidar 1, Nazlet Khater 2, Brno 2, and Cro-Magnon 1 (Dastugue, 1967; Walker et al., 1982; Crubézy and Trinkaus, 1992; Oliva, 1996; Fennell and Trinkaus, 1997; Bräuer et al., 2003; Crevecoeur and Villotte, 2006; Kappleman et al., 2008; Trinkaus et al., 2008).

Dietary Implications

In addition to behavioral implications of the Tianyuan 1 human remains from its population affinities, functional morphology, and paleopathology, indications of its dietary spectrum have broader implications for human behavior at this time period in eastern Eurasia. In particular, its stable isotopes indicating a significant diet of freshwater aquatic resources (Appendix 2), as well as of terrestrial herbivores indicated by the humanly processed faunal remains (Appendix 1), suggest a broad spectrum of animal exploitation.

From western Eurasia there have been suggestions that there was a shift to a habitual exploitation of a broader range of dietary sources around the time of modern human emergence (Stiner et al., 1999; Richards et al., 2001; Trinkaus, Soficaru, et al., 2009; Richards and Trinkaus, 2009), even though there is evidence for the full range of faunal exploitation among earlier Middle Paleolithic humans in the region (Stiner, 1994; Muñoz and Casadevall, 1998; Stiner et al., 2000; Richards et al., 2001; Hardy et al., 2001; Grayson and Delpech, 2008). At the same time, zoo-archeological analyses of large terrestrial mammals in eastern Eurasia (Norton and Gao, 2008a,b) suggest a similar continuity of human predatory abilities through this time period, although fish remains first appear in Zhoukoudian–Upper Cave (Gao and Norton, 2008a) and there are inferences of a broader dietary spectrum from Niah Cave in Sarawak (Barker et al., 2007).

The Tianyuan 1 stable isotopes in particular document, for the first time, that aquatic resources were a substantial part of the diet of at least this early modern human in eastern Eurasia, even though similar data for more early modern human individuals and for late archaic humans in the same region are needed to assess whether this marks a shift or a continuation of such a habitual aquatic resource exploitation. If the former case holds, then it may well correlate with the expansion of modern human biology, by whatever population processes, documented by Tianyuan 1. Any such population expansion would have involved population increase and thus augmented pressure on available resources. The evidence of a broader habitual dietary range may well reflect the need by these early modern humans to access small package, harder-to-capture, prey to support their populations.

Conclusions

The Late Pleistocene human fossil record for eastern Eurasia consists of scattered remains from a variety of sites, a situation that continues to make it difficult to document the nature of the human biological and populational changes through this time period. Since this is the portion of the Pleistocene that saw the emergence and establishment of anatomically modern humans and the evolution of cultural systems into ones similar to those of ethnohistorically known forging human groups, discovering, documenting, and describing the available evidence for this time period across the eastern Old World becomes an important paleoanthropological endeavor.

The 2001 discovery of human remains in Tianyuan Cave (Tianyuandong) in northern China is an important piece of this puzzle. This human partial skeleton, despite preservation gaps, provides a unique window into the biology and behavior of the earliest modern humans to occupy this portion of the Old World. In many ways it fits the emerging pattern for these human populations. It shows a minority of archaic traits in the context of an otherwise modern human skeleton, implying only subtle behavioral differences with preceding human populations. As an older individual at death, it shows the accumulation of the wear and tear of a Pleistocene foraging existence, combined with a persistence of mobility despite lower limb abnormalities. Survival, despite tooth loss and muscular alterations, depended on that mobility, mobility that was apparently aided by the earliest known evidence for habitual footwear. At the same time, there was some reduction in the strength of the arms and the use of the teeth for manipulation relative to earlier humans, implying increased mechanical effectiveness of the technology, a phenomenon poorly documented in the eastern Asian archeological record.

It is hoped that this is but part of the beginning of a renaissance in the comprehension of early modern human biology and behavior in the eastern Old World—and that the details and insights provided here will serve in that process.

Appendix 1

Taphonomy of the Human Remains from Tianyuandong

YOLANDA FERNÁNDEZ-JALVO AND PETER ANDREWS

THIS APPENDIX focuses on the differences in preservation and taphonomy between the human remains from Tianyuandong and the animal bones. A full report of the site taphonomy is being prepared.

We visited the Institute of Vertebrate Paleontology and Paleoanthropology (IVPP) in April–May 2008 and taphonomically analyzed 748 animal bones from Tianyuandong and the 32 skeletal elements of the Tianyuan 1 human fossil (Chapter 3). The animal bones are all from Layer III, and several of the human bones were recovered from excavations at the same horizon. The remainder of the human fossils from Tianyuan Cave came from the initial digging at the site by the landowners, and they appear to represent the skeleton of a single individual (Chapters 2 and 3). A large quantity of animal bones were recovered with the human skeleton. The bones from Layer III, including the few human fossils found in the course of the excavation, show no trace of porcupine gnawing. The absence of porcupine gnawing contrasts with the fossils from Layer I (N = 116), the preservation of which is distinctive; many of the bones have been damaged by porcupine gnawing and by weathering (Tong et al., 2008). The other levels in the cave had very few fossils (see Table 2-1).

Excavations by members of the IVPP were undertaken in 2003–2004 to recover additional remains of both animals and the human (Chapter 2, especially Figure 2-3), and in the process several human bones were found in stratigraphic context. They were found in the blind cave extension behind the limestone wall bordering the entrance chamber within Layer III, but no stratigraphic evidence is available to show whether the human fossils were intrusive (i.e., part of a burial in a grave cut into the deposits) or whether they were from the surface of this bed. Too few bones were recovered in situ, with no information available as to their deposition, which might enable one to distinguish between burial and surface deposition. The human fossils found in the course of the excavation have the preservation typical of the lower level, with no evidence of the porcupine gnawing that characterizes the fossils from the upper level. In addition, the direct radiocarbon date on the right human femur (see Table 2-2), which was among the disturbed bones, has the same age as the animals fossils from Layer III. Whether the human individual was buried or left on the surface is not known, but it is evident that it was contemporaneous with the animal fossils.

Methods

Length, breadth, and thickness of all of the bones were measured, and surface areas and volumes were calculated. There are few fragments smaller than 2 cm, and it is likely that such bones were not collected. Surface modifications were counted and measured. Percussion marks (Blumenschein and Selvaggio, 1988) were the most common modification, and breakage was analyzed following Villa and Mahieu (1991) on long bones larger than 4 cm. We have added a "category 4" when the fragment has the complete circumference at the distal/proximal end and the shaft has less than half of the circumference (Figure A1-1 bottom). The whole collection was examined with a light microscope (10x to 40x magnification), and a sample was examined with the scanning electron microscope at the IVPP, a Jeol JSM 6100. These samples could not be prepared or coated, and the SEM observations were performed at 2 KV at a working distance of 32 mm.

Results

The animal remains from Layer III are extremely fragmented, and initial descriptions of them did not include any reference to cultural or carnivore modification (Tong, Shang, et al., 2004; Tong, Liu, and Zhang 2004; Shang et al., 2007). This result was consistent with the complete absence of lithic artifacts in the deposits. We confirmed the absence of carnivore chewing marks, a remarkable finding given the small size of the cave and the presence of so many bones within a few meters of the cave entrance. The human remains are much less broken, and they, too, show no evidence of carnivore action (Chapter 3).

There is a substantial difference in fragment size and skeletal element completeness between the human and nonhuman remains of Tianyuan Cave. Nearly 60% of the human fossils (N = 19) are complete (Table A1-1; Figure A1-1; see also Table 3-1) compared with 8.4% of the animal fossils (N = 63). Excluding these complete fossils, the comparison of fossil fragment sizes in Table A1-1 shows that only 6.3% of the human fragments are less than 50 mm in length (N = 2) compared with 45% of the animal bones (N = 337). This effect is emphasized particularly in the volume dimension, as shown in Figure A1-2, where the human size distribution is strongly skewed to the right whereas the animal distribution is skewed to the left.

Table A1-1. Distribution of breakage patterns of animal and human bones from Layer III of Tianyuan Cave

	Total		*Human fragments*		*Nonhuman fragments*	
Category	*N*	*%*	*N*	*%*	*N*	*%*
Length (mm)						
Complete	38	4.9	19	59.4	19	2.5
A: 0.1–20.0	47	6.0	0	0.0	47	6.3
B: 20.1–50.0	274	35.1	2	6.3	272	36.4
C: 50.1–100.0	368	47.2	6	18.8	362	48.4
D: 100.1–200.0	51	6.5	3	9.4	48	6.4
E: >200	2	0.3	2	6.3	0	0.0
	780		32		748	
Surface (mm^2)						
A: 1–500	39	5.2	0	0.0	39	5.5
B: 501–1,000	91	12.2	1	3.1	90	12.6
C: 1,001–2,500	183	24.6	4	12.5	179	25.1
D: 2,501–5,000	394	53.0	2	6.3	392	55.1
E: 5,001–10,000	17	2.3	6	18.8	11	1.5
>10,000	1	0.1	0	0.0	1	0.1
	744		32		712	
Volume (mm^3)						
A: 1–5,000	76	19.5	0	0.0	76	21.3
B: 5,001–10,000	85	21.9	0	0.0	85	23.8
C: 10,001–25,000	105	27.0	3	9.4	102	28.6
D: 25,001–50,000	51	13.1	2	6.3	49	13.7
E: 50,001–500,000	52	13.4	7	21.9	45	12.6
F: >500,000	1	0.3	1	3.1	0	0.0
	389		32		357	

The shaft circumferences of the animal bone fragments (Villa and Mahieu, 1991) are also less complete than the human bones. Almost all of the nonhuman bones had half or less than half of the circumference preserved, whereas the human bones had half or more preserved (Table A1-2; Figure A1-1). Preservation of the circumference outlines is shown in Figure A1-1 compared with the lengths of the fragments, and the contrast between human and nonhuman preservation is clear. The animal bones are strongly fragmented, whereas the human ones are not (Figure A1-2).

In general, the human bones have little surface damage except for postdepositional modifications such as manganese staining. There are some humidity cracks, and the manganese staining is concentrated along these cracks. Manganese staining is also more abundant on one side of the bone than the other, an indication that the bones were stabilized in the deposits, with one side in contact with wet substrate. The human bones have a few impact marks scattered on the surfaces,

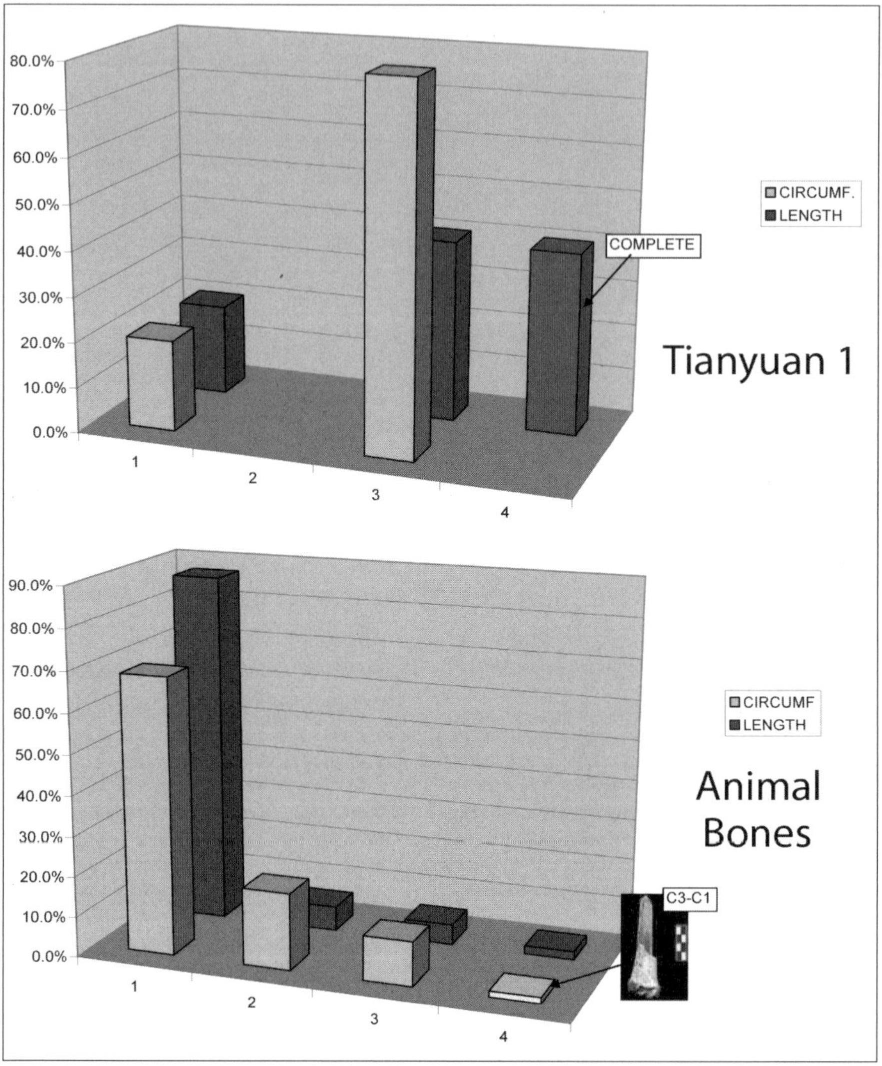

Figure A1-1. Distributions of the preservation of the circumference and the length of the long bone fragments (>4 cm), following Villa and Mahieu (1991), comparing Tianyuan 1 (*top*) and animal bones (*bottom*) from Tianyuan Cave. See Table A1-2 for categories of degrees of completeness.

without any concentration at points where the bones might have been opened up to extract the marrow; the marks are inferred to be the result of rock falls. This pattern could indicate that the bones were exposed on the surface of the cave or were not buried immediately after death, but such impacts could also occur in a shallow burial. This pattern therefore does not resolve the burial issue.

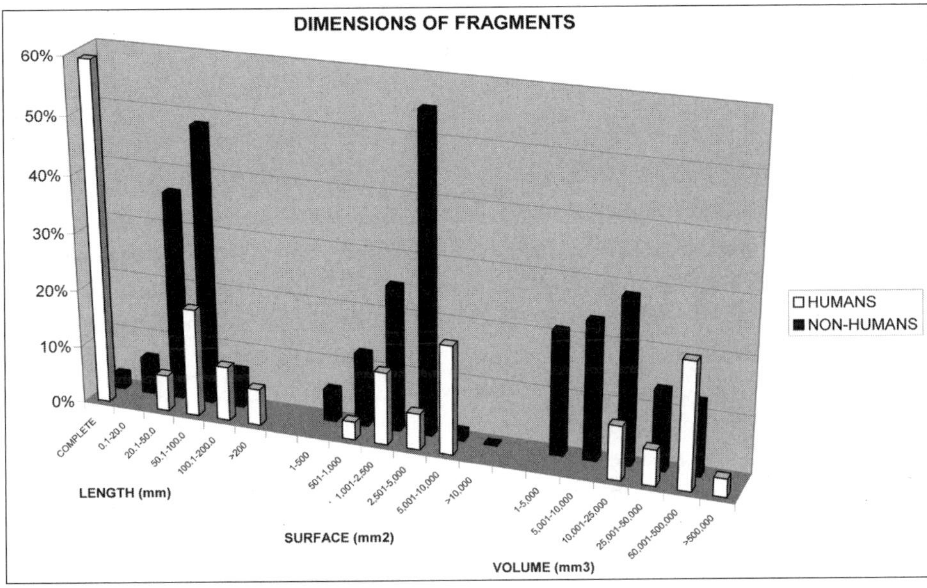

Figure A1-2. Percentages of the animal bone assemblage and the Tianyuan 1 skeleton shown in Table A1-1 that fall into length, surface area, and bone volume intervals.

Table A1-2. Distributions of the preservation of the circumference and the length of the long bone fragments (>4cm)

Category	*1 (%)*	*2 (%)*	*3 (%)*	*4 (%)*
Circumference				
Tianyuan 1	20.0	0.0	80.0	0.0
Animal bones	68.5	19.0	11.0	1.5
Length				
Tianyuan 1	20.0	0.0	40.0	40.0
Animal bones	87.0	6.0	5.0	2.0

Source: Following Villa and Mahieu (1991).
Note: The length categories refer to: category 1: one-quarter; category 2: one-half; category 3: three-quarters; category 4: complete. The circumference categories are: category 1: less than one-half; category 2: greater than one-half; category 3: complete; (new) category 4: complete epiphyseal circumference but less than half of the adjacent diaphyseal circumference.

The scarcity of impact marks on the human bones contrasts with the more heavily damaged animal bones, which have extensive impact and percussion marks. Percussion marks on the surfaces of the animal bones and impacts on broken edges are evidence that the bones were broken for access to the marrow. The greater breakage of the animal bones noted above was undoubtedly the product of the extensive percussion damage, which was observed on the great majority of the animal fossils. The combination of a high proportion of animal bones bearing percussion marks and the absence of evidence of carnivore chewing show that the animal assemblage is the product of human butchery and that all or most of the assemblage was accumulated by human action.

Summary

In conclusion, the human skeleton has a taphonomic history distinct from the animal remains. The human bones are more complete and lack surface modifications, whereas the animal fossils are extensively broken, have multiple percussion marks on their surfaces, and have impact marks on their broken edges. The faunal remains were clearly processed by people, which is inferred to be butchery, but the human remains themselves were not processed by humans. With the evidence available, we cannot determine whether the single human individual entered the cave and died, with the bones of the skeleton remaining on the surface for a period of time, or if the body was carried into the cave and placed in a shallow grave.

Acknowledgments

This taphonomic study has been supported by project CGL-2007 66231-BTE of the Spanish Ministry of Science and Innovation. We are grateful to Erik Trinkaus and Paola Villa for proposing that we do this taphonomic study. Thanks to Dr. Haowen Tong and Prof. Xinzhi Wu for providing access to both faunal and human collections of Tianyuandong site. We are also very grateful to Dr. Tong for his kindness and great help during our stay in the IVPP in Beijing, and for accompanying us to the site and explaining previous studies and excavation procedures. Thanks to Zhang Wending from the Electronic Microscopy Unit for his expert assistance and good work.

Appendix 2

Isotopic Analysis of Tianyuan 1 and Associated Faunal Remains

YAOWU HU AND MICHAEL P. RICHARDS

AN UNDERSTANDING of changing human adaptations to the environment, specifically their subsistence strategies, is a key part of determining the processes and nature of human evolution. The position of the Tianyuan 1 human remains as one of the oldest marine isotope stage (MIS) 3 modern humans in Eurasia poses the question of whether there might have been documentable shifts in human dietary spectra and emphasis associated with the spread of modern human biology. There have been suggestions, based on European faunal assemblages and inferred from technological changes associated with the emergence of the Upper Paleolithic, that there was a shift in human predatory abilities and associated changes in diet. At the same time, carbon and nitrogen stable isotope analyses of both late archaic humans (Neandertals) and Upper Paleolithic early modern humans in Europe (Richards, Pettitt, et al., 2001; Trinkaus et al., 2009; Richards and Trinkaus, 2009) and analyses of smaller faunal remains (Stiner et al., 1999) have suggested that there was a shift to a broader dietary spectrum with the spread of modern humans, probably including greater emphasis on aquatic resources. But analyses of European large mammal faunal remains (Grayson and Delpech, 2002, 2003, 2008) and organic residues (Hardy et al., 2001), as well as inferences from human functional anatomy (Trinkaus, 2008a), have suggested little change in the human dietary spectrum or predation prior to the Mid Upper Paleolithic.

In eastern Eurasia, the nature of human dietary changes that might have been associated with the emergence of modern humans remains unclear. There is evidence at Xujiayao prior to the human biological transition of effective faunal predation (principally cervids) (Norton and Gao, 2008b), and similar evidence is present for Zhoukoudian–Upper Cave (Norton and Gao, 2008a) and Layer III of Tianyuan Cave (Appendix 1). In addition, there is evidence for fish (freshwater carp [*Cyprinus* and *Ctenopharygodon*], plus *Arca* shells) at Zhoukoudian–Upper Cave (Norton and Gao, 2008a). There have been some preliminary discussions of Late Pleistocene dietary changes in China based on the changes in lithic technology (Chen, 2006; Gao and Pei, 2006), and there are suggestions of dietary breadth expansion in peninsular southeast Asia from the same time horizon as Tianyuan Cave Layer III (Barker et al., 2007).

It is in this context that we present a carbon, nitrogen, and sulfur stable isotopic analysis of the Tianyuan 1 human remains and associated fauna from Tianyuan Cave. Stable isotope analysis has been proved to be useful for dietary reconstruction, since it provides direct (if general) evidence for human diets over the last few years of life (Richards, 2002; Sponheimer et al., 2006; Sponheimer and Lee-Thorp, 2007). In addition to the more commonly employed carbon and nitrogen stable isotopes, recent studies have also shown that sulfur isotope values have the potential to reveal whether the principal foods were from terrestrial or freshwater ecosystems (Richards, Fuller, and Hedges, 2001; Richards, Fuller, et al., 2003; Privat et al., 2007). Therefore, sulfur isotope ratios were also analyzed to assess whether Tianyuan 1 consumed significant aquatic resources.

Materials and Methods

Stable Isotope Analysis

All of the samples were prepared using a modified Longin method (Richards and Hedges, 1999) with the addition of an ultrafiltration step (Brown et al., 1988). The procedure for collagen extraction was as follows. The samples were cleaned mechanically by powder abrasion, rinsed in 0.5M HCl at 4°C and refreshed for several days to completely decalcify the bones. The samples were then rinsed to neutrality in deionized water and then gelatinized in HCl (pH: 3) at 70°C for 48 hours. Then the residues were ultrafiltered to isolate the fraction with a molecular weight >30 kDa. Finally, the residues were frozen and lyophilized.

Carbon, nitrogen, and sulfur contents and $\delta^{13}C$, $\delta^{15}N$, and $\delta^{34}S$ values of collagen were determined through EA-IRMS (Thermo-Finnigan Delta Plus V coupled with a Heka elemental analyzer). Stable isotope ratios are expressed relative to the vPDB standard for carbon, atmospheric N (AIR) for nitrogen, and vCDT for sulfur respectively, using the delta (δ) notation in parts per thousand (‰). Each sample was run in duplicate, and an internal standard was measured with each set of ten samples. The contents of carbon, nitrogen, and sulfur, and the values of $\delta^{13}C$, $\delta^{15}N$, and $\delta^{34}S$ of all samples, are listed in Table A2-1. All samples discussed below have an average C content of 42.3 ± 0.9%, an average N content of 15.3 ± 0.5%, and atomic C/N ratios in the range of 3.2–3.3, similar to those in modern bones (41% C content, 15% N content and 2.9–3.6 C/N ratio) (DeNiro, 1985; Ambrose, 1990), suggesting that all samples retained their in vivo isotopic signatures. In addition, the atomic C/S and N/S ratios of four samples (one human and three from animals) are also within the range of modern collagen.

Table A2-1: Isotope samples and resultant data

	Layer	C%	N%	S%	C/N	C/S	N/S	$\delta^{13}C$	$\delta^{15}N$	$\delta^{34}S$
Tianyuan 1	3	43.3	15.5	0.19	3.3	623	223	−17.6	11.1	4.1
Wild cat (*Felis microtis*)	1	43.7	16.0	0.20	3.2	587	215	−17.8	8.7	7.6
Sika deer (*Cervus nippon*)	3	41.2	14.7		3.3			−19.2	6.6	
Sika deer (*Cervus nippon*)	3	40.9	14.3		3.3			−19.1	5.2	
Unidentified herbivore	1	42.1	15.4		3.2			−18.8	3.8	
Unidentified herbivore	1	42.1	15.3		3.2			−18.0	4.8	
Unidentified herbivore	1	43.1	15.9	0.18	3.2	639	235	−21.2	6.4	7.2
Unidentified herbivore	2	41.8	15.0		3.2			−19.2	3.9	
Unidentified herbivore	2	43.6	16.0	0.20	3.2	572	209	−20.6	3.8	7.9
Unidentified herbivore	2	41.6	15.0		3.2			−20.3	3.9	
Unidentified herbivore	3	42.0	15.3		3.2			−19.0	5.1	
Unidentified herbivore	3	42.3	15.4		3.2			−19.6	4.4	
Unidentified herbivore	3	42.0	15.2		3.2			−19.1	4.2	
Donghulin fish	n/a	26.0	8.5	0.13	3.6	385	108	n/a	n/a	5.6
Donghulin fish	n/a	39.9	12.9	0.32	3.6	336	93	n/a	n/a	5.3

Note: Data for human and terrestrial fauna from Tianyuan Cave; data for fish from the nearby Early Neolithic site of Donghulin. "Unidentified herbivore" samples are postcranial fragments of cervids, most likely *Cervus nippon*.

Materials

Although most of the Tianyuan Cave animal bones are very fragmentary (Appendix 1), 36 animal species could be identified in the faunal assemblages with a possible six additional species (see Table 2-1). However, 20 of these species are rodents, hares, or bats. Most of the macromammal species are herbivores; carnivores are rare and extremely fragmentary (see Table 2-1; Tong et al., 2004; Tong, 2005). As a result, 13 bones were selected for stable isotopic analysis (Table A2-1). A sample of bone, which had been removed from the diaphysis of the right humerus (Chapter 3) for radiocarbon dating, was used for Tianyuan 1. For the faunal remains two samples of sika deer (*Cervus nippon*) from Layer III were included, plus nine samples of unidentified herbivores (cervids indeterminate as to species, but mostly likely *C. nippon* based on size and species representation in the diagnostic remains), three from each of Layers I, II, and III. In addition, a sample was taken from a Layer I wild cat (*Felis microtis*) given the dearth of carnivores in Layer III.

There are no fish remains preserved from Tianyuandong. However, the sulfur isotope signals from a region relate to the local geology and so remain constant through time. Therefore, in order to establish a local sulfur signal for freshwater ecosystems in this area, we sampled fish from the nearby Early Neolithic site of Donghulin dating to 11,000–9,000 BP (School of Archaeology and Museology of Beijing University, 2006). This site is much later than Tianyuan Cave, but the sulfur values from these fish can be used as a proxy for the local freshwater sulfur isotope values.

Results and Discussion

The Isotopic Food Web at Tianyuan Cave

The $\delta^{13}C$ and $\delta^{15}N$ values of all the samples from Tianyuan Cave are plotted in Figure A2-1. The two sika deer and the nine other herbivores have $\delta^{15}N$ values of 4.7 ± 1.6 ‰, plotting in the expected range for herbivores. The herbivore mean $\delta^{13}C$ value of −19.5 ± 0.9 ‰ suggests that C_3 plants were generally dominant in their diets. However, the large ranges of $\delta^{13}C$ values (−18.0 ‰ to −21.2 ‰) and $\delta^{15}N$ values (3.8 ‰ to 6.6 ‰) in the herbivores show that their diets were quite varied. The individual with highest (least negative) $\delta^{13}C$ value (−18.0 ‰) likely consumed some C_4 plants, while the individual with the lowest $\delta^{13}C$ value (−21.2 ‰) and highest $\delta^{15}N$ value (6.4 ‰) may have lived in a more closed habitat and consumed relatively ^{15}N-enriched C_3 plants. The sika deer have higher $\delta^{15}N$ values than most of the other herbivores, suggesting that ^{15}N-enriched C_3 plants were included in their diets.

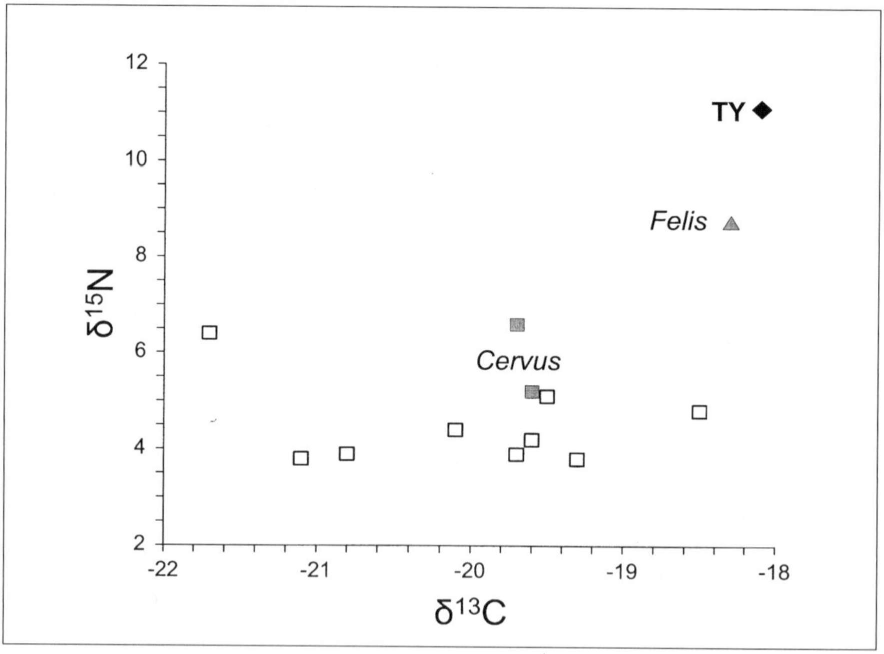

Figure A2-1. Scatter plot of $\delta^{13}C$ and $\delta^{15}N$ values for the human skeleton and faunal remains from Tianyuan Cave. *TY:* Tianyuan 1.

Numerous studies indicate that bone collagen $\delta^{13}C$ and $\delta^{15}N$ values are enriched by 0.8 ‰ ~ 1.3 ‰ and by 3 ‰ ~ 5 ‰ respectively from herbivores to carnivores in the same food web (DeNiro and Epstein, 1981; Schoeninger et al., 1983; Bocherens and Drucker, 2003; Hedges and Reynard, 2007). A 4 ‰ enrichment of $\delta^{15}N$ value between the mean herbivore $\delta^{15}N$ value and the carnivore (wild cat) (Table A2-1) is well within the expected enrichment factor for predator-prey species. However, the difference of $\delta^{13}C$ value (1.7 ‰) between the wild cat and the mean herbivore $\delta^{13}C$ value is beyond the range of carbon isotopic fractionation in the food web, indicating that the wild cat sometimes preyed on herbivores which consumed C_4 plants.

The Carbon and Nitrogen Values of Tianyuan 1

The Tianyuan 1 has a $\delta^{13}C$ value close to that of the wild cat, suggesting that some C_4 foods or animals that ate C_4 plants were also included in the diet. However, the human $\delta^{15}N$ value (11.1 ‰) is much higher than that of the wild cat (8.7 ‰) and all of the herbivores. It is 4.5 ‰ higher than the highest herbivore $\delta^{15}N$ and 6.4 ‰ above the overall herbivore $\delta^{15}N$ average value of 4.7 ‰, differences that are at or above the normal predator-prey enrichment in $\delta^{15}N$. It is therefore unlikely that herbivores were the sole source of protein in the diet of this human.

The $\delta^{15}N$ values of animals in aquatic ecosystems (marine and freshwater) are generally higher than those in terrestrial ecosystems due to their longer food chains (Richards, Pettitt, et al., 2001). Therefore, the human nitrogen isotope value may well be explained by the consumption of freshwater foods and perhaps terrestrial animals that consumed C_4 vegetation. No freshwater fish bones were found in Tianyuan Cave (Tong et al., 2004); however, fish remains have been found at sites near Tianyuan Cave, including Zhoukoudian–Upper Cave and the Donghulin Early Neolithic site (School of Archaeology and Museology of Beijing University, 2006; Norton and Gao, 2008a), strongly suggesting that there must have been freshwater resources available around the Zhoukoudian area.

The Sulfur Isotope Values of Tianyuan 1 and the Fauna

To further evaluate this inference of freshwater fish in the diet of Tianyuan 1, the sulfur isotope value of Tianyuan 1 was compared to those of a subset of the terrestrial animals from Tianyuan Cave and of fish from Donghulin (Table A2-1). The terrestrial animal sulfur isotope values are similar and average 7.6 ± 0.4 ‰ (7.2 ‰–7.9 ‰). As expected for freshwater fish, the values for the Donghulin fish are lower, at 5.6 ‰ and 5.3 ‰. The sulfur isotope value of Tianyuan 1 (4.1 ‰) is even lower and substantially below those of the few terrestrial animals providing data.

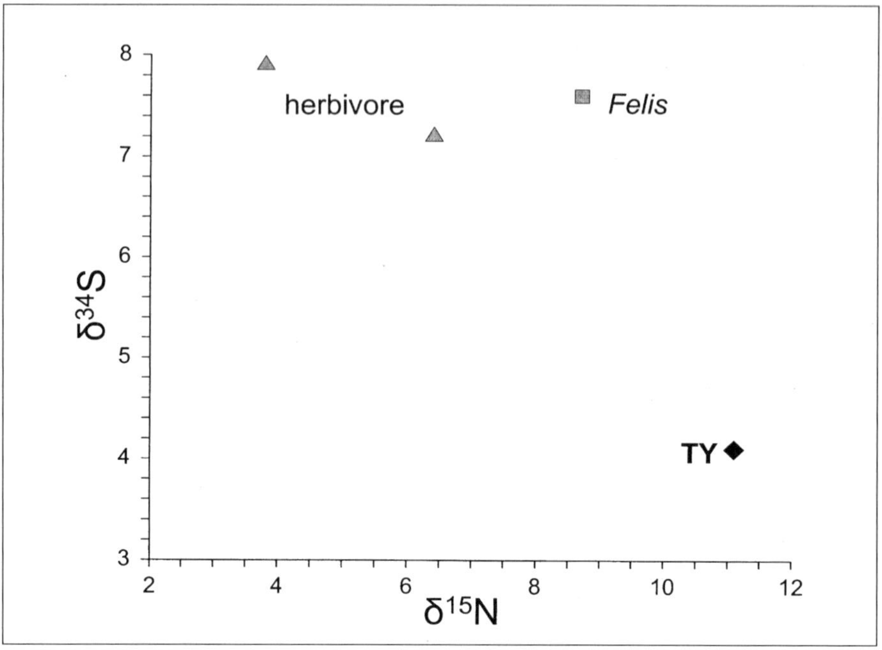

Figure A2-2. Nitrogen and sulfer isotopes of the human skeleton and three terrestrial animals from Tianyuan Cave.

Conclusions

Carbon, nitrogen, and sulfur isotope analyses of bones from the Tianyuan 1 early modern human and the associated animals in Tianyuan Cave indicated that the human most likely obtained a substantial portion of its protein from a freshwater ecosystem, probably from freshwater fish. This result provides the earliest direct evidence of significant freshwater food exploitation by modern humans in eastern Eurasia.

Acknowledgments

This research has been funded through a partnership between the Max Planck Society and the Chinese Academy of Sciences (CAS), and by the Knowledge Innovative Project of CAS (KJCX3.SYW.N12) and the National Science Foundation in China (40702003). Many archeologists and paleoanthropologists permitted sampling and provided expertise. We thank all of them.

References

Al Quran FA, Al-Dwairi ZN. 2006. Torus palatinus and torus mandibularis in edentulous patients. *Journal of Contemporary Dental Practice* 7:112–119.

Ambrose SH. 1990. Preparation and characterization of bone and tooth collagen for stable carbon and nitrogen isotope analysis. *Journal of Archaeological Science* 17:431–451.

Anderson JY. 2000. Regional cranial stability: constancy of the mid-mandibulo-facial region as defined by foraminal determinants (abstract). *American Journal of Physical Anthropology Supplement* 30:96–97.

Anderson JY, Trinkaus E. 1998. Patterns of sexual, bilateral, and inter-populational variation in human femoral neck-shaft angles. *Journal of Anatomy* 192:279–285.

Arensburg B, Kaffe I, Littner MM. 1989. The anterior buccal mandibular depressions: ontogeny and phylogeny. *American Journal of Physical Anthropology* 78:431–437.

Arora AK, Gupta SC, Gupta CD, Jeyasingh P. 1979. Variations in calcanean facets in Indian tali. *Anatomischer Anzeiger* 146:377–380.

Arsuaga JL, Lorenzo C, Carretero JM, Gracia A, Martínez I, García N, Bermúdez de Castro JM, Carbonell E. 1999. A complete human pelvis from the Middle Pleistocene of Spain. *Nature* 399:255–258.

Auerbach BM. 2007. Human skeletal variation in the New World during the Holocene: effects of climate and subsistence across geography and time. PhD dissertation, Johns Hopkins University.

Auerbach BM, Ruff CB. 2004. Human body mass estimation: a comparison of "morphometric" and "mechanical" methods. *American Journal of Physical Anthropology* 125: 331–342.

Auerbach BM, Ruff CB. 2006. Limb bone bilateral asymmetry: variability and commonality among modern humans. *Journal of Human Evolution* 50:203–218.

Aufderheide AC, Rodríguez-Martin C. 1998. *The Cambridge Encyclopedia of Human Paleopathology.* Cambridge UK: Cambridge University Press.

Baba H. 1970. On some morphological characters of Japanese lower-limb bones from the viewpoint of squatting and other sitting postures in Jomon, Edo and Modern periods (in Japanese with English summary). *Journal of the Anthropological Society of Nippon* 78:213–234.

Baba H, Endo B. 1982. Postcranial skeleton of the Minatogawa man. In Suzuki H, Hanihara K (eds) *The Minatogawa Man: The Upper Pleistocene Man from the Island of Okinawa. Bulletin of the University Museum, University of Tokyo* 19:61–195.

Barker G, Barton H, Bird M, Daly P, Datan I, Dykes A, Farr L, Gilbertson D, Harrisson B, Hunt C, Higham T, Kealhofer L, Krigbaum J, Lewis H, McLaren S, Paz V, Pike A, Piper P, Pyatt B, Rabett R, Reynolds T, Rose J, Rushworth G, Stephens M, Stringer C, Thompson J, Turney C. 2007. The "human revolution" in lowland tropical Southeast Asia: the antiquity and behavior of anatomically modern humans at Niah Cave (Sarawak, Borneo). *Journal of Human Evolution* 52:243–261.

Barnett CH. 1954. Squatting facets on the European talus. *Journal of Anatomy* 88:509–513.

Blumenschine RJ, Selvaggio MM. 1988. Percussion marks on bone surfaces as a new diagnostic of hominid behaviour. *Nature* 333:763–765.

Bocherens H, Drucker D. 2003. Trophic level isotopic enrichment of carbon and nitrogen in bone collagen: case studies from recent and ancient terrestrial ecosystems. *International Journal of Osteoarchaeology* 13:46–53.

Boule M. 1911–1913. L'homme fossile de La Chapelle-aux-Saints. *Annales de Paléontologie* 6:111–172; 7:21–56, 85–192; 8:1–70.

Bowler JM, Thorne AG. 1976. Human remains from Lake Mungo: discovery and excavation of Lake Mungo III. In Kirk RL, Thorne AG (eds) *The Origin of the Australians.* Canberra: Australian Institute of Aboriginal Studies, 127–138.

Bräuer G. 1988. Osteometrie. In Knussman R (ed) *Anthropologie I.* Stuttgart: Fischer Verlag, 160–232.

Bräuer G. 1992. The origins of modern Asians: by regional evolution or replacement? In Akazawa T, Aoki K, Kimura T (eds) *The Evolution and Dispersal of Modern Humans in Asia.* Tokyo: Hokusen-Sha, 401–413.

Bräuer G. 2008. The origin of modern anatomy: by speciation or intraspecific evolution? *Evolutionary Anthropology* 17:22–37.

Bräuer G, Groden C, Delling G, Kupczik K, Mbua E, Schultz M. 2003. Pathological alterations in the archaic *Homo sapiens* cranium from Eliye Springs, Kenya. *American Journal of Physical Anthropology* 120:200–204.

Bräuer G, Mehlman MJ. 1988. Hominid molars from a Middle Stone Age level at the Mumba Rock Shelter, Tanzania. *American Journal of Physical Anthropology* 75:69–76.

Bräuer G, Mímisson K. 2004. Morphological affinities of early modern crania from China. In Rubio S (ed) *Miscalánea en Homenaje a Emiliano Aguirre III: Paleoantropología.* Alcalá de Henares: Museo Arqueológico Regional, 58–70.

Brennan MU. 1991. Health and disease in the Middle and Upper Paleolithic of southwestern France: a bioarcheological study. PhD dissertation, New York University.

Bridges PS. 1994. Vertebral arthritis and physical activities in the prehistoric Southeastern United States. *American Journal of Physical Anthropology* 93:83–93.

Brothwell DR. 1960. Upper Pleistocene human skull from Niah Caves, Sarawak. *Sarawak Museum Journal* 9:323–349.

Brothwell DR 1963. *Digging Up Bones.* London: British Museum.

Brown P. 1993. Recent human evolution in east Asia and Australasia. In Aitken MJ, Stringer CB, Mellars PA (eds) *The Origin of Modern Humans and the Impact of Chronometric Dating.* Princeton: Princeton University Press, 217–233.

Brown TA, Nelson DE, Vogel JS, Southon JR. 1988. Improved collagen extraction by modified Longin method. *Radiocarbon* 30:171–177.

Brůžek J. 2002. A method for visual determination of sex, using the human hip bone. *American Journal of Physical Anthropology* 117:157–168.

Brůžek J, Franciscus RG, Novotný V, Trinkaus E. 2006. The assessment of sex. In Trinkaus E, Svoboda JA (eds) *Early Modern Human Evolution in Central Europe: The People of Dolní Věstonice and Pavlov.* New York: Oxford University Press, 46–62.

Buzhilova AP. 2000. The analysis of anomalies and indicators of physiological stress in non-mature Sunghir individuals (in Russian with English summary). In Alexeeva TI, Bader NO, Munchaev RM, Buzhilova AP, Kozlovskaya MV, Mednikova MB (eds) *Homo Sungirensis.* Moscow: Scientific World, 302–314.

Carretero JM, Arsuaga JL, Lorenzo C. 1997. Clavicles, scapulae and humeri from the Sima de los Huesos site (Sierra de Atapuerca, Spain). *Journal of Human Evolution* 33: 357–408.

Carretero JM, Arsuaga JL, Martínez I, Quam RM, Lorenzo C, Gracia A, Ortega AI. 2004. Los humanos de la Sima de los Huesos (Sierra de Atapuerca) y la evolución del cuerpo en el género *Homo.* In Rubio S (ed) *Miscalánea en Homenaje a Emiliano Aguirre III: Paleoantropología.* Alcalá de Henares: Museo Arqueológico Regional, 120–135.

Carretero JM, Hailie-Selassie Y, Rodriguez L, Arsuaga JL. 2009. A partial distal humerus from the Middle Pleistocene deposits at Bodo, Middle Awash, Ethiopia. *Anthropological Science.*117:19–31.

Carter JT. 1928. The teeth of Rhodesian man. In Bather FA (ed) *Rhodesian Man and Associated Remains.* London: British Museum, 64–65.

Cartmill M, Smith FH. 2009. *The Human Lineage.* New York: Wiley-Blackwell.

Chen S. 2006. Environment, adaptation, and diffusion of north China humans in late Pleistocene (in Chinese). *Quaternary Sciences* 26:522–533.

Chen T, Hedges REM, Yuan Z.1992. The second batch of accelerator radiocarbon dates for the Upper Cave site of Zhoukoudian (in Chinese with English abstract). *Acta Anthropologica Sinica* 11:112–116.

Chia LP. 1957. Notes on the human and some other mammalian remains from Changyang, Hupei (in Chinese with English summary). *Vertebrata PalAsiatica* 1:247–257.

Chia LP, Wei Q, Li CR. 1979. Report on the excavation of Hsuchiayao man site in 1976 (in Chinese with English summary). *Vertebrata PalAsiatica* 17:277–293.

Chilardi S, Frayer DW, Giola P, Macchiarelli R, Mussi M. 1996. Fontana Nuova di Ragusa (Sicily, Italy): southernmost Aurignacian site in Europe. *Antiquity* 70:553–563.

Churchill SE. 1994. Human upper body evolution in the Eurasian later Pleistocene. PhD dissertation, University of New Mexico.

Churchill SE, Berger LR, Parkington JE. 2000. A Middle Pleistocene human tibia from Hoedjiespunt, Western Cape, South Africa. *South African Journal of Science* 96:367–368.

Churchill SE, Formicola V. 1997. A case of marked bilateral asymmetry in the upper limbs of an Upper Palaeolithic male from Barma Grande (Liguria), Italy. *International Journal of Osteoarchaeology* 7:18–38.

Churchill, S.E., Pearson, O.M., Grine, F.E., Trinkaus, E., and Holliday, T.W. 1996. Morphological affinities of the proximal ulna from Klasies River Mouth Main Site: archaic or modern? *Journal of Human Evolution* 31:213–237.

Churchill SE, Rhodes JA. 2009. The evolution of the human capacity for "killing at a distance": the human fossil evidence for the evolution of projectile weaponry. In Hublin JJ, Richards MP (eds). *The Evolution of Hominid Diets: Integrating Approaches to the Study of Palaeolithic Subsistence.* Springer, Dordrecht, 201–210.

Churchill SE, Trinkaus E. 1990. Neandertal scapular glenoid morphology. *American Journal of Physical Anthropology.* 83:147–160.

Constantinesco A. 1974. Étude biomécanique des mouvements du sternum chez l'homme adult. *Archives d'Anatomie, d'Histologie et d'Embryologie, Normales et Experimentales* 57:153–196.

Coppens Y, Tseveendorj D, Demeter F, Turbat T, Giscard PH. 2008. Discovery of an archaic *Homo sapiens* skullcap in Northeast Mongolia. *Comptes Rendus Palévol* 7:51–60.

Corruccini RS, Jacobi KP, Handler JS, Aufderheide AC. 1987. Implications of tooth root hypercementosis in a Barbados slave skeletal collection. *American Journal of Physical Anthropology* 74:179–184.

Costa RL Jr. 1980. Age, sex, and antemortem loss of teeth in prehistoric Eskimo samples from Point Hope and Kodiak Island, Alaska. *American Journal of Physical Anthropology* 53:579–587.

Costa RL Jr. 1982. Periodontal disease in the prehistoric Ipiutak and Tigara skeletal remains from Point Hope, Alaska. *American Journal of Physical Anthropology* 59:97–110.

Courtaud P. 1996. Anthropologie de sauvetage: vers une optimisation des méthodes d'enregistrement: présentation d'une fiche anthropologique. *Bulletins et Mémoires de la Société d'Anthropologie de Paris* 8:157–167.

Cowgill LW. 2008. The ontogeny of Recent and Late Pleistocene human postcranial robusticity. PhD dissertation, Washington University in Saint Louis.

Cowgill LW, Trinkaus E, Zeder MA. 2007. Shanidar 10: a Middle Paleolithic immature distal lower limb from Shanidar Cave, Iraqi Kurdistan. *Journal of Human Evolution* 53: 213–223.

Crevecoeur I. 2002. Étude Morphologique et Biométrique des Ossements de Mains Retrouvés dans la Couche Castelperronienne du Gisement de Saint-Césaire (Charente-Maritime). D.E.A. d'Anthropologie Biologique, Université de Bordeaux 1.

Crevecoeur I. 2008. *Étude Anthropologique du Squelette du Paléolithique supérieur de Nazlet Khater, Egyptian Prehistory Monographs* (Leuven University Press) 8:1–318.

Crevecoeur I, Rougier H, Grine F, Froment A. 2009. Modern human cranial diversity in the Late Pleistocene of Africa and Eurasia: evidence from Nazlet Khater, Peştera cu Oase and Hofmeyr. *American Journal of Physical Anthropology.*

Crevecoeur I, Trinkaus E. 2004. From the Nile to the Danube: a comparison of the Nazlet Khater 2 and Oase 1 early modern human mandibles. *Anthropologie (Brno)* 42:229–239.

Crevecoeur I, Villotte S. 2006. Atteintes pathologiques de Nazlet Khater 2 et activité minière au début du Paléolithique supérieur en Égypte. *Bulletins et Mémoires de la Société d'Anthropologie de Paris* 18:165–175.

Crubézy E, Trinkaus E. 1992. Shanidar 1: a case of hyperostotic disease (DISH) in the Middle Paleolithic. *American Journal of Physical Anthropology* 89:411–420.

Čuk T, Leben-Seljak P, Štefančič M. 2001. Lateral asymmetry of human long bones. *Variability and Evolution* 9:19–32.

Cunningham DJ. 1886. The neural spines of the cervical vertebrae as a race-character. *Journal of Anatomy and Physiology* 20:637–640.

Daegling DJ. 1993. Functional morphology of the human chin. *Evolutionary Anthropology* 1:170–177.

Daegling DJ. 2001. Biomechanical scaling of the hominoid mandibular symphysis. *Journal of Morphology* 250:12–23.

Daegling DJ. 2002. Estimation of torsional rigidity in primate long bones. *Journal of Human Evolution* 43:229–239.

Darroch JN, Mosimann JE. 1985. Canonical and principal components of shape. *Biometrika* 72:241–252.

Dastugue J. 1967. Pathologie des hommes fossiles de l'Abri de Cro-Magnon. *L'Anthropologie* 71:479–492.

Davies TGH, Pedersen PO. 1955. The degree of attrition of the deciduous teeth and the first permanent molars of primitive and urbanised Greenland natives. *British Dental Journal* 99:35–43.

Dawson JE, Trinkaus E. 1997. Vertebral osteoarthritis of the La Chapelle-aux-Saints 1 Neandertal. *Journal of Archaeological Science* 24:1015–1021.

Day MH. 1969. Omo human skeletal remains. *Nature* 222:1135–1138.

Day MH, Stringer CB. 1982. A reconsideration of the Omo Kibish remains and the *erectus/sapiens* transition. In Lumley H de (ed) *L'*Homo erectus *et la Place de l'Homme de Tautavel Parmi les Hominidés Fossiles.* Paris: C.N.R.S, 814–846.

Day MH, Twist MHC, Ward S. 1991. Les vestiges post-crâniens d'Omo I (Kibish). *L'Anthropologie* 95:595–610.

Demes B, Qin YX, Stern JT, Larson SG, Rubin CT. 2001. Patterns of strain in the macaque tibia during functional activity. *American Journal of Physical Anthropology* 116: 257–265.

DeNiro MJ. 1985. Post-mortem preservation of alteration of *in vivo* bone collagen isotope ratios in relation to paleodietary reconstruction. *Nature* 317:806–809.

DeNiro MJ, Epstein S. 1981. Influence of diet on the distribution of nitrogen isotopes in animals. *Geochimica et Cosmochimica Acta* 45:341–351.

Dobson SD, Trinkaus E. 2002. Cross-sectional geometry and morphology of the mandibular symphysis in Middle and Late Pleistocene *Homo. Journal of Human Evolution* 43:67–87.

Dočkalová M, Vančata V. 2005. Gravettian human remains Brno II: postcranial skeleton. *Anthropologie (Brno)* 43:295–304.

Drayer-Verhagen F. 1993. Arthritis of the subtalar joint associated with sustentaculum tali facet configuration. *Journal of Anatomy* 183:631–634.

Duday H, Arensburg B. 1991. La pathologie. In BarYosef O, Vandermeersch B (eds) *Le Squelette Moustérien de Kébara 2.* Paris: Éditions du CNRS, 179–193.

Eickstedt EF von. 1925. Variationen am Axillarrand der Scapula. *Anthropologischer Anzeiger* 2:217–228.

Einwögerer T, Friesinger H, Händel M, Neugebauer-Maresch C, Simon U, Teschler-Nicola M. 2006. Upper Palaeolithic infant burials. *Nature* 444:285.

El-Najjar MY, DeSanti MV, Ozebek L 1978. Prevalence and possible etiology of dental enamel hypoplasia. *American Journal of Physical Anthropology* 48:185–192.

Eschman PN. 1992. *SLCOMM Version 1.6.* Albuquerque: Eschman Archeological Services.

Feldesman MR, Kleckner JG, Lundy JK 1990. Femur/stature ratio and estimates of stature in mid- and late-Pleistocene fossil hominids. *American Journal of Physical Anthropology* 83:359–372.

Fennell KJ, Trinkaus E. 1997. Bilateral femoral and tibial periostitis in the La Ferrassie 1 Neandertal. *Journal of Archaeological Science* 24:985–995.

Ferembach D. 1976. Les restes humains de la grotte de Dar-es-Soltane 2 (Maroc) Campagne 1975. *Bulletins et Mémoires de la Société d'Anthropologie de Paris* Série XIII, 3:183–193.

Ferembach D. 1998. Le crâne Atérien de Témara (Maroc Atlantique). *Bulletin d'Archéologie Marocaine* 18:19–66.

Flemming DJ, Murphey MD, Shekitka KM, Temple HT, Jelinek JJ, Kransdorf MJ. 2003. Osseus involvement in calcific tendinitis: a retrospective review of 50 cases. *American Journal of Roentgenology* 181:965–972.

Formicola V, Buzhilova AP. 2004. Double child burial from Sunghir (Russia): pathology and inferences for Upper Paleolithic funerary practices. *American Journal of Physical Anthropology* 124:189–198.

Franciscus RG. 1989. Neandertal mesosterna and noses: implications for activity and biogeographical patterning (abstract). *American Journal of Physical Anthropology* 78:223.

Franciscus RG, Trinkaus E. 1995. Determinants of retromolar space presence in Pleistocene *Homo* mandibles. *Journal of Human Evolution* 28:577–595.

Frayer DW, Jelínek J, Oliva M, Wolpoff MH. 2006. Aurignacian male crania, jaws and teeth from the Mladeč Caves, Moravia, Czech Republic. In Teschler-Nicola M (ed) *Early Modern Humans at the Moravian Gate: The Mladeč Caves and their Remains.* Vienna: Springer Verlag, 185–272.

Fritz P, Bardin T, Laredo JD, Ziza JM, D'Anglejan G, Lansaman J, Bucki B, Forest M, Kuntz D. 1994. Paradiaphyseal calcific tendinitis with cortical bone erosion. *Arthritis and Rheumatology* 37:718–723.

Galtés I, Jordana X, Cos M, Malgosa A, Manyosa J. 2008. Biomechanical model of pronation efficiency: new insight into skeletal adaptation of the hominoid upper limb. *American Journal of Physical Anthropology* 135:293–300.

Galtés I, Jordana X, Manyosa J, Malgosa A. 2009. Functional implications of radial diaphyseal curvature. *American Journal of Physical Anthropology* 138:286–292.

Gao X, Pei S. 2006. An archaeological interpretation of ancient human lithic technology and adaptive strategies in China (in Chinese). *Quaternary Sciences* 26:504–513.

Gerasimov MM. 1935. Excavations of the Paleolithic site in the village of Mal'ta (in Russian). *Paleolit SSSR* 118:78–124.

Gindhart PS. 1964. The frequency of appearance of transverse lines in the tibia in relation to childhood illnesses. *American Journal of Physical Anthropology* 31:17–22.

Goodman AH, Rose JC. 1991. Dental enamel hypoplasias as indicators of nutritional status. In Kelly MA, Larsen CS (eds) *Advances in Dental Anthropology.* New York: Wiley-Liss, 279–293.

Gorlin, RJ, Goldman, HM. 1970. *Thomas's Oral Pathology.* St. Louis: Mosby.

Grayson DK, Delpech F, 2002. Specialized Early Upper Palaeolithic hunters in southwestern France? *Journal of Archaeological Science* 29:1439–1449.

Grayson DK, Delpech F. 2003. Ungulates and the Middle-to-Upper Paleolithic transition at Grotte XVI (Dordogne, France). *Journal of Archaeological Science* 30:1633–1648.

Grayson DK, Delpech F. 2008. The large mammals of Roc de Combe (Lot, France): the Châtelperronian and Aurignacian assemblages. *Journal of Anthropological Archaeology* 27:338–362.

Grine FE. 1998. Additional human fossils from the Middle Stone Age of Die Kelders Cave 1, South Africa, 1995 excavation. *South African Journal of Science* 94:229–235.

Grine FE, Bailey RM, Harvati K, Nathan RP, Morris AG, Henderson GM, Ribot I, Pike AWG. 2007. Late Pleistocene human skull from Hofmeyr, South Africa, and modern human origins. *Science* 315:226–229.

Grine F, Jungers WL, Tobias PV, Pearson OM. 1995. Fossil *Homo* femur from Berg Aukas, northern Namibia. *American Journal of Physical Anthropology* 97:151–185.

Grine FE, Klein RG. 1985. Pleistocene and Holocene human remains from Equus Cave, South Africa. *Anthropology* 8:55–98.

Grine FE, Klein RG. 1993. Late Pleistocene human remains from the Sea Harvest site, Saldanha Bay, South Africa. *South African Journal of Science* 89:145–152.

Gu Y. 1978. New cave man of Zhoukoudian and his living environment (in Chinese). In Institute of Vertebrate Paleontology and Paleoanthropology (ed) *Collected Papers of Paleoanthropology.* Beijing: Science Press, 158–174.

Guatelli-Steinberg D, Larsen CS, Hutchinson DL. 2004. Prevalence and the duration of linear enamel hypoplasia: a comparative study of Neandertals and Inuit foragers. *Journal of Human Evolution* 47:65–84.

Habgood PJ. 1992. The origin of anatomically modern humans in east Asia. In Bräuer G (ed) *Continuity or Replacement. Controversies in* Homo sapiens *Evolution.* Rotterdam: Balkema, 273–288.

Habgood PJ, Franklin NR. 2008. The revolution that didn't arrive: a review of Pleistocene Sahul. *Journal of Human Evolution* 55:187–222.

Hailie-Selassie Y, Asfaw B, White TD. 2004. Hominid cranial remains from Upper Pleistocene deposits at Aduma, Middle Awash, Ethiopia. *American Journal of Physical Anthropology* 123:1–10.

Hambücken A. 1993. Variabilité Morphologique et Métrique de l'Humérus, du Radius et de l'Ulna des Néandertaliens. Comparaison avec l'Homme Moderne. Thèse de Doctorat, Université de Bordeaux 1.

Hanihara K, Ueda H. 1982. Dentition of the Minatogawa man. In Suzuki H, Hanihara K (eds) *The Minatogawa Man: The Upper Pleistocene Man from the Island of Okinawa. Bulletin of the University Museum, University of Tokyo* 19:51–59.

Hardy BL, Kay M, Marks AE, Monigal K. 2001. Stone tool function at the Paleolithic sites of Starosele and Buran Kaya III, Crimea: behavioral implications. *Proceedings of the National Academy of Sciences USA* 98:10972–10977.

Hasebe K. 1912. De Wirbelsäule der Japaner. *Zeitschrift für Morphologie und Anthropologie* 15:259–380.

Hedges REM, Reynard LM. 2007. Nitrogen isotopes and the trophic level of humans in archaeology. *Journal of Archaeological Science* 34:1240–1251.

Heim JL. 1982. Les hommes fossiles de La Ferrassie II: les squelettes adultes (squelettes des membres). *Archives de l'Institut de Paléontologie Humaine* 38:1–272.

Henry-Gambier D, Maureille B, White R. 2004.Vestiges humains des niveaux de l'Aurignacien ancien du site de Brassempouy (Landes). *Bulletins et Mémoires de la Société d'Anthropologie de Paris* 16:49–87.

Henry-Gambier D, Sacchi D. 2008. La Crouzade V–VI (Aude, France): un des plus anciens fossiles d'anatomie moderne en Europe occidentale. *Bulletins et Mémoires de la Société d'Anthropologie de Paris* 20:79–104.

Hillson SW. 1996. *Dental Anthropology.* Cambridge UK: Cambridge University Press.

Hillson SW, Franciscus RG, Holliday TW, Trinkaus E. 2006. The ages-at-death. In Trinkaus E, Svoboda JA (eds) *Early Modern Human Evolution in Central Europe: The People of Dolní Věstonice and Pavlov.* New York: Oxford University Press, 31–45.

Hintze J. 2007. *NCSS, PASS, and GESS.* Kaysville UT: NCSS.

Holliday TW. 1995. Body size and proportions in the Late Pleistocene Western Old World and the origins of Modern humans. PhD dissertation, University of New Mexico.

Holliday TW, 1997a. Body proportions in Late Pleistocene Europe and modern human origins. *Journal of Human Evolution* 32:423–447.

Holliday TW. 1997b. Postcranial evidence of cold adaptation in European Neandertals. *American Journal of Physical Anthropology* 104:245–258.

Holliday TW. 2000. Evolution at the crossroads: modern human emergence in western Asia. *American Anthropologist.* 102:54–68.

Holliday TW. 2006. The costal skeletons. In Trinkaus E, Svoboda JA (eds) *Early Modern Human Evolution in Central Europe: The People of Dolní Věstonice and Pavlov.* New York: Oxford University Press, 295–326.

Holliday TW, Falsetti AB. 1995. Lower limb length of European early modern humans in relation to mobility and climate. *Journal of Human Evolution* 29:141–153.

Holliday TW, Ruff CB. 1997. Ecogeographic patterning and stature prediction in fossil hominids. *American Journal of Physical Anthropology* 103:137–140.

Holt B. 1999. Biomechanical evidence for decreased mobility in Upper Paleolithic and Mesolithic Europe. PhD dissertation, University of Missouri–Columbia.

Houston CS, Zaleski WA. 1967. The shape of vertebral bodies and femoral necks in relation to activity. *Radiology* 89:59–66.

Hrdlička A. 1937. The gluteal ridge and the gluteal tuberosities (3rd trochanters). *American Journal of Physical Anthropology* 23:127–198.

Huang WP, Fang DS, Ye XX. 1982. Preliminary study on the fossil hominid skull and fauna of Hexian, Anhui. *Vertebrata PalAsiatica* 20:248–256.

Hublin JJ. 1991. L'émergence des *Homo sapiens* archaïques: Afrique du Nord-Ouest et Europe Occidentale. Thèse de Doctorat d'État, Université de Bordeaux 1.

Hublin JJ. 1992. Recent human evolution in northwestern Africa. In Aitken MJ, Stringer CB, Mellars PA (eds) *The Origin of Modern Humans and the Impact of Chronometric Dating.* Princeton: Princeton University Press, 118–131.

Ihunwo AO, Phukubye P. 2006. The frequency and anatomical features of torus mandibularis in a Black South African population. *Homo* 57:253–262.

Jablonski NJ, Chaplin G. 2000. The evolution of human skin coloration. *Journal of Human Evolution* 39:57–106.

Jans MME, Nielsen-Marsh CM, Smith CI, Collins MJ, Kars H. 2004. Characterisation of microbial attack on archaeological bone. *Journal of Archaeological Science* 31:87–95.

Jelínek J. 1959. Der fossile Mensch Brno II. *Anthropos* 9:17–22.

Jidoi K, Nara T, Dodo Y. 2000. Bony bridging of the mylohyoid groove of the human mandible. *Anthropological Science* 108:345–370.

Jolicoeur P. 1963. Bilateral symmetry and asymmetry in the limb bones of *Martes americana* and man. *Revue Canadien de Biologie* 22:409–432.

Jungers WL, Falsetti AB, Wall CE. 1995. Shape, relative size, and size-adjustments in morphometrics. *Yearbook of Physical Anthropology* 38:137–161.

Kallay J. 1970. Komparativne napomene o čeljustima Krapinskih praljudi obzirom na položaj među Hominidima (Comparative observations on the mandible of a Krapina Neanderthal man in respect to its position among Hominids). In Malez M (ed) *Krapina, 1899–1969.* Zagreb: Jugoslavenske Akademije Znanosti I Umjetmosti, 153–164.

Kamminga J. 1992. New interpretations of the Upper Cave, Zhoukoudian. In Akazawa T, Aoki K, Kimura T (eds) *The Evolution and Dispersal of Modern Humans in Asia.* Tokyo: Hokusen-Sha, 279–400.

Kappleman J, Alçiçek MC, Kazanci N, Schultz M, Özkul M, Şen Ş. 2008. First *Homo erectus* from Turkey and implications for migrations into temperate Eurasia. *American Journal of Physical Anthropology* 135:110–116.

Kate BR. 1966. A study of regional variation of the Indian femur—the third trochanter. *Eastern Anthropologist* 19:69–74.

Kennedy KAR, Deraniyagala SU. 1989. Fossil remains of 28,000-year-old hominids from Sri Lanka. *Current Anthropology* 30:394–399.

Kennedy KAR, Zahorsky JL. 1995. Trends in prehistoric technology and biological adaptations: new evidence from Pleistocene deposits at Fa Hien Cave, Sri Lanka. In Allchin R, Allchin B (eds) *South Asian Archaeology 1995.* New Delhi: Oxford and IBH Publishing. 2:839–853.

Kimura T, Takahashi H. 1992. Cross sectional geometry of the Minatogawa limb bones. In Akazawa T, Aoki K, Kimura T (eds) *The Evolution and Dispersal of Modern Humans in Asia.* Tokyo: Hokusen-Sha, 305–320.

Klein RG, Cruz-Uribe K. 1984. *The Analysis of Animal Bones from Archeological Sites.* Chicago: University of Chicago Press.

Korbmacher H, Koch L, Eggers-Stroeder G, Kahl-Niele B. 2007. Associations between orthopaedic disturbances and unilateral crossbite in children with asymmetry of the upper cervical spine. *European Journal of Orthodontics* 29:100–104.

Kozlovskaya MV, Mednikova MB 2000. Catalogue of photos and tables on materials from Sunghirian graves 1 and 2 (in Russian with English summary). In Alexeeva TI, Bader NO, Munchaev RM, Buzhilova AP, Kozlovskaya MV, Mednikova MB (eds) *Homo Sungirensis.* Moscow: Scientific World, 85–144.

Kreshover SJ. 1960. Metabolic disturbances in tooth formation. *Annals of the New York Academy of Science* 85:161–167.

Kuzmin YV, Orlova LA. 1998. Radiocarbon chronology of the Siberian Paleolithic. *Journal of World Prehistory* 12:1–53.

Lam YM, Pearson OM, Smith CM. 1996. Chin morphology and sexual dimorphism in the fossil hominid mandible sample from Klasies River Mouth. *American Journal of Physical Anthropology* 100:545–557.

Lanier RR Jr. 1939. The presacral vertebrae of American white and Negro males. *American Journal of Physical Anthropology* 25:341–420.

Lebel S, Trinkaus E. 2002. Middle Pleistocene human remains from the Bau de l'Aubesier. *Journal of Human Evolution* 43:659–685.

Lewis M, Roberts C. 1997. Growing pains: the interpretation of stress indicators. *International Journal of Osteoarchaeology* 7:581–586.

Li Q. 2007. Faunal analysis of large and middle-sized mammals at Tianyuan Cave (in Chinese). Master's thesis, Institute of Vertebrate Paleontology and Paleoanthropology, Chinese Academy of Sciences.

Li X, Gao L, Zhang S. 2008. Preliminary study of cave development and filling sequence at Zhoukoudian Tianyuan Cave (in Chinese with English summary). *Quaternary Science* 28:1098–1105.

Lieberman DE, Polk JD, Demes B. 2004. Predicting long bone loading from cross-sectional geometry. *American Journal of Physical Anthropology* 123:156–171.

Lisowski FP. 1967. Angular growth changes and comparisons in the primate talus. *Folia Primatologica* 7:81–97.

Lordkipanidze D, Vekua A, Ferring R, Rightmire GP, Agusti J, Kiladze G, Mouskhelishvili A, Nioradze M, Ponce de león M, Tappen M, Zollikofer CPE. 2005. The earliest toothless hominin skull. *Nature* 434:717–718.

Lü Z. 1989. Date of Jinniushan man and his position in human evolution (in Chinese). *Liaohai Wenwu Xuekan* 1:44–55.

Lü Z. 1990. La découverte de l'homme fossile de Jing-niu-shan. Première étude. *L'Anthropologie* 94:899–902.

Macchiarelli R, Bondioli L, Censi L, Hernaez MK, Salvadei L, Sperduti A. 1994. Intra- and interobserver concordance in scoring Harris lines: a test of bone sections and radiographs. *American Journal of Physical Anthropology* 95:77–83.

Madhavi C, Madhuri V, George VM, Antonisamy B. 2008. South Indian calcaneal talar facet configurations and osteoarthritic changes. *Clinical Anatomy* 21:581–586.

Mafart B, Guipert G, Alliez-Philip C, Brau JJ. 2007. Virtual reconstitution and new palaeopathological study of the Magdalenian child's skull of Rochereil. *Comptes Rendus Palévol* 6:569–579.

Mallegni F, Segre-Naldini E. 1992. A human maxilla (Fossellone 1) and scapula (Fossellone 2) recovered in the Pleistocene layers of the Fossellone Cave, Mt. Circeo, Italy. *Quaternaria Nova* 2:211–225.

Mann A, Monge J. 2006. A Neandertal parietal fragment from Krapina (Croatia) with a serious cranial trauma. *Periodicum Biologorum* 108:495–502.

Marean CW, Nilssen PJ, Brown K, Jerardina A, Stynder D. 2004. Paleoanthropological investigations of Middle Stone Age sites at Pinnacle Point, Mossel Bay (South Africa): archaeology and hominid remains from the 2000 field season. *PaleoAnthropology 2004:*14–83.

Marzke MW. 1983. Joint functions and grips of the *Australopithecus afarensis* hand, with special reference to the region of the capitate. *Journal of Human Evolution* 12:197–211.

Marzke MW, Marzke RF. 1987. The third metacarpal styloid process in humans: origins and functions. *American Journal of Physical Anthropology* 73:415–431.

Matiegka J. 1934. *Homo předmostensis. Fosilní člověk z Předmostí na Moravě I. Lebky.* Prague: Česká Akademie Věd a Umění.

Matiegka J. 1938. *Homo předmostensis. Fosilní člověk z Předmostí na Moravě II. Ostatní Části Kostrové.* Prague: Česká Akademie Věd a Umění.

Matsumura H, Pookajorn S. 2005. A morphometric analysis of the Late Pleistocene human skeleton from the Moh Khiew Cave in Thailand. *Homo* 56:93–118.

Mays S, Brickley M, Ives R. 2006. Skeletal manifestations of rickets in infants and young children in a historic population from England. *American Journal of Physical Anthropology* 129:362–374.

McCown TD, Keith A. 1939. *The Stone Age of Mount Carmel II: The Fossil Human Remains from the Levalloiso-Mousterian.* Oxford: Clarendon Press.

McDougall I, Brown FH, Fleagle JG. 2005. Stratigraphic placement and age of modern humans from Kibish, Ethiopia. *Nature* 433:733–736.

McHenry HM. 1992. Body size and proportions in early hominids. *American Journal of Physical Anthropology* 87:407–431.

McKern TW, Stewart TD. 1957. Skeletal age changes in young American males analyzed from the standpoint of age identification. *Headquarters Quartermaster Research and Development Command Technical Report* EP-45:1–179.

Ménard J. 1998. Odontologie des dents de la Grotte de Témara (Maroc). *Bulletin d'Archéologie Marocaine* 18:67–97.

Merbs CF. 1983. Patterns of activity-induced pathology in a Canadian Inuit population. *Archaeological Survey of Canada Paper* 119:1–199.

Meseke CA, Duray SM, Brillon SR. 2008. Principal components analysis of the atlas vertebra. *Journal of Manipulative and Physiological Therapeutics* 31:212–216.

Mitra ES, Smith MF, Lemelin P, Jungers WL. 2007. Comparative morphometrics of the primate apical tuft. *American Journal of Physical Anthropology* 134:449–459.

Molnar S. 1971. Human tooth wear, tooth function and cultural variability. *American Journal of Physical Anthropology* 34:175–190.

Moorrees CFA. 1957. *The Aleut Dentition.* Cambridge MA: Harvard University Press.

Muñoz M, Casadevall M. 1998. Fish remains from Arbreda Cave (Serinyà, Girona), northeast Spain, and their palaeoecological significance. *Journal of Quaternary Sciences* 12:111–115.

Musgrave JH. 1970. An anatomical study of the hands of Pleistocene and Recent man. PhD dissertation, University of Cambridge.

Musgrave JH. 1973. The phalanges of Neanderthal and Upper Palaeolithic hands. In Day MH (ed) *Human Evolution.* London: Taylor and Francis, 59–85.

Nagurka ML, Hayes WC. 1980. An interactive graphics package for calculating cross-sectional properties of complex shapes. *Journal of Biomechanics* 13:59–64.

Nara T, Hanihara T, Dodo Y, Vandermeersch B. 1998. Influence of the interproximal attrition of teeth on the formation of Neandertal retromolar space. *Anthropological Science* 106:297–309.

Newman H. 1998. Do not confuse the bone loss of periodontitis with root exposure due to supereruption. Factors affecting the lifespan of the human dentition in Britain prior to the seventeenth century. *British Dental Journal* 184:229–229.

Nielsen-Marsh CM, Smith CI, Jans MME, Nord A, Kars H, Collins MJ. 2007. Bone diagenesis in the European Holocene II: taphonomic and environmental considerations. *Journal of Archaeological Science* 34:1523–1531.

Niewoehner WA 2001. Behavioral inferences from the Skhul/Qafzeh early modern human hand remains. *Proceedings of the National Academy of Sciences USA* 98:2979–2984.

Niewoehner WA, Weaver AH, Trinkaus E. 1997. Neandertal capitate-metacarpal articular morphology. *American Journal of Physical Anthropology* 103:219–233.

Norton CJ, Gao X. 2008a. Zhoukoudian Upper Cave revisited. *Current Anthropology* 49: 732–745.

Norton CJ, Gao X. 2008b. Hominin-carnivore interactions during the Chinese Early Paleolithic: taphonomic perspectives from Xujiayao. *Journal of Human Evolution* 55:164–178.

Oakley KP, Campbell BG, Molleson TI. 1975. *Catalogue of Fossil Hominids, III: Americas, Asia, Australasia.* London: British Museum (Natural History).

Ogilvie MD, Curran BK, Trinkaus E. 1989. Incidence and patterning of dental enamel hypoplasia among the Neandertals. *American Journal of Physical Anthropology* 79:25–41.

Ohzeki S, Igarashi Y. 2007. Frequency of mandibular torus in Jomon people, medieval Japanese and modern Japanese (in Japanese with English abstract). *Anthropological Science* 115:97–107.

Oliva M. 1996. Mladopaleolitický hrob Brno II jako příspěvek k počátkům šamanismu. *Archeologické rozhledy* 48:353–383, 537–542.

Olivier G. 1960. *Pratique Anthropologique.* Paris: Vigot Frères.

O'Neill MC, Ruff CB. 2004. Estimating human long bone cross-sectional geometric properties: a comparison of noninvasive methods. *Journal of Human Evolution* 47:221–235.

Ortner DJ, Mays S. 1998. Dry-bone manifestations of rickets in infancy and early childhood. *International Journal of Osteoarchaeology* 8:45–55.

Ortner DJ, Putschar WGJ. 1981. Identification of Pathological Conditions in Human Skeletal Remains. *Smithsonian Contributions to Anthropology* 28:1–479.

Pearson OM, Fleagle JG, Grine FE, Royer DF. 2008. Further new hominin fossils from the Kibish Formation, southwestern Ethiopia. *Journal of Human Evolution* 55:444–447.

Pearson OM, Grine FE. 1997. Re-analysis of the hominid radii from Cave of Hearths and Klasies River Mouth, South Africa. *Journal of Human Evolution* 32:577–592.

Pearson OM, Lieberman DE. 2004. The aging of Wolff's "law": ontogeny and responses to mechanical loading in cortical bone. *Yearbook of Physical Anthropology* 47:63–99.

Pearson OM, Royer DF, Grine FE, Fleagle JG. 2008. A description of the Omo 1 postcranial skeleton, including newly discovered fossils. *Journal of Human Evolution* 55:421–437.

Pei WC, Woo JK. 1957. *Tzeyang Paleolithic Man* (in Chinese). Beijing: Science Press.

Polk JD, Demes B, Jungers WL, Biknevicius AR, Heinrich RE, Runestad JA. 2000. A comparison of primate, carnivoran and rodent limb bone cross-sectional properties: are primates really unique? *Journal of Human Evolution* 39:297–325.

Pope GG. 1992. Replacement versus regionally continuous models: the paleobehavioral and fossil evidence from east Asia. In Akazawa T, Aoki K, Kimura T (eds) *The Evolution and Dispersal of Modern Humans in Asia.* Tokyo: Hokusen-Sha, 3–14.

Privat KL, O'Connell TC, Hedges REM. 2007. The distinction between freshwater- and terrestrial-based diets: methodological concerns and archaeological applications of sulphur stable isotope analysis. *Journal of Archaeological Science* 34:1197–1204.

Pycraft WP. 1928. Description of the human remains. In Bather FA (ed) *Rhodesian Man and Associated Remains.* London: British Museum, 1–51.

Qiu ZL, Gu Y, Zhang Y, Zhang S. 1973. New discovery of the fossils and cultural remains of *Sinanthropus pekinensis* from Zhoukoudian (in Chinese). *Vertebrata PalAsiatica* 11:109–131.

Ramirez Rossi FV, d'Errico F, Vanhaeren M, Grootes PM, Kerautret B, Dujardin V. 2009. Cut-marked human remains bearing Neandertal features and modern human remains associated with the Aurignacian at Les Rois. *Journal of Anthropological Sciences* 87:1–30.

Rhoads JG, Trinkaus E. 1977. Morphometrics of the Neandertal talus. *American Journal of Physical Anthropology* 46:29–44.

Richards GD, Jabbour RS, Anderson JY. 2003. Medial mandibular ramus. Ontogenetic, idiosyncratic, and geographic variation in recent *Homo,* great apes and fossil hominids. *British Archaeological Reports: International Series* 1138:1–113.

Richards MP. 2002. A brief review of the archaeological evidence for Palaeolithic and Neolithic subsistence. *European Journal of Clinical Nutrition* 56:1270–1278.

Richards MP, Fuller BT, Hedges REM. 2001. Sulphur isotopic variation in ancient bone collagen from Europe: implications for human palaeodiet, residence mobility, and modern pollutant studies. *Earth and Planetary Science Letters* 191:185–190.

Richards MP, Fuller BT, Sponheimer M, Robinson T, Ayliffe L. 2003. Sulphur isotopes in palaeodietary studies: a review and results from a controlled feeding experiment. *International Journal of Osteoarchaeology* 13:37–45.

Richards MP, Hedges REM. 1999. Stable isotope evidence for similarities in the types of marine foods used by late Mesolithic humans at sites along the Atlantic coast of Europe. *Journal of Archaeological Science* 26:717–722.

Richards MP, Pettitt PB, Stiner MC, Trinkaus E. 2001. Stable isotope evidence for increasing dietary breadth in the European mid-Upper Paleolithic. *Proceedings of the National Academy of Sciences USA* 98:6528–6532.

Richards MP, Trinkaus E. 2009. Isotopic evidence for the diets of European Neandertals and early modern humans. *Proceedings of the National Academy of Sciences USA* (doi:10.1073/pnas.0903821106).

Rosas A. 1985. Falanges humanas de la Sima des los Huesos (Cueva Mayor), Sierra de Atapuerca (Burgos). Estudio anatomico y comparativo. *Actas del IV Congreso Español de Antropologia Biologica* 4:557–566.

Rosas A. 2001. Occurrence of Neanderthal features in mandibles from the Atapuerca-SH site. *American Journal of Physical Anthropology* 114:74–91.

Rosen PS, Forman D. 1999. The role of orthognathic surgery in the treatment of severe dentoalveolar extrusion. *Journal of the American Dental Association* 130:1619–1622.

Rosenberg KR, Lü Z, Ruff CB. 2006. Body size, body proportions, and encephalization in a Middle Pleistocene archaic human from northern China. *Proceedings of the National Academy of Sciences USA* 103:3552–3556.

Rougier H, Milota Ş, Rodrigo R, Gherase M, Sarcină, L, Moldovan O, Zilhão J, Constantin S, Franciscus RG, Zollikofer CPE, Ponce-de-León M, Trinkaus E. 2007. Peştera cu Oase 2 and the cranial morphology of early modern Europeans. *Proceedings of the National Academy of Sciences USA* 104:1164–1170.

Ruff CB. 2000. Body size, body shape, and long bone strength in modern humans. *Journal of Human Evolution* 38:269–290.

Ruff CB. 2006. Gracilization of the modern human skeleton. *American Scientist* 94: 508–514.

Ruff CB, Hayes WC. 1983a. Cross-sectional geometry of Pecos Pueblo femora and tibiae—a biomechanical investigation, I: method and general patterns of variation. *American Journal of Physical Anthropology* 60:259–381.

Ruff CB, Hayes WC. 1983b. Cross-sectional geometry of Pecos Pueblo femora and tibiae—a biomechanical investigation, II: sex, age, and side differences. *American Journal of Physical Anthropology* 60:383–400.

Ruff CB, Holt B, Trinkaus E. 2006. Who's afraid of the big bad Wolff? "Wolff's Law" and bone functional adaptation. *American Journal of Physical Anthropology* 129:484–498.

Ruff CB, Scott WW, Liu AYC. 1991 Articular and diaphyseal remodeling of the proximal femur with changes in body mass in adults. *American Journal of Physical Anthropology* 86:397–413.

Ruff CB, Trinkaus E, Holliday TW. 1997. Body mass and encephalization in Pleistocene *Homo. Nature* 387:173–176.

Ruff CB, Trinkaus E, Walker A, Larsen CS. 1993. Postcranial robusticity in *Homo,* I: temporal trends and mechanical interpretations. *American Journal of Physical Anthropology* 91:21–53.

Ruff CB, Walker A. 1993. Body size and body shape. In Walker A, Leakey R (ed) *The Nariokotome* Homo erectus *Skeleton.* Cambridge MA: Harvard University Press, 234–265.

Ruff CB, Walker A, Trinkaus E. 1994. Postcranial robusticity in *Homo,* III: ontogeny. *American Journal of Physical Anthropology* 93:35–54.

Sakura H. 1981. Pleistocene human bones found at Pinza-Abu (Goat Cave), Miyako Island—a short report. *Bulletin of the National Museum of Science* 7D:1–6.

Sakura H. 1985. Pleistocene human fossil remains from Pinza-Abu (Goat Cave), Miyako Island, Okinawa, Japan (in Japanese). *Reports on Excavation of the Pinza-Abu Cave.* Naha: Department of Education, Okinawa Prefectural Government, 161–173.

Scheuer L, Black S. 2000. *Developmental Juvenile Osteology.* London: Academic Press.

Schoeninger MJ, DeNiro MJ, Tauber H. 1983. Stable nitrogen isotope ratios of bone collagen reflect marine and terrestrial components of prehistoric human diet. *Science* 220:1381–1383.

School of Archaeology and Museology of Beijing University. 2006. Center for the Study of Chinese Archaeology of Beijing University, Institute of Cultural Relics and Archaeology of Beijing. Donghulin site during Early Neolithic in Mentougou, Beijing (in Chinese with English abstract). *Archaeology* 7:3–8.

Seeger LL, Butler DI, Eckardt JJ, Layfield L, Adams JS. 1990. Tumoral calcinosis-like lesion of the proximal linea aspera. *Skeletal Radiology* 19:579–583.

Senut B. 1981. *L'Humérus et ses Articulations chez les Hominidés Plio-Pleistocènes.* Paris: CNRS.

Sergi S. 1974. *Il Cranio Neandertaliano del Monte Circeo (Circeo I).* Rome: Accademia Naziolane dei Lincei.

Shackelford LL. 2005. Patterns of geographic variation in the postcranial robusticity of Late Upper Paleolithic humans. PhD dissertation, Washington University in St. Louis.

Shackelford LL. 2007. Regional variation in the postcranial robusticity of Late Upper Paleolithic humans. *American Journal of Physical Anthropology* 133:655–668.

Shackelford LL, Trinkaus E. 2002. Late Pleistocene human femoral diaphyseal curvature. *American Journal of Physical Anthropology* 118:359–370.

Shang H, Tong H, Zhang S, Chen F, Trinkaus E. 2007. An early modern human from Tianyuan Cave, Zhoukoudian, China. *Proceedings of the National Academy of Sciences USA* 104:6573–6578.

Shang H, Trinkaus E. 2008. An ectocranial lesion on the Middle Pleistocene human cranium from Hulu Cave, Nanjing, China. *American Journal of Physical Anthropology* 135: 431–437.

Shang H, Wei Q, Wu X, 2006 An issue on the date of fossil human remains from Salawusu, Inner Mongolia (in Chinese with English abstract). *Acta Anthropologica Sinica* 25: 82–86.

Shaw CN, Stock JT. 2009a. Intensity, repetitiveness, and directionality of habitual adolescent mobility patterns influence the tibial diaphysis morphology of athletes. *American Journal of Physical Anthropology* 140:149–159.

Shaw CN, Stock JT. 2009b. Habitual throwing and swimming correspond with upper limb diaphyseal strength and shape in modern human athletes. *American Journal of Physical Anthropology* 140:160–172.

Shen G, Wang W, Wang Q, Zhao J, Collerson K, Zhou C, Tobias PV. 2002. U-series dating of Liujiang hominid site in Guangxi, South China. *Journal of Human Evolution* 43:817–829.

Shiraishi N, Matsumura G, Uehara A. 2007. Morphological classification of the anterior and middle talocalcaneal joints in terms of biomechanical stress dispersion. *Okajimas Folia Anatomica Japonica* 84:43–48.

Shore LR. 1933. A report on the spinous processes of the cervical vertebrae in a series of Egyptian skeletons. *Journal of Anatomy* 67:422–442.

Shpakova EG. 2005. Paleolithic human dental remains from Siberia. In Derevianko AP (ed) *The Middle to Upper Paleolithic Transition in Eurasia. Hypotheses and Facts.* Novosibirsk: Institute of Archaeology and Ethnography Press, 414–426.

Sigmon BA. 1982. Comparative morphology of the locomotor skeleton of *Homo erectus* and the other fossil hominids, with special reference to the Tautavel innominate and femora. In Lumley H de (ed) *L'Homo* erectus *et la Place de l'Homme de Tautavel Parmi les Hominidés Fossiles.* Paris: CNRS, 422–446.

Simpson SW, Quade J, Levin NE, Butler R, Dupont-Nivet G, Everett M, Semaw S. 2008. A female *Homo erectus* pelvis from Gona, Ethiopia. *Science* 322:1089–1092.

Singer R, Wymer J. 1982. *The Middle Stone Age at Klasies River Mouth in South Africa.* Chicago: University of Chicago Press.

Singh I. 1959. Squatting facets on the talus and tibia in Indians. *Journal of Anatomy* 93: 540–550.

Sinitsyn A. 2004. Les sépultures de Kostenki: chronologie, attribution culturelle, rite funéraire. In Otte M (ed) *La Spiritualité. Etudes et Recherches Archéologiques de l'Université de Liège* 106:237–244.

Smith BH. 1991. Standards of human tooth formation and dental age assessment. In Kelley MA, Larsen CS (eds) *Advances in Dental Anthropology.* New York: Wiley-Liss, 143–168.

Smith FH. 1978. Evolutionary significance of the mandibular foramen area in Neandertals. *American Journal of Physical Anthropology* 48:523–532.

Smith FH, Smith MO, Schmitz RW. 2006. Human skeletal remains from the 1997 and 2000 excavations of cave deposits derived from Kleine Feldhofer Grotte in the Neander Valley, Germany. In Schmitz RW (ed) *Neanderthal 1856–2006.* Mainz: Philipp von Zabern, 187–246.

Soficaru A, Doboş A, Trinkaus E. 2006. Early modern humans from the Peştera Muierii, Baia de Fier, Romania. *Proceedings of the National Academy of Sciences USA* 103:17196–17201.

Soficaru A, Petrea C, Doboş A, Trinkaus E. 2007. The human cranium from the Peştera Cioclovina Uscată, Romania: context, age, taphonomy, morphology, and paleopathology. *Current Anthropology* 48:611–619.

Solow B, Sandham A. 2002. Cranio-cervical posture: a factor in the development and function of the dentofacial structures. *European Journal of Orthodontics* 24:447–456.

Sponheimer M, Lee-Thorp J. 2007. Hominin paleodiets: the contribution of stable isotopes. In Henke W, Tattersall I, Hardt T (eds) *Handbook of Paleoanthropology.* New York: Springer Verlag, 555–585.

Sponheimer M, Passey BH, de Ruiter DJ, Guatelli-Steinberg D, Cerling TE, Lee-Thorp JA. 2006. Isotopic evidence for dietary variability in the early hominin *Paranthropus robustus. Science* 314:980–982.

Spoor F, Stringer CB, Zonneveld F. 1998. Rare temporal bone pathology of the Singa calvaria from Sudan. *American Journal of Physical Anthropology* 107:41–50.

Steele DG, Bramblett CA. 1988. *The Anatomy and Biology of the Human Skeleton.* College Station: Texas A&M University Press.

Steele J, Mays S. 1995. Handedness and directional asymmetry in the long bones of the human upper limb. *International Journal of Osteoarchaeology* 5:39–49.

Stefan VH, Trinkaus E. 1998a. La Quina 9 and Neandertal mandibular variability. *Bulletins et Mémoires de la Société d'Anthropologie de Paris* ns 10:293–324.

Stefan VH, Trinkaus E. 1998b. Discrete trait and dental morphometric affinities of the Tabun 2 mandible. *Journal of Human Evolution* 34:443–468.

Stevenson PH. 1929. On racial differences in stature long bone regression formulae, with special reference to stature reconstruction formulae for the Chinese. *Biometrika* 21:303–318.

Stewart TD. 1962a. Neanderthal cervical vertebrae with special attention to the Shanidar Neanderthals from Iraq. *Bibliotheca Primatologica* 1:130–154.

Stewart TD. 1962b. Neanderthal scapulae with special reference to the Shanidar Neanderthals from Iraq. *Anthropos* 57:779–800.

Stiner MC. 1994. *Honor among Thieves: A Zooarchaeological Study of Neandertal Ecology.* Princeton: Princeton University Press.

Stiner MC, Munro ND, Surovell TA. 2000. The tortoise and the hare. Small-game use, the Broad-Spectrum Revolution, and Paleolithic demography. *Current Anthropology* 41:39–73.

Stiner MC, Munro ND, Surovell TA, Tchernov E, Bar-Yosef O. 1999. Paleolithic population growth pulses evidence by small animal exploitation. *Science* 283:190–194.

Stringer CB, Trinkaus E, Roberts MB, Parfitt SA, Macphail RI. 1998. The Middle Pleistocene human tibia from Boxgrove. *Journal of Human Evolution* 34:509–547.

Suzuki H. 1982. Skulls of the Minatogawa man. In Suzuki H, Hanihara K (eds) *The Minatogawa Man: The Upper Pleistocene Man from the Island of Okinawa. Bulletin of the University Museum, University of Tokyo* 19:7–49.

Suzuki H. 1983. The Yamashita-cho Man: a late Pleistocene infantile skeleton from the Yamashita-cho Cave (Okinawa). *Bulletins et Mémoires de la Société d'Anthropologie de Paris* Série 13, 10:81–87.

Suzuki H, Hanihara K (eds). 1982. *The Minatogawa Man: The Upper Pleistocene Man from the Island of Okinawa. Bulletin of the University Museum,* University of Tokyo, 19.

Svoboda JA. 2000. The depositional contexts of the Early Upper Paleolithic human fossils from the Koněprusy (Zlatý kůň) and Mladeč Caves, Czech Republic. *Journal of Human Evolution* 38:523–536.

Sweeney EA, Saffir AJ, Leon RD. 1971. Linear hypoplasia of deciduous incisor teeth in malnourished children. *American Journal of Clinical Nutrition* 24:29–31.

Tangtrakulwanich B, Chongsuvivatwong V, Geater AF. 2007. Habitual floor activities increases risk of knee osteoarthritis. *Clinical Orthopedics and Related Research* 454:147–154.

Teschler-Nicola M (ed). 2006. *Early Modern Humans at the Moravian Gate: The Mladeč Caves and their Remains.* Vienna: Springer Verlag.

Tillier AM. 1999. *Les Enfants Moustériens de Qafzeh. Interprétation Phylogénétique et Paléoauxologique.* Paris: CNRS Éditions.

Tobias PV. 1967. The hominid skeletal remains of Haua Fteah. In McBurney CBM (ed) *The Haua Fteah (Cyrenaica) and the Stone Age of the South-East Mediterranean.* Cambridge UK: Cambridge University Press, 338–352.

Tong H. 2005. *Hystrix subcristata* (Mammalia, Rodentia) from Tianyuan Cave, a human fossil site newly discovered near Zhoukoudian (Choukoutien) (in Chinese with English summary). *Vertebrata PalAsiatica* 43:135–150.

Tong H. 2007. *Aeretes melanopterus* (Pteromyinae, Rodentia) from Tianyuan Cave near Zhoukoudian (Choukoutien) in China. *Geobios* 40:219–230.

Tong H. 2008. Quaternary *Hystrix* (Rodentia, Mammalia) from North China: taxonomy, stratigraphy, and zoogeography, with discussions on the distribution of *Hystrix* in Palearctic Eurasia. *Quaternary International* 179:126–134.

Tong H, Liu J, Zhang S. 2004. Large mammal fossils from the Tianyuan Cave, a human fossil site of end-Pleistocene age recently discovered near Zhoukoudian (in Chinese with English abstract). *Acta Anthropologica Sinica* 24:41–51.

Tong H, Shang H, Zhang S, Chen F. 2004. A preliminary report on the newly found Tianyuan Cave, a Late Pleistocene human fossil site near Zhoukoudian. *Chinese Science Bulletin* 49:853–857.

Tong H, Shang H, Zhang S, Liu J, Chen FY, Wu X, Li Q. 2006. Mammalian biostratigraphy of Tianyuan Cave, compared with that of Upper Cave at Zhoukoudian (Choukoutien) (in Chinese with English summary). *Acta Anthropologica Sinica* 25:68–81.

Tong H, Zhang S, Chen F, Li Q. 2008. Rongements sélectifs des os par les porcs-épics et autres rongeurs: cas de la grotte Tianyuan, un site avec des restes humains fossiles récemment découvert près de Zhoukoudian (Choukoutien). *L'Anthropologie* 112:353–369.

Trinkaus E. 1975a. A functional analysis of the Neandertal foot. PhD dissertation, University of Pennsylvania.

Trinkaus E. 1975b. Squatting among the Neandertals: a problem in the behavioral interpretation of skeletal morphology. *Journal of Archaeological Science* 2:327–351.

Trinkaus E. 1978. Bilateral asymmetry of human skeletal non-metric traits. *American Journal of Physical Anthropology* 49:315–318.

Trinkaus E. 1981. Neanderthal limb proportions and cold adaptation. In Stringer CB (ed) *Aspects of Human Evolution*. London: Taylor and Francis, 187–224.

Trinkaus E. 1983a. *The Shanidar Neandertals*. New York: Academic Press.

Trinkaus E. 1983b. Functional aspects of Neandertal pedal remains. *Foot and Ankle* 3: 377–390.

Trinkaus E. 1984. Western Asia. In Smith FH, Spencer F (eds) *The Origins of Modern Humans: A World Survey of the Fossil Evidence*. New York: Alan R. Liss, 251–293.

Trinkaus E. 1985. Pathology and the posture of the La Chapelle-aux-Saints Neandertal. *American Journal of Physical Anthropology* 67:19–41.

Trinkaus E. 1992. Morphological contrasts between the Near Eastern Qafzeh-Skhul and late archaic human samples: grounds for a behavioral difference? In Akazawa T, Aoki K, Kimura T (eds) *The Evolution and Dispersal of Modern Humans in Asia*. Tokyo: Hokusen-Sha, 277–294.

Trinkaus E. 1993a. Variability in the position of the mandibular mental foramen and the identification of Neandertal apomorphies. *Rivista di Antropologia* 71:259–274.

Trinkaus E. 1993b. Femoral neck-shaft angles of the Qafzeh-Skhul early modern humans, and activity levels among immature Near Eastern Middle Paleolithic hominids. *Journal of Human Evolution* 25:393–416.

Trinkaus E. 1993c. A note on the KNM-ER 999 hominid femur. *Journal of Human Evolution* 24:493–504.

Trinkaus E. 2000. The "Robusticity Transition" revisited. In Stringer C, Barton RNE, Finlayson C (eds) *Neanderthals on the Edge*. Oxford: Oxbow Books, 227–236.

Trinkaus E. 2003. Neandertal faces were not long; modern human faces are short. *Proceedings of the National Academy of Sciences USA* 100:8142–8145.

Trinkaus E. 2004. Dental crown dimensions of Middle and Late Pleistocene European humans. In Rubio S (ed) *Miscalánea en Homenaje a Emiliano Aguirre III: Paleoantropología*. Alcalá de Henares: Museo Arqueológico Regional, 393–398.

Trinkaus E. 2005a. Early modern humans. *Annual Review of Anthropology* 34:207–230.

Trinkaus E. 2005b. Anatomical evidence for the antiquity of human footwear use. *Journal of Archaeological Science* 32:1515–1526.

Trinkaus E. 2006a. Modern human versus Neandertal evolutionary distinctiveness. *Current Anthropology* 47:597–620.

Trinkaus E. 2006b. The upper limb remains. In Trinkaus E, Svoboda JA (eds) *Early Modern Human Evolution in Central Europe: The People of Dolní Věstonice and Pavlov*. New York: Oxford University Press, 327–372.

Trinkaus E. 2006c. The Krapina scapulae. *Periodicum Biologorum* 108:341–351.

Trinkaus E. 2006d. The lower limb remains. In Trinkaus E, Svoboda JA (eds) *Early Modern Human Evolution in Central Europe: The People of Dolní Věstonice and Pavlov.* New York: Oxford University Press, 380–418.

Trinkaus E. 2007a. European early modern humans and the fate of the Neandertals. *Proceedings of the National Academy of Sciences USA* 104:7367–7372.

Trinkaus E. 2007b. Activité, stress et survie chez les Néandertaliens. In Vandermeersch B, Maureille B (eds) *Les Néandertaliens: Biologie et Cultures.* Paris: Comité des Travaux Historiques et Scientifiques (Documents préhistoriques), 131–137.

Trinkaus E. 2008a. Behavioral implications of the Muierii 1 early modern human scapula. *Annuaire Roumain d'Anthropologie* 45:27–41.

Trinkaus E. 2008b. Kiik-Koba 2 and Neandertal axillary border ontogeny. *Anthropological Science* 116:231–236.

Trinkaus E. 2009. The human tibia from Broken Hill, Kabwe, Zambia. *PaleoAnthropology* 2009:145–165.

Trinkaus E, Biglari F. 2006. Middle Paleolithic human remains from Bisitun Cave, Iran. *Paléorient* 32:105–111.

Trinkaus E, Churchill SE. 1988. Neandertal radial tuberosity orientation. *American Journal of Physical Anthropology* 75:15–21.

Trinkaus E, Churchill SE, Ruff CB. 1994. Postcranial robusticity in *Homo*, II: humeral bilateral asymmetry and bone plasticity. *American Journal of Physical Anthropology* 93:1–34.

Trinkaus E, Churchill SE, Ruff CB, Vandermeersch B. 1999. Long bone shaft robusticity and body proportions of the Saint-Césaire 1 Châtelperronian Neandertal. *Journal of Archaeological Science.* 26:753–773.

Trinkaus E, Formicola V, Svoboda J, Hillson SW, Holliday TW. 2001. Dolní Věstonice 15: Pathology and persistence in the Pavlovian. *Journal of Archaeological Science* 28:1291–1308.

Trinkaus E, Hillson SW, Franciscus RG, Holliday TW. 2006. Skeletal and dental paleopathology. In Trinkaus E, Svoboda JA (eds) *Early Modern Human Evolution in Central Europe: The People of Dolní Věstonice and Pavlov.* New York: Oxford University Press, 419–458.

Trinkaus E, Hillson SW, Santos Coelho JM. 2002. Paleopathology. In Zilhão J, Trinkaus E (eds) *Portrait of the Artist as a Child. The Gravettian Human Skeleton from the Abrigo do Lagar Velho and its Archeological Context. Trabalhos de Arqueologia* 22:489–495.

Trinkaus E, Hilton CE. 1996. Neandertal pedal proximal phalanges: diaphyseal loading patterns. *Journal of Human Evolution* 30:399–425.

Trinkaus E, Jelínek J. 1997. Human remains from the Moravian Gravettian: the Dolní Věstonice 3 postcrania. *Journal of Human Evolution* 33:33–82.

Trinkaus E, Maki J, Zilhão J. 2007. Middle Paleolithic human remains from the Gruta da Oliveira (Torres Novas), Portugal. *American Journal of Physical Anthropology* 134: 263–273.

Trinkaus E, Maley B, Buzhilova AP. 2008. Paleopathology of the Kiik-Koba 1 Neandertal. *American Journal of Physical Anthropology* 137:106–112.

Trinkaus E, Moldovan O, Milota Ş, Bîlgăr A, Sarcină L, Athreya S, Bailey SE, Rodrigo R,

Gherase M, Higham T, Bronk Ramsey C, van der Plicht J. 2003. An early modern human from the Peştera cu Oase, Romania. *Proceedings of the National Academy of Sciences USA* 100:11231–11236.

Trinkaus E, Ruff CB. 1989. Diaphyseal cross-sectional geometry and biomechanics of the Fond-de-Forêt 1 femur and the Spy 2 femur and tibia. *Bulletin de la Société royale belge d'Anthropologie et de Préhistoire* 100:33–42.

Trinkaus E, Ruff CB. 1996. Early modern human remains from eastern Asia: the Yamashita-cho 1 immature postcrania. *Journal of Human Evolution* 30:299–314.

Trinkaus E, Ruff CB. 1999. Diaphyseal cross-sectional geometry of Near Eastern Middle Paleolithic humans: the tibia. *Journal of Archaeological Science* 26:1289–1300.

Trinkaus E, Ruff CB. 2000. Comment on O. M. Pearson, "Activity, climate, and postcranial robusticity. Implications for modern human origins and scenarios of adaptive change." *Current Anthropology* 41:598.

Trinkaus E, Ruff CB, Conroy GC. 1999. The anomalous archaic *Homo* femur from Berg Aukas, Namibia: a biomechanical assessment. *American Journal of Physical Anthropology* 110:379–391.

Trinkaus E, Shang H. 2008. Anatomical evidence for the antiquity of human footwear: Tianyuan and Sunghir. *Journal of Archaeological Science* 35:1928–1933.

Trinkaus E, Smith FH, Stockton TC, Shackelford LL. 2006. The human postcranial remains from Mladeč. In Teschler-Nicola M (ed) *Early Modern Humans at the Moravian Gate: The Mladeč Caves and their Remains.* Vienna: Springer Verlag, 385–445.

Trinkaus E, Soficaru A, Doboş A, Constantin S, Zilhão J, Richards M. 2009. Stable isotope evidence for early modern human diet in southeastern Europe: Peştera cu Oase, Peştera Muierii and Peştera Cioclovina Uscată. *Materiale şi Cercetări Arheologice* 5:5–14.

Trinkaus E, Stringer CB, Ruff CB, Hennessy RJ, Roberts MB, Parfitt SA. 1999. Diaphyseal cross-sectional geometry of the Boxgrove 1 Middle Pleistocene human tibia. *Journal of Human Evolution* 37:1–25.

Trinkaus E, Svoboda J, Wojtal P, Nývltová Fišákova M, Wilczyński J. 2010. Human remains from the Moravian Gravettian: morphology and taphonomy of additional elements from Dolní Věstonice II and Pavlov I. *International Journal of Osteoarchaeology.*

Trinkaus E, Thompson DD. 1987. Femoral diaphyseal histomorphometric age determinations for the Shanidar 3, 4, 5 and 6 Neandertals and Neandertal longevity. *American Journal of Physical Anthropology.* 72:123–129.

Trotter M. 1934. Synostosis between manubrium and body of the sternum in whites and Negroes. *American Journal of Physical Anthropology* 18:439–442.

Trotter M, Gleser GC. 1952. Estimation of stature from long bones of American whites and Negroes. *American Journal of Physical Anthropology* 10:463–514.

Twiesselmann F. 1961. Le fémur Néanderthalien de Fond-de-Forêt (Province de Liège). *Mémoire de l'Institut Royal des Sciences Naturelles de Belgique* 148:1–164.

Twiesselmann F. 1973. Évolution des dimensions et de la forme de la mandibule, du palais et des dents de l'homme. *Annales de Paléontologie (Vertébrés)* 59:173–277.

Twiesselmann F, Brabant H. 1967. Les dents et les maxillaires de la population d'âge Franc de Coxyde (Belgique). *Bulletin du Groupement International pour la Recherche Scientifique en Stomatologie* 10:5–180.

Valladas H, Reyes JL, Joron JL, Valladas G, Bar-Yosef O, Vandermeersch B. 1988. Thermoluminescence dating of Mousterian "Proto-Cro-Magnon" remains from Israel and the origin of modern man. *Nature* 331:614–616.

Vallois HV. 1958. Le gisement aurignacien des Rois à Mouthiers (Charente): les restes humains. *Gallia Supplément* 9:118–137.

Vallois HV, Roche J. 1958. La mandibule acheuléenne de Témara, Maroc. *Comptes rendus de l'Académie des Sciences Paris* 246:3113–3116.

Vandermeersch B. 1981. *Les Hommes Fossiles de Qafzeh (Israël).* Paris: CNRS.

Vandermeersch B, Cleyet-Merle JJ, Jaubert J, Maureille B, Turq A (eds). (2008) *Première Humanité: Gestes Funéraires des Néandertaliens.* Paris: Réunion des Musées Nationaux.

Vandermeersch B, Trinkaus E. 1995. The postcranial remains of the Régourdou 1 Neandertal: the shoulder and arm remains. *Journal of Human Evolution* 28:439–476.

Veleminská J, Brůžek J (eds). 2008. *Early Modern Humans from Předmostí near Přerov, Czech Republic. A New Reading of Old Documentation.* Prague: Academia.

Vermeersch PM. 2002. Two Upper Palaeolithic burials at Nazlet Khater. In Vermeersch PM (ed) *Palaeolithic Quarrying Sites in Upper and Middle Egypt.* Leuven: Leuven University Press, 273–282.

Villa P, Mahieu E. 1991. Breakage patterns of human long bones. *Journal of Human Evolution* 21:27–48.

Villemeur I. 1994. *La Main des Néandertaliens: Comparaison avec la Main des Hommes de Type Moderne: Morphologie et Mécanique.* Paris: CNRS Éditions.

Walker A, Leakey R (eds). 1993. *The Nariokotome* Homo erectus *Skeleton.* Cambridge MA: Harvard University Press.

Walker A, Zimmerman MR, Leakey REF. 1982. A possible case of hypervitaminosis A in *Homo erectus. Nature* 296:248–250.

Walker PL. 1994. Documentation of sex differences and age changes in adults. In Buikstra JE, Ubelaker DH (eds) *Standards for Data Collection from Human Skeletal Remains. Arkansas Archeological Survey Research Series* 44:15–38.

Walker PL. 2008. Sexing skulls using discriminant function analysis of visually assessed traits. *American Journal of Physical Anthropology* 136:39–50.

Weaver TD, Steudel-Numbers K. 2007. Does climate or mobility explain the differences in body proportions between Neandertals and their Upper Paleolithic successors? *Evolutionary Anthropology* 14:218–223.

Webb SG. 1989. *The Willandra Lakes Hominids.* Canberra: Research School of Pacific Studies.

Wei C. 1979. Searching for descendants of "Peking man." *Anthropogischer Anzeiger* 37:61–67.

Weidenreich F. 1936. The mandibles of *Sinanthropus pekinensis:* a comparative study. *Palaeontologia Sinica* 7D:1–162.

Weidenreich F. 1939a. On the earliest representatives of modern mankind recovered on the soil of east Asia. *Peking Natural History Bulletin* 13:161–173.

Weidenreich F. 1939b The duration of life of fossil man in China and the pathological lesions found in his skeleton. *Chinese Medical Journal* 55:34–44.

Weidenreich F. 1941. The extremity bones of *Sinanthropus pekinensis. Palaeontologia Sinica* 5D:1–82.

Weidenreich F. 1943. The skull of *Sinanthropus pekinensis:* a comparative study on a primitive hominid skull. *Palaeontologia Sinica* 10D:1–485.

Weidenreich F. 1947. Facts and speculations concerning the origin of *Homo sapiens. American Anthropologist* 49:187–203.

Weninger B, Jöris O, Danzeglocke U. 2007. *CalPal 2007. Cologne Radiocarbon Calibration and Palaeoclimate Research Package* (www.calpal.de).

White TD, Asfaw B, DeGusta D, Gibert H, Richards GD, Suwa G, Howell FC. 2003. Pleistocene *Homo sapiens* from Middle Awash, Ethiopia. *Nature* 423:742–747.

White TD, Folkens PA. 1991. *Human Osteology.* San Diego: Academic Press.

Whittaker DK, Parker JH, Jenkins C. 1982. Tooth attrition and continuing eruption in a Romano-British population. *Archives of Oral Biology* 27:405–409.

Wild EM, Teschler-Nicola M, Kutschera W, Steier P, Trinkaus E, Wanek W. 2005. First direct dating of Early Upper Paleolithic human remains from Mladeč. *Nature* 435:332–335.

Wolpoff MH, Frayer DW, Jelínek J. 2006. Aurignacian female crania and teeth from the Mladeč Caves, Moravia, Czech Republic. In Teschler-Nicola M (ed) *Early Modern Humans at the Moravian Gate: The Mladeč Caves and Their Remains.* Vienna: Springer Verlag, 273–340.

Wolpoff MH, Hawks J, Frayer DW, Hunley K. 2001. Modern human ancestry at the peripheries: a test of the replacement theory. *Science* 291:293–297.

Wolpoff MH, Wu X, Thorne AG. 1984. Modern *Homo sapiens* origins: a general theory of hominid evolution involving the fossil evidence from east Asia. In Smith FH, Spencer F (eds) *The Origins of Modern Humans: A World Survey of the Fossil Evidence.* New York: Alan R. Liss, 411–483.

Woo JK. 1958. Investigation of human teeth (in Chinese with English abstract). In Pei W (ed) *Report on the Excavation of the Paleolithic Site at Tingtsun, Hsiangfenhsien, Shansi Province, China.* Beijing: Science Press, 15–20.

Woo JK. 1959. Fossil human skull of early Paleoanthropic stage found at Mapa, Shaoquan, Kwangtung Province. *Vertebrata PalAsiatica* 3:176–182.

Woo JK. 1964. A newly discovered mandible of the *Sinanthropus* type—*Sinanthropus lantianensis. Scientia Sinica* 13:801–811.

Woo JK. 1965. The hominid skull of Lantian, Shensi. *Vertebrata PalAsiatica* 10:1–22.

Woo JK, Chia L, 1954. New discoveries of *Sinanthropus pekinensis* in Choukoutien (in Chinese with English abstract). *Acta Paleontologica Sinica* 2:267–288.

Wood NK, Goaz PW. 1997. *Differential Diagnosis of Oral and Maxillofacial Lesions.* 5th ed. St. Louis: Mosby.

Wu R, Li X, Wu X (eds). 2002. Homo erectus *from Nanjing* (in Chinese with English summaries). Nanjing: Jiangsu Science and Technology Press.

Wu X. 1961 Study on the Upper Cave man of Choukoutien (in Chinese with English summary). *Vertebrata PalAsiatica* 3:181–211.

Wu X. 1981. A well-preserved cranium of an archaic type of early *Homo sapiens* from Dali, China. *Scienta Sinica* 24:530–539.

Wu X. 1992. The origin and dispersal of anatomically modern humans in east and southeast Asia. In Akazawa T, Aoki K, Kimura T (eds) *The Evolution and Dispersal of Modern Humans in Asia.* Tokyo: Hokusen-Sha, 373–378.

Wu X. 2004. On the origin of modern humans in China. *Quaternary International* 117: 131–140.

Wu X. 2006. Evidence of multiregional human evolution hypothesis from China (in Chinese with English summary). *Quaternary Sciences* 26:702–709.

Wu X, Poirier FE. 1995. *Human Evolution in China: A Metric Description of the Fossils and a Review of the Sites.* Oxford: Oxford University Press.

Yamano S, Yamaguchi B. 1976. On the mylohyoid canal in the human mandible. *Bulletin of the National Science Museum* 2D:37–44.

Yang HS. 1995. Torus palatinus and torus mandibularis: a review of the literature. *Australian Dental Journal* 40:318–321.

Yokley TR, Churchill SE. 2006. Archaic and modern human distal humeral morphology. *Journal of Human Evolution* 51:603–616.

Yuan SX, Chen TM, Gao SJ. 1986. Uranium series chronological sequence of some Paleolithic sites in South China (in Chinese with English abstract). *Acta Anthropologica Sinica* 5:179–190.

Zhang Y, Hunter DJ, Nevitt MC, Xu L, Niu J, Liu LY, Yu W, Aliabadi P, Felson DT. 2004. Association of squatting with increased prevalence of radiographic tibiofemoral knee osteoarthritis: the Beijing Osteoarthritis study. *Arthritis and Rheumatology* 50:1187–1192.

Zilhão J, Trinkaus E. 2002. Social implications. In Zilhão J, Trinkaus E (eds) *Portrait of the Artist as a Child: The Gravettian Human Skeleton from the Abrigo do Lagar Velho and Its Archeological Context. Trabalhos de Arqueologia* 22:519–541, Lisbon: Instituto Português de Arqueologia.

Zollikofer CPE, Ponce de León M, Vandermeersch B, Lévêque F. 2002. Evidence for interpersonal violence in the St. Césaire Neanderthal. *Proceedings of the National Academy of Sciences USA* 99:6444–6448.

Index

Texas A&M University Anthropology Series

Number One	*Making Faces: Using Forensic and Archaeological Evidence*, John Prag and Richard Neave
Number Two	*Culture of Stone: Sacred and Profane Uses of Stone among the Dani*, O. W. "Bud" Hampton
Number Three	*The Archaeology of Death and Burial*, Mike Parker Pearson
Number Four	*The Archaeology of Animal Bones*, Terry O'Connor, Terry
Number Five	*Stories That Float from Afar: Ancestral Folklore of the San of Southern Africa*, J. D. Lewis-Williams
Number Six	*Archaeological Conservation Using Polymers: Practical Applications for Organic Artifact Stabilization*, C. Wayne Smith
Number Seven	*Rock Art of the Lower Pecos*, Carolyn E. Boyd
Number Eight	*The Proto-Neolithic Cemetery in Shanidar Cave*, Ralph S. Solecki, Rose L. Solecki, and Anagnostis P. Agelarakis
Number Nine	*The Prehistory of Texas*, Timothy K. Perttula
Number Ten	*How War Began*, Keith Otterbein
Number Eleven	*In Search of Maya Sea Traders*, Heather McKillop
Number Twelve	*The Osteology of Infants and Children*, Brenda J. Baker, Tosha L. Dupras, and Matthew W. Tocheri
Number Thirteen	*Human Origins: From Bones to Genomes*, Rob DeSalle and Ian Tattersall

ISBN-13: 978-1-60344-177-3
ISBN-10: 1-60344-177-8